Studies in Systems, Decision and Control

Volume 23

Series editor

Janusz Kacprzyk, Polish Academy of Sciences, Warsaw, Poland
e-mail: kacprzyk@ibspan.waw.pl

About this Series

The series "Studies in Systems, Decision and Control" (SSDC) covers both new developments and advances, as well as the state of the art, in the various areas of broadly perceived systems, decision making and control- quickly, up to date and with a high quality. The intent is to cover the theory, applications, and perspectives on the state of the art and future developments relevant to systems, decision making, control, complex processes and related areas, as embedded in the fields of engineering, computer science, physics, economics, social and life sciences, as well as the paradigms and methodologies behind them. The series contains monographs, textbooks, lecture notes and edited volumes in systems, decision making and control spanning the areas of Cyber-Physical Systems, Autonomous Systems, Sensor Networks, Control Systems, Energy Systems, Automotive Systems, Biological Systems, Vehicular Networking and Connected Vehicles, Aerospace Systems, Automation, Manufacturing, Smart Grids, Nonlinear Systems, Power Systems, Robotics, Social Systems, Economic Systems and other. Of particular value to both the contributors and the readership are the short publication timeframe and the world-wide distribution and exposure which enable both a wide and rapid dissemination of research output.

More information about this series at http://www.springer.com/series/13304

Khalid Abidi · Jian-Xin Xu

Advanced Discrete-Time Control

Designs and Applications

Springer

Khalid Abidi
Newcastle University
Ang Mo Kio
Singapore

Jian-Xin Xu
Department of Electrical and Computer Engineering
National University of Singapore
Kent Ridge Crescent
Singapore

ISSN 2198-4182 ISSN 2198-4190 (electronic)
Studies in Systems, Decision and Control
ISBN 978-981-10-1275-4 ISBN 978-981-287-478-8 (eBook)
DOI 10.1007/978-981-287-478-8

Springer Singapore Heidelberg New York Dordrecht London

Softcover reprint of the hardcover 1st edition 2015

Printed on acid-free paper

Springer Science+Business Media Singapore Pte Ltd. is part of Springer Science+Business Media
(www.springer.com)

To my wife Burcu and daughter Nora

Khalid Abidi

To my wife Iris Chen and daughter Elizabeth Xu

Jian-Xin Xu

Preface

This monograph aims to cover a wide spectrum of systems such as linear and nonlinear multivariable systems as well as control problems such as disturbance, uncertainty, and time-delays. The purpose is to provide researchers and practitioners a manual for the design and application of advanced discrete-time controllers.

The monograph presents six different control approaches depending on the type of system and control problem. The first and second approaches are based on Sliding Mode control (SMC) theory and are intended for linear systems with exogenous disturbances. The third and fourth approaches are based on adaptive control theory and are aimed at linear/nonlinear systems with periodically varying parametric uncertainty or systems with input delay. The fifth approach is based on learning control (ILC) theory and is aimed at uncertain linear/nonlinear systems with repeatable tasks and the final approach is based on fuzzy logic control (FLC) and is intended for highly uncertain systems with heuristic control knowledge.

In the presentation of the above control approaches, it is worthwhile highlighting that, unlike in continuous-time problems, robust control approaches characterized by high feedback gain are no longer suitable in discrete-time implementations due to the inherent stability property. As a consequence, low gain profiles are essential in discrete-time control. To meet the control requirement such as precision tracking when model uncertainties are present, it is necessary to explore more subtle or smart control approaches that are based on the underlying characters of system uncertainties.

In this monograph, we first present a disturbance estimation approach together with SMC. By making full use of discrete-time or sampled-data properties, a time-delay-based estimator is constructed to perform disturbance estimation, where the disturbance can be any exogenous uncertain factors. By virtue of the disturbance estimation that works as a kind of universal feed-forward compensation, low gain feedback is sufficient to warrant a precise tracking control performance.

Next, we present adaptive control approaches that can deal with parametric uncertainties. Both time invariant and time varying unknown parameters can be adaptively estimated so that a low feedback gain control can be employed to achieve generic tracking tasks.

Then, we present an iterative learning control approach that can significantly enhance tracking performance as far as the control task repeats. Learning not only from previous control tracking error profiles but also from previous control signals, the effect of system uncertainties, either parameterized or lumped, can be nullified completely when learning repeats iteratively. We particularly present ILC design and property analysis, which assures a monotonic tracking convergence along the iterative horizon. Monotonic tracking convergence is highly desired in practical control problems.

When heuristic control knowledge is available, fuzzy logic control is a suitable approach because it can easily incorporate heuristic knowledge through constructing an appropriate rule base. In this monograph, we present a fuzzy PID controller with a parallel structure, and implement an autotuning scheme according to classical gain and phase margins. In this way, the classical control design is well connected to the advanced FLC.

For each approach presented in the monograph, real-world-based examples are used as case studies to demonstrate the effectiveness and ease of implementation of the designed controllers. First, when introducing each control approach in a chapter, numerical examples are provided to illustrate the controller design guidelines, and the effectiveness of the control approach is compared with classical control methods. Next, we present a benchmark control task: precision control of a piezomotor-driven linear stage. Experimentally, SMC and ILC show excellent tracking performance with the achieved precision at micrometer or sub-micrometer scale. Finally, we apply advanced discrete-time control approaches to four representative engineering control problems, (1) Speed control of PM synchronous motor, a common engineering problem; (2) position control of ball and beam system, a typical motion control task; (3) level control of a coupled tank system, a process control task; and (4) ramp metering control of a freeway traffic system, a large-scale traffic control task.

Singapore, February 2015

Khalid Abidi
Jian-Xin Xu

Contents

Chapter 1
Introduction

1.1 Background

In recent years there has been a rapid increase in the use of digital controllers in control systems. Digital controls are used for achieving optimal performance, e.g., in the form of maximum productivity, maximum profit, minimum cost, or minimum energy use.

Most recently, the application of computer control has made possible "intelligent" motion in industrial robots, the optimization of fuel economy in automobiles, and the refinements in the operation of household appliances and machines such as microwaves and sewing machines, among others. Decision-making capability and flexibility in the control program are major advantages of digital control systems.

The current trend toward digital rather than analog control of dynamic systems is mainly due to the availability of low-cost digital computers and the advantages found in working with digital signals rather than continuous-time signals, [17, 59].

It is well known that most, if not all, engineering systems are continuous in nature. Owing to the capacity of digital computers to process discrete data, the continuous-time systems are controlled using sampled observations taken at discrete-time instants. Thus, the resulting control systems are a hybrid, consisting of interacting discrete and continuous components as depicted in Fig. 1.1. These hybrid systems, in which the system to be controlled evolves in continuous-time and the controller evolves in discrete-time, are called sampled-data systems.

The significant feature of sampled-data system design that distinguishes it from standard techniques for control system design is that it must contend with plant models and control laws lying in different domains. There are three major methodologies for design and analysis of sampled-data systems which are pictorially represented in Fig. 1.2 where G is a continuous-time process and K_{d} is a discrete-time control law. All three methods begin with the principle continuous-time model G and aim to design the discrete-time controller K_{d} and analyze its performance.

The two well known approaches follow the paths around the perimeter of the diagram. The first is to conduct all analysis and design in continuous-time domain

K. Abidi and J.-X. Xu, *Advanced Discrete-Time Control*, Studies in Systems, Decision and Control 23, DOI 10.1007/978-981-287-478-8_1

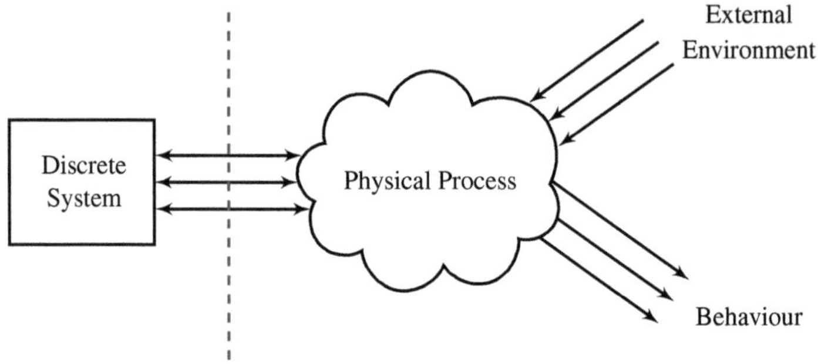

Fig. 1.1 General sampled-data arrangement

Fig. 1.2 Design approaches

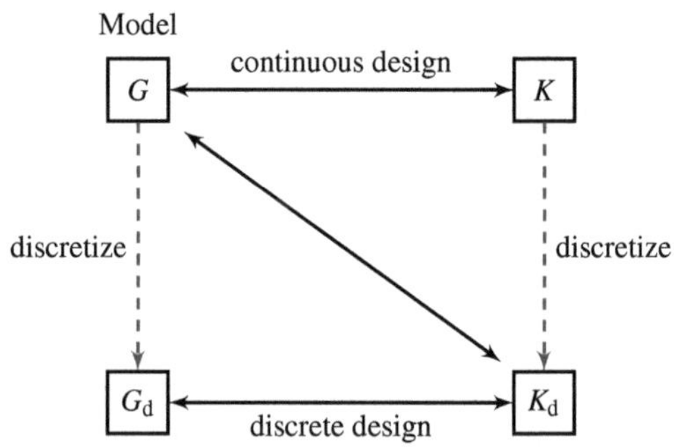

using a system that is believed to be a close approximation to the sampled-data system. This is accomplished by associating every continuous-time controller K with a discrete-time approximation K_d via discretization method; synthesis and analysis of the controller are then performed in continuous-time, with the underlying assumption that the closed-loop system behavior obtained controller K closely reflects that achieved with the sampled-data implementation K_d. Thus, this method does not directly address the issue of implementation in the design stage. The second approach starts instead by discretizing the continuous-time system G, giving a discrete-time approximation G_d, thus, ignoring intersample behavior. Then the controller K_d is designed directly in discrete-time using G_d, with the belief that the performance of this purely discrete-time system approximates that of the sampled-data system. The third approach has attracted considerable research activity. In this approach the system G and the controller K_d interconnection is treated directly and exactly. In this monograph we will, as much as possible, focus on this approach while in some cases use the second approach in order to more simply explain the controller design approach.

In the first study we focus on sliding mode control for sampled-data systems. Sliding mode control is well known in continuous-time control where it is characterized by high frequency switching which gives sliding mode control its very good robustness properties. This, however, is hard to achieve in sampled-data systems due

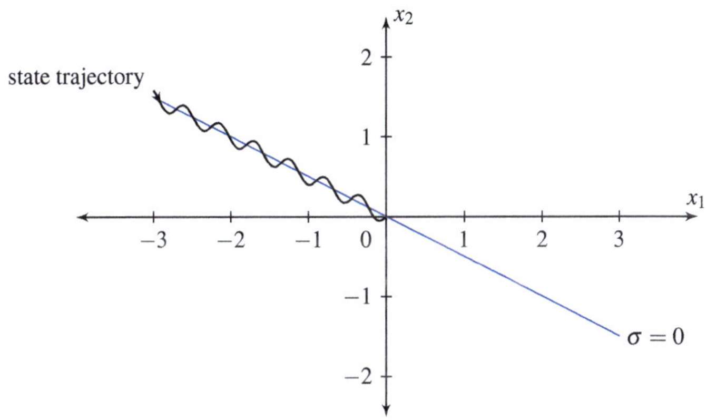

Fig. 1.3 Chattering phenomenon with switching sliding mode control

to hardware limitations such as processor speed, A/D and D/A conversion delays, etc. The use of discontinuous control under these circumstances would lead to the well known chattering phenomenon around the sliding manifold (Fig.1.3), leading to a boundary of order $O(T)$, [150]. In order to avoid this problem, in [150, 154] a discrete-time control equivalent in the prescribed boundary is proposed, whose size is defined by the restriction to the control variables. This approach results in motion within an $O(T^2)$ boundary around the sliding manifold. In this monograph we explore two different modifications to the classical sliding manifold that achieve better tracking performance than that in [150, 154].

In the second study we focus on discrete-time adaptive control for systems with uncertain periodically varying parameters. In [160] the author asks the following question: "Within the current framework of adaptive control, can we deal with time-varying parametric uncertainties?" This is a challenging problem to the control community. Adaptive algorithms have been reported for systems with slow time-varying parametric uncertainties, [9, 66, 115], etc., with arbitrarily rapid time-varying parameters in a known compact set, [152], and with rapid time-varying parameters which converge asymptotically to constants, [110]. However, as indicated in [110], no adaptive control algorithms developed hitherto can solve unknown parameters with arbitrarily fast and nonvanishing variations. Considering the fact that, as a function of time, the classes of time varying parameters are in essence infinite, it would be extremely difficult to find a general solution to such a broad control problem. A more realistic way is first to classify the time-varying parametric uncertainties into subclasses, and then look for an appropriate adaptive control approach for each subclass. Instead of classifying parameters into slow versus rapid time-varying, in this work we classify parameters into periodic versus nonperiodic ones. When the periodicity of system parameters is known a priori, a new adaptive controller with periodic updating can be constructed by means of a pointwise integral mechanism. This method is proposed in [160] for continuous-time systems. As a natural extension to this we propose a similar methodology for discrete-time systems.

In the third study we focus on adaptive posicast control for uncertain time-delay systems. The basic premise behind adaptive posicast controller (APC), [170–175], is the use of the plant model to predict the future states and, thereby, cancel the effects of the time-delay. APC has been implemented successfully with considerable performance improvements, [52]. However, the premise of time-delay compensation using future output prediction, as proven by the theory, had to be approximately realized in these applications. The main reason being that the controller had to be implemented using a microprocessor and, therefore, all the terms in the control law had to be digitally approximated. This is a standard approach to many controller implementations and in most of the cases works perfectly well as long as the sampling is fast enough. One exception to this rule is the implementation of the finite spectrum assignment (FSA) controller. It is shown in [156] that, as the sampling frequency increases, the phase margin of the FSA controller decreases. A remedy to this problem is provided in [116]. Since APC is based on FSA controller, fast sampling to achieve good approximation of the continuous control laws may degrade the system performance. To eliminate the need for approximation and, therefore, to exploit the full benefits of APC, a fully discrete time APC design is presented in this monograph.

In the fourth study we focus on iterative learning control for sampled-data systems. Iterative learning control (ILC) is based on the idea that the performance of a system that executes the same task multiple times can be improved by learning from previous executions (trials, iterations, passes). When letting a machine do the same task repeatedly it is, at least from an engineering point of view, very sound to use knowledge from previous iterations of the same task to try to reduce the error next time the task is performed. The first academic contribution to what today is called ILC appears to be a paper by Uchiyama [153]. Since it was published in Japanese only, the ideas did not become widely spread. What is a bit remarkable, however, is that an application for a US patent on 'Learning control of actuators in control systems', [64], was already done in 1967 and that it was accepted as a patent in 1971. The idea in the patent is to store a 'command signal' in a computer memory and iteratively update this command signal using the error between the actual response and the desired response of the actuator. This is clearly an implementation of ILC, although the actual ILC updating equation was not explicitly formulated in the patent. From an academic perspective it was not until 1984 that ILC started to become an active research area. In this study we present a framework for linear iterative control, which enables several results from linear control theory to be applied.

In the final study we focus on fuzzy PID controllers. Conventional proportional-integral-derivative (PID) controllers have been well developed and applied for many decades, and are the most extensively used controllers in industrial automation and process control. The main reason being their simplicity of implementation, ease of design, and effectiveness for most linear systems. Motivated by the rapidly developing advanced microelectronics and digital processors technologies, conventional PID controllers have undergone a technological evolution, from pneumatic controllers via analog electronics to microprocessors via digital circuits. However, it is well known that conventional PID controllers generally do not perform very well for nonlinear systems, higher order and time-delayed linear systems, and particularly complex and uncertain systems that do not have precise mathematical models. To overcome these difficulties, various types of modified conventional PID controllers such as autotuning and adaptive PID controllers were developed lately, [13, 14]. In this study we present a discrete-time fuzzy PID controller design with the aim of highlighting the ease of implementation and the superior performance with respect to conventional PID.

1.2 Contributions

The contributions of this monograph can be summarized as follows:

(1) Discrete-Time Sliding Mode Control

In this study, two different approaches to the sliding surface design are presented. First, a discrete-time integral sliding mode (ISM) control scheme for sampled-data systems is discussed. The control scheme is characterized by a discrete-time integral switching surface which inherits the desired properties of the continuous-time integral switching surface, such as full order sliding manifold with eigenvalue assignment, and elimination of the reaching phase. In particular, comparing with existing discrete-time sliding mode control, the scheme is able to achieve more precise tracking performance. It will be shown that, the control scheme achieves $O(T^2)$ steady-state error for state regulation and reference tracking with the widely adopted delay-based disturbance estimation. Another desirable feature is that the Discrete-time ISM control prevents the generation of overlarge control actions which are usually inevitable due to the deadbeat poles of a reduced order sliding manifold designed for sampled data systems.

Second, a terminal sliding mode control scheme is discussed. Terminal Sliding Mode (TSM) control is known for its high gain property nearby the vicinity of the equilibrium while retaining reasonably low gain elsewhere. This is desirable in digital implementation where the limited sampling frequency may incur chattering if the controller gain is overly high. The overall sliding surface integrates a linear switching surface with a terminal switching surface. The switching surface can be designed according to the precision requirement. The design is implemented on a specific SISO system example, but, the approach can be used in exactly the same way for any other system as long as it is SISO. The analysis and experimental investigation show that the TSM controller design outperforms the linear SM control.

(2) Discrete-Time Periodic Adaptive Control

In this study a periodic adaptive control approach is discussed for a class of nonlinear discrete-time systems with time-varying parametric uncertainties which are periodic, and the only prior knowledge is the periodicity. The adaptive controller updates the parameters and the control signal periodically in a pointwise manner over one entire period, in the sequel achieves the asymptotic tracking convergence. The result is further extended to a scenario with mixed time-varying and time-invariant parameters, and a hybrid classical and periodic adaptation law is proposed to handle the scenario more appropriately. Extension of the periodic adaptation to systems with unknown input gain, higher order dynamics, and tracking problems are also discussed.

(3) Discrete-Time Adaptive Posicast Control

In this study, we discuss the discrete version of the Adaptive Posicast Controller (APC) that deals with parametric uncertainties in systems with input time-delays. The continuous-time APC is based on the Smith Predictor and Finite Spectrum Assignment with time-varying parameters adjusted online. Although the continuous-time APC showed dramatic performance improvements in experimental studies with internal combustion engines, the full benefits could not be realized since the finite integral term in the control law had to be approximated in computer implementation. It is shown in the literature that integral approximation in time-delay compensating controllers degrades the performance if care is not taken. In this study, we discuss a development of the APC in the discrete-time domain, eliminating the need for approximation. Rigorous and complete derivation is provided with a Lyapunov stability proof. The discussed discrete-time APC is developed in State Space to easily accomodate multivariable systems and also allow for the extension to nonlinear systems. In essence, this study presents a unified development of the discrete-time APC for systems that are linear/nonlinear with known input time-delays or linear systems with unknown but upper-bounded time-delays. Performances of the continuous-time and discrete-time APC, as well as conventional Model Reference Adaptive Controller (MRAC) for linear systems with known time-delay are compared in simulation studies. It is shown that discrete-time APC outperforms it's continuous-time counterpart and MRAC. Further simulations studies are also presented to show the performance of the design for nonlinear systems and also for systems with unknown time-delay.

(4) Discrete-Time Iterative Learning Control

In this study the convergence properties of iterative learning control (ILC) algorithms are discussed. The analysis is carried out in a framework using linear iterative systems, which enables several results from the theory of linear systems to be applied. This makes it possible to analyse both first-order and high-order ILC algorithms in both the time and frequency domains. The time and frequency domain results can also be tied together in a clear way. Illustrative examples are presented to support the analytical results.

(5) Discrete-Time Fuzzy PID Control

In this study, a parallel structure of fuzzy PID control systems is presented. It is associated with a new tuning method which, based on gain margin and phase margin specifications, determines the parameters of the fuzzy PID controller. In comparison with conventional PID controllers, the presented fuzzy PID controller shows higher control gains when system states are away from equilibrium and, at the same time, retains lower profile of control signals. Consequently better control performance is achieved. With the presented formula, the weighting factors of a fuzzy logic controller can be systematically selected according to the plant under control. By virtue of using the simplest structure of fuzzy logic control, the stability of the nonlinear control system is able to be analyzed and a sufficient BIBO stability condition is given. The superior performance of the controller is demonstrated through both numerical and experimental examples.

1.3 Organization

The monograph is organized as follows.

In Chap. 2, we discuss the Discrete-Time Integral Sliding Mode Control and Terminal Sliding Mode Control for Sampled-Data systems. Section 2.2 gives the problem formulation. Section 2.3 revisits the existing SMC properties in sampled-data systems. Section 2.4 discusses the appropriate discrete-time integral. Section 2.5 discusses the discrete-time terminal sliding mode manifold design for a 2nd order SISO system.

In Chap. 3, we discuss the Discrete-Time Periodic Adaptive Control Approach for Time-Varying Parameters with Known Periodicity. In Sect. 3.2, we discuss the new periodic adaptive control approach and give complete analysis. To clearly demonstrate the underlying idea and method, we consider the simplest nonlinear dynamics with a single time-varying parameter. In Sect. 3.3, the extension to more general cases is discussed. The first extension considers multiple time-varying parameters and time-varying gain of the system input. The second extension considers a mixture of time-varying and time-invariant parameters, and a new hybrid adaptive control scheme is developed. The third extension considers a general tracking control problem. The fourth extension considers a higher order system in canonical form. In Sect. 3.4, an illustrative example is provided.

In Chap. 4, we discuss the Discrete-Time Adaptive Posicast Control for Time-Delay Systems. Section 4.2 gives the problem statement. Section 4.3 gives the Discrete-Time Adaptive Posicast Controller Design. Section 4.4 gives the Extension to More General Cases. Section 4.5 gives the Illustrative Examples.

In Chap. 5, we discuss Iterative Learning Control for Sampled-Data systems. In Sect. 5.3, we discuss the time domain analysis of different ILC. In Sect. 5.4, we analyze the same ILC laws in the frequency domain and highlight the connection between the time domain and frequency domain results. In Sect. 5.5 a special case is presented where ILC is combined with Multirate technique. In Sects. 5.6 and 5.7, illustrative examples are provided to support the results in each domain.

In Chap. 6, we discuss the Discrete-Time Fuzzy PID Control. In Sect. 6.2 the fuzzy PID controller constructed by the parallel combination of fuzzy PI and fuzzy PD controllers and its tuning formula is first discussed. In Sect. 6.3 the stability condition and the property of the fuzzy PID controller are then studied. In Sect. 6.4 a number of illustrative examples are presented which demonstrate better control performance of the proposed fuzzy PID controller.

In Chap. 7, we present a practical application for the discussed control laws. The aim is to design control laws that would achieve high-precision motion of a piezo-motor driven linear stage. In Sect. 7.2 we describe the model of the piezo-motor. In Sect. 7.3 we present the ISM design and in Sect. 7.4 we present the TSM design. Finally, in Sect. 7.5 we present the ILC design.

In Chap. 8, we present further engineering applications. In Sect. 8.2 the Periodic Adaptive Control of a PM Synchronous Motor is presented. In Sect. 8.3 the Multirate ILC of a Ball and Beam System is presented. In Sect. 8.4 the discrete-time Fuzzy PID of a Coupled Tank System is presented. Finally, in Sect. 8.5 the ILC for Freeway Traffic Control is presented.

Throughout this monograph, $\| \cdot \|$ denotes the Euclidean Norm. For notational convenience, in mathematical expressions f_k represents $f(k)$.

Chapter 2
Discrete-Time Sliding Mode Control

Abstract In this study, two different approaches to the sliding surface design are presented. First, a discrete-time integral sliding mode (ISM) control scheme for sampled-data systems is discussed. The control scheme is characterized by a discrete-time integral switching surface which inherits the desired properties of the continuous-time integral switching surface, such as full order sliding manifold with eigenvalue assignment, and elimination of the reaching phase. In particular, comparing with existing discrete-time sliding mode control, the scheme is able to achieve more precise tracking performance. It will be shown that, the control scheme achieves $O(T^2)$ steady-state error for state regulation and reference tracking with the widely adopted delay-based disturbance estimation. Another desirable feature is that the Discrete-time ISM control prevents the generation of overlarge control actions which are usually inevitable due to the deadbeat poles of a reduced order sliding manifold designed for sampled-data systems. Second, a terminal sliding mode control scheme is discussed. Terminal Sliding Mode (TSM) control is known for its high gain property nearby the vicinity of the equilibrium while retaining reasonably low gain elsewhere. This is desirable in digital implementation where the limited sampling frequency may incur chattering if the controller gain is overly high. The overall sliding surface integrates a linear switching surface with a terminal switching surface. The switching surface can be designed according to the precision requirement. The design is implemented on a specific SISO system example, but, the approach can be used in exactly the same way for any other system as long as it is SISO. The analysis and experimental investigation show that the TSM controller design outperforms the linear SM control.

2.1 Introduction

Research in discrete-time control has been intensified in recent years. A primary reason is that most control strategies nowadays are implemented in discrete-time. This also necessitated a rework in the sliding mode (SM) control strategy for sampled-data systems, [150, 154]. In such systems, the switching frequency in control variables is limited by T^{-1}; where T is the sampling period. This has led researchers to approach

K. Abidi and J.-X. Xu, *Advanced Discrete-Time Control*, Studies in Systems, Decision and Control 23, DOI 10.1007/978-981-287-478-8_2

discrete-time sliding mode control from two directions. The first is the emulation that focuses on how to map continuous-time sliding mode control to discrete-time, and the switching term can be preserved, [62, 67]. The second is based on the equivalent control design and disturbance observer, [150, 154]. In the former, although high-frequency switching is theoretically desirable from the robustness point of view, it is usually hard to achieve in practice because of physical constraints, such as processor computational speed, A/D and D/A conversion delays, actuator bandwidth, etc. The use of a discontinuous control law in a sampled-data system will bring about chattering phenomenon in the vicinity of the sliding manifold, hence lead to a boundary layer with thickness $O(T)$, [150].

The effort to eliminate the chattering has been paid over 30 years. In continuous-time SM control, smoothing schemes such as boundary layer (saturator) are widely used, which in fact results in a continuous nonlinear feedback instead of switching control. Nevertheless, it is widely accepted by the community that this class of controllers can still be regarded as SM control. Similarly, in discrete-time SM control, by introducing a continuous control law, chattering can be eliminated. In such circumstance, the central issue is to guarantee the precision bound or the smallness of the error.

In [154] a discrete-time equivalent control was proposed. This approach results in the motion in $O(T^2)$ vicinity of the sliding manifold. The main difficulty in the implementation of this control law is that we need to know the disturbances for calculating the equivalent control. Lack of such information leads to an $O(T)$ error boundary.

The control proposed in [150] drives the sliding mode to $O(T^2)$ in one-step owing to the incorporation of deadbeat poles in the closed-loop system. State regulation was not considered in [150]. In fact, as far as the state regulation is concerned, the same SM controller design will produce an accuracy in $O(T)$ instead of $O(T^2)$ boundary. Moreover, the SM control with deadbeat poles requires large control efforts that might be undesirable in practice. Introducing saturation in the control input endangers the global stability or accuracy of the closed-loop system.

In this chapter we begin by discussing a discrete-time integral sliding manifold (ISM). With the full control of the system closed-loop poles and the elimination of the reaching phase, like the continuous-time integral sliding mode control, [40, 60, 155], the closed-loop system can achieve the desired control performance while avoiding the generation of overly large control inputs. It is worth highlighting that the discrete-time ISM control does not only drive the sliding mode into the $O(T^2)$ boundary, but also achieve the $O(T^2)$ boundary for state regulation.

After focusing on state feedback based ISM regulation, we consider the situation where output tracking and output feedback is required. Based on output feedback two approaches arose—design based on obervers to construct the missing states, [55, 177], or design based on the output measurement only, [54, 146]. Recently integral sliding-mode control has been developed to improve controller design and consequently the control performance, [40, 60, 155], which use full state information. The first objective of this work is to extend ISM control to output-tracking problems.

We present three ISM control design approaches associated with state feedback, output feedback, and output feedback with state estimation, respectively.

After the discussion on ISM is concluded a second approach to discrete-time Sliding Mode control is presented. The approach is called terminal sliding mode control and has been developed in [150] to achieve finite time convergence of the system dynamics in the terminal sliding mode. In [3, 43, 60], the first-order terminal sliding mode control technique is developed for the control of a simple second-order nonlinear system and an 4th order nonlinear rigid robotic manipulator system with the result that the output tracking error can converge to zero in finite time.

Most of the Terminal Sliding Mode (TSM) approaches have been developed from the continuous-time point of view, [40, 150, 154, 155], however less work exists in the design from the discrete-time point of view. In [67] a continuous-time terminal sliding mode controller is first discretized and then applied to a sampled-data system. While it is possible to achieve acceptable result via this approach it makes more sense if the design was tackled entirely from the discrete-time point of view. This would allow us to gain more insight on the performance and stability issues and, thereby, achieve the best possible performance. In this paper, a revised terminal sliding mode control law is developed from the discrete-time point of view. It is shown that the new method can achieve better performance than with the linear SM control owing to the high gain property of the Terminal Sliding Mode in the vicinity of the origin.

In each of the above approaches, robustness is enhanced by the use of disturbance observers or estimators. Different disturbance observer approaches will be presented depending on the control problem. When the system states are accessible, the disturbance can be directly estimated using state and control signals delayed by one sampling period. The resulting control can perform arbitrary trajectory tracking or state regulation with $O(T^2)$ accuracy. When only outputs are accessible, the delayed disturbance estimation cannot be performed. In that case a dynamic disturbance observer design based on an unconventional Sliding Mode approach will be used. It will be shown that the SM observer can quickly and effectively estimate the disturbance and avoid the undesirable deadbeat response inherent in conventional SM based designs for sampled-data systems; in the sequel, avoid the generation of overly large estimation signals in the controller.

2.2 Problem Formulation

Consider the following continuous-time system with a nominal linear time invariant model and matched disturbance

$$\begin{aligned}\dot{\mathbf{x}}(t) &= A\mathbf{x}(t) + B\mathbf{u}(t) + B\mathbf{f}(\mathbf{x}, t)\\ \mathbf{y}(t) &= C\mathbf{x}(t)\end{aligned} \tag{2.1}$$

where the state $\mathbf{x}(t) \in \Re^n$, the output $\mathbf{y}(t) \in \Re^m$, the control $\mathbf{u}(t) \in \Re^m$, and the disturbance $\mathbf{f}(\mathbf{x}, t) \in \Re^m$. The state matrix is $A \in \Re^{n\times n}$ and the control matrix is $B \in \Re^{n\times m}$ and the output matrix is $C \in \Re^{m\times n}$. For the system (2.1) the following assumptions are made:

Assumption 2.1 The pair (A, B) is Controllable and the pair (A, C) is Observable.

Assumption 2.2 Controllability and Observability is not lost upon sampling.

Assumption 2.3 The disturbance $\mathbf{f}(\mathbf{x}, t) \equiv \mathbf{f}(t)$ is smooth and uniformly bounded.

Proceeding further, the discretized counterpart of (2.1) can be given by

$$\begin{aligned} \mathbf{x}_{k+1} &= \Phi\mathbf{x}_k + \Gamma\mathbf{u}_k + \mathbf{d}_k, \quad \mathbf{x}_0 = \mathbf{x}(0) \\ \mathbf{y}_k &= C\mathbf{x}_k \end{aligned} \tag{2.2}$$

where

$$\Phi = e^{AT}, \quad \Gamma = \int_0^T e^{A\tau} d\tau B$$
$$\mathbf{d}_k = \int_0^T e^{A\tau} B\mathbf{f}((k+1)T - \tau)\, d\tau$$

and T is the sampling period. Here the disturbance $\mathbf{d}_k$ represents the influence accumulated from kT to $(k+1)T$, in the sequel it shall directly link to $\mathbf{x}_{k+1} = \mathbf{x}((k+1)T)$. From the definition of Γ it can be shown that

$$\Gamma = BT + \frac{1}{2!}ABT^2 + \cdots = BT + MT^2 + O\left(T^3\right) \Rightarrow BT = \Gamma - MT^2 + O\left(T^3\right) \tag{2.3}$$

where M is a constant matrix because T is fixed.

The control objective is to design a discrete-time sliding manifold and a discrete-time SM control law for the sampled-data system (2.2), hence achieve as precisely as possible state regulation or output tracking.

Remark 2.1 The smoothness assumption made on the disturbance is to ensure that the disturbance bandwidth is sufficiently lower than the controller bandwidth, or the ignorance of high frequency components does not significantly affect the control performance. Indeed, if a disturbance has frequencies nearby or higher than the Nyquist frequency, for instance a non-smooth disturbance, a discrete-time SM control will not be able to handle it.

In order to proceed further, the following definition is necessary:

Definition 2.1 The magnitude of a variable v is said to be $O(T^r)$ if, and only if, there is a $C > 0$ such that for any sufficiently small T the following inequality holds

$$|v| \leq CT^r$$

where r is an integer. Denote $O(T^0) = O(1)$.

Remark 2.2 Note that $O(T^r)$ can be a scalar function or a vector valued function.

Associated with the above definition if there exists two variables v_1 and v_2 such that $v_1 \in O(T^r)$ and $v_2 \in O(T^{r+1})$ then $v_1 \gg v_2$ and, therefore, the following relations hold

$$O\left(T^{r+1}\right) + O\left(T^r\right) = O\left(T^r\right) \quad \forall r \in Z$$

$$O\left(T^r\right) \cdot O(1) = O\left(T^r\right) \quad \forall r \in Z$$

$$O\left(T^r\right) \cdot O\left(T^{-s}\right) = O\left(T^{r-s}\right) \quad \forall r, s \in Z$$

where Z is the set of integers.

Based on (2.3) and the Definition, the magnitude of Γ is $O(T)$. Note that, as a consequence of sampling, the disturbance originally matched in continuous-time will contain mismatched components in the sampled-data system. This is summarized in the following lemma:

Lemma 2.1 *If the disturbance* $\mathbf{f}(t)$ *in (2.1) is bounded and smooth, then*

$$\mathbf{d}_k = \int_0^T e^{A\tau} B\mathbf{f}\left((k+1)T - \tau\right) d\tau = \Gamma \mathbf{f}_k + \frac{1}{2}\Gamma \mathbf{v}_k T + O\left(T^3\right) \tag{2.4}$$

where $\mathbf{v}_k = \mathbf{v}(kT)$, $\mathbf{v}(t) = \frac{d}{dt}\mathbf{f}(t)$, $\mathbf{d}_k - \mathbf{d}_{k-1} \in O\left(T^2\right)$, and $\mathbf{d}_k - 2\mathbf{d}_{k-1} + \mathbf{d}_{k-2} \in O\left(T^3\right)$.

Proof Consider the Taylor's series expansion of $\mathbf{f}((k+1)T - \tau)$

$$\mathbf{f}(kT + T - \tau) = \mathbf{f}_k + \mathbf{v}_k(T-\tau) + \frac{1}{2!}\mathbf{w}_k(T-\tau)^2 + \cdots = \mathbf{f}_k + \mathbf{v}_k(T-\tau) + \boldsymbol{\xi}(T-\tau)^2 \tag{2.5}$$

where $\mathbf{v}(t) = \frac{d}{dt}\mathbf{f}(t)$, $\mathbf{w}(t) = \frac{d^2}{dt^2}\mathbf{f}(t)$ and $\boldsymbol{\xi} = \frac{1}{2!}\mathbf{w}(\mu)$ and μ is a time value between kT and $(k+1)T$, [43]. Substituting (2.5) into the expression of $\mathbf{d}_k$

$$\mathbf{d}_k = \int_0^T e^{A\tau}\mathbf{B}\mathbf{f}_k d\tau + \int_0^T e^{A\tau} B\mathbf{v}_k(T-\tau)d\tau + \int_0^T e^{A\tau} B\boldsymbol{\xi}(T-\tau)^2 d\tau. \tag{2.6}$$

For clarity, each integral will be analyzed separately. Since $\mathbf{f}_k$ is independent of τ it can be taken out of the first integral

$$\int_0^T e^{A\tau} B\mathbf{f}_k d\tau = \int_0^T e^{A\tau} B d\tau \mathbf{f}_k = \Gamma \mathbf{f}_k. \tag{2.7}$$

In order to solve the second integral term, it is necessary to expand $e^{A\tau}$ into series form. Thus,

$$\int_0^T e^{A\tau} B\mathbf{v}_k(T-\tau)d\tau = \int_0^T \left[e^{A\tau} B - \left(B + AB\tau + \frac{1}{2!}A^2B\tau^2 + \cdots \right)\tau \right] d\tau \mathbf{v}_k. \tag{2.8}$$

Solving the integral leads to

$$\int_0^T e^{A\tau} B\mathbf{v}_k(T-\tau)d\tau = \left[\Gamma - \left(\frac{1}{2!}BT + \frac{1}{3!}ABT^2 + \cdots \right) \right] T\mathbf{v}_k. \tag{2.9}$$

Simplifying the result with the aid of (2.3)

$$\int_0^T e^{A\tau} B\mathbf{v}_k(T-\tau)d\tau = \left[\Gamma - \frac{1}{2}\Gamma + \frac{1}{2}MT^2 - \left(\frac{1}{3!}ABT^2 + \frac{1}{4!}A^2BT^2 + \cdots \right) \right] T\mathbf{v}_k. \tag{2.10}$$

Simplifying the above expression further

$$\int_0^T e^{A\tau} B\mathbf{v}_k(T-\tau)d\tau = \frac{1}{2}\Gamma \mathbf{v}_k T + \hat{M}\mathbf{v}_k T^3 \tag{2.11}$$

where $\hat{M}$ is a constant matrix. Finally, note that in (2.6) the third integral is $O(T^3)$, since, the term inside the integral is already $O(T^2)$, therefore

$$\int_0^T e^{A\tau} B\boldsymbol{\xi}(T-\tau)^2 d\tau = O(T^3). \tag{2.12}$$

Thus, combining (2.7), (2.10) and (2.12) leads to

$$\mathbf{d}_k = \Gamma \mathbf{f}_k + \frac{1}{2}\Gamma \mathbf{v}_k T + \hat{M}T^3 \mathbf{v}_k + O(T^3) = \Gamma \mathbf{f}_k + \frac{1}{2}\Gamma \mathbf{v}_k T + O(T^3). \tag{2.13}$$

Now evaluate

$$\mathbf{d}_k - \mathbf{d}_{k-1} = \Gamma(\mathbf{f}_k - \mathbf{f}_{k-1}) + \frac{1}{2}\Gamma(\mathbf{v}_k - \mathbf{v}_{k-1})T + O(T^3). \tag{2.14}$$

From (2.5) and letting $\tau = 0, \mathbf{f}_k - \mathbf{f}_{k-1} \in O(T)$. From (2.3), $\Gamma \in O(T)$. In the sequel $\mathbf{d}_k - \mathbf{d}_{k-1} \in O(T^2)$, if the assumptions on the boundedness and smoothness of $\mathbf{f}(t)$

hold. Finally, we notice that (2.14) is the difference of the first order approximation, whereas

$$\mathbf{d}_k - 2\mathbf{d}_{k-1} + \mathbf{d}_{k-2} = \Gamma(\mathbf{f}_k - 2\mathbf{f}_{k-1} + \mathbf{f}_{k-2}) + \frac{1}{2}\Gamma(\mathbf{v}_k - 2\mathbf{v}_{k-1} + \mathbf{v}_{k-2}) + O(T^3) \tag{2.15}$$

is the difference of the second order approximation. Accordingly the magnitude of $\mathbf{d}_k - 2\mathbf{d}_{k-1} + \mathbf{d}_{k-2}$ is $O\left(T^3\right)$, [43].

Note that the magnitude of the mismatched part in the disturbance $\mathbf{d}_k$ is of the order $O\left(T^3\right)$.

2.3 Classical Discrete-Time Sliding Mode Control Revisited

2.3.1 State Regulation

Consider the well established discrete-time sliding surface, [150, 154], shown below

$$\boldsymbol{\sigma}_k = D\mathbf{x}_k \tag{2.16}$$

where $\boldsymbol{\sigma}_k \in \Re^m$ and D is a constant matrix of rank m. The objective is to steer the states towards and force them to stay on the sliding manifold $\boldsymbol{\sigma}_k = 0$ at every sampling instant. The control accuracy of this class of sampled-data SM control is given by the following lemma:

Lemma 2.2 *With $\boldsymbol{\sigma}_k = D\mathbf{x}_k$ and equivalent control based on a disturbance estimate*

$$\hat{\mathbf{d}}_k = \mathbf{x}_k - \Phi\mathbf{x}_{k-1} - \Gamma\mathbf{u}_{k-1}$$

then there exists a matrix D such that the control accuracy of the closed-loop system is

$$\lim_{k\to\infty} \|\mathbf{x}_k\| \le O\left(T\right).$$

Proof Discrete-time equivalent control is defined by solving $\boldsymbol{\sigma}_{k+1} = 0$, [150]. This leads to

$$\mathbf{u}_k^{\mathrm{eq}} = -\left(D\Gamma\right)^{-1} D\left(\Phi\mathbf{x}_k + \mathbf{d}_k\right) \tag{2.17}$$

with D selected such that the closed-loop system achieves desired performance and $D\Gamma$ is invertible, [3]. Under practical considerations, the control cannot be implemented in the same form as in (2.33) because of the lack of prior knowledge regarding the discretized disturbance $\mathbf{d}_k$. However, with some continuity assumptions on the disturbance, $\mathbf{d}_k$ can be estimated by its previous value $\mathbf{d}_{k-1}$, [150]. The substitution of $\mathbf{d}_k$ by $\mathbf{d}_{k-1}$ will at most result in an error of $O\left(T^2\right)$. With reasonably small sampling period as in motion control or mechatronics, such a substitution will be effective. Let

$$\hat{\mathbf{d}}_k = \mathbf{d}_{k-1} = \mathbf{x}_k - \Phi \mathbf{x}_{k-1} - \Gamma \mathbf{u}_{k-1} \tag{2.18}$$

where $\hat{\mathbf{d}}_k$ is the estimate of $\mathbf{d}_k$. Thus, analogous to the equivalent control law (2.33), the practical control law is

$$\mathbf{u}_k = -\left(D\Gamma\right)^{-1} D\left(\Phi \mathbf{x}_k + \mathbf{d}_{k-1}\right). \tag{2.19}$$

Substituting the sampled-data dynamics (2.2), applying the above control law, and using the conclusions in *Lemma* 2.1, yield

$$\boldsymbol{\sigma}_{k+1} = D\left(\Phi \mathbf{x}_k + \Gamma \mathbf{u}_k + \mathbf{d}_k\right) = D\left(\mathbf{d}_k - \mathbf{d}_{k-1}\right) = O\left(T^2\right) \tag{2.20}$$

which is the result shown in [150]. The closed-loop dynamics is

$$\mathbf{x}_{k+1} = \left(\Phi - \Gamma\left(D\Gamma\right)^{-1} D\Phi\right)\mathbf{x}_k + \left(I - \Gamma\left(D\Gamma\right)^{-1} D\right)\mathbf{d}_{k-1} + \mathbf{d}_k - \mathbf{d}_{k-1} \tag{2.21}$$

where the matrix $\left(\Phi - \Gamma\left(D\Gamma\right)^{-1} D\Phi\right)$ has m zero eigenvalues and $n-m$ eigenvalues to be assigned inside the unit circle in the complex *z-plane*. It is possible to simplify (2.21) further to

$$\mathbf{x}_{k+1} = \left(\Phi - \Gamma\left(D\Gamma\right)^{-1} D\Phi\right)\mathbf{x}_k + \boldsymbol{\delta}_k \tag{2.22}$$

where $\boldsymbol{\delta}_k = \left(I - \Gamma\left(D\Gamma\right)^{-1} D\right)\mathbf{d}_{k-1} + \mathbf{d}_k - \mathbf{d}_{k-1}$. From *Lemma* 2.1,

$$\begin{aligned}\boldsymbol{\delta}_k &= \mathbf{d}_k - \mathbf{d}_{k-1} + \left(I - \Gamma\left(D\Gamma\right)^{-1} D\right)\left(\Gamma \mathbf{f}_{k-1} + \frac{1}{2}\Gamma \mathbf{v}_{k-1} T + O\left(T^3\right)\right)\\ &= O\left(T^2\right) + \left(I - \Gamma\left(D\Gamma\right)^{-1} D\right) O\left(T^3\right) = O\left(T^2\right).\end{aligned} \tag{2.23}$$

In the preceeding derivation, we use the relations $\left(I - \Gamma\left(D\Gamma\right)^{-1} D\right)\Gamma = 0$, $\|I - \Gamma\left(D\Gamma\right)^{-1} D\| \le 1$ and $O\left(1\right) \cdot O\left(T^3\right) = O\left(T^3\right)$. Note that since m eigenvalues of the matrix $\left(\Phi - \Gamma\left(D\Gamma\right)^{-1} D\Phi\right)$ are deadbeat, it can be written as

$$\left(\Phi - \Gamma\left(D\Gamma\right)^{-1} D\Phi\right) = PJP^{-1} \tag{2.24}$$

where P is a transformation matrix and J is the Jordan matrix of the eigenvalues of $\left(\Phi - \Gamma (D\Gamma)^{-1} D\Phi\right)$. The matrix J can be written as

$$J = \begin{bmatrix} J_1 & \mathbf{0} \\ \mathbf{0} & J_2 \end{bmatrix} \tag{2.25}$$

where $J_1 \in \Re^{m\times m}$ and $J_2 \in \Re^{(n-m)\times(n-m)}$ and are given by

$$J_1 = \begin{bmatrix} \mathbf{0} & I_{m-1} \\ 0 & \mathbf{0} \end{bmatrix} \quad \& \quad J_2 = \begin{bmatrix} \lambda_{m+1} & 0 & \cdots & 0 \\ 0 & \ddots & \ddots & \vdots \\ \vdots & \ddots & \ddots & 0 \\ 0 & \cdots & 0 & \lambda_n \end{bmatrix}$$

where λ_j are the eigenvalues of $\left(\Phi - \Gamma (D\Gamma)^{-1} D\Phi\right)$. For simplicity it is assumed that the non-zero eigenvalues are designed to be distinct and that their continuous time counterparts are of order $O(1)$. Then the solution of (2.22) is

$$\mathbf{x}_k = PJ^kP^{-1}\mathbf{x}_0 + P\left(\sum_{i=0}^{k-1} J^iP^{-1}\delta_{k-i-1}\right). \tag{2.26}$$

Rewriting (2.26) as

$$\mathbf{x}_k = PJ^kP^{-1}\mathbf{x}_0 + P\left(\sum_{i=0}^{k-1}\begin{bmatrix} J_1^i & \mathbf{0} \\ \mathbf{0} & \mathbf{0} \end{bmatrix} P^{-1}\delta_{k-i-1}\right) + P\left(\sum_{i=0}^{k-1}\begin{bmatrix} \mathbf{0} & \mathbf{0} \\ \mathbf{0} & J_2^i \end{bmatrix} P^{-1}\delta_{k-i-1}\right) \tag{2.27}$$

it is easy to verify that $J_1^i = \mathbf{0}$ for $i \geq m$. Thus, (2.27) becomes (for $k \geq m$)

$$\mathbf{x}_k = PJ^kP^{-1}\mathbf{x}_0 + P\left(\sum_{i=0}^{m}\begin{bmatrix} J_1^i & \mathbf{0} \\ \mathbf{0} & \mathbf{0} \end{bmatrix} P^{-1}\delta_{k-i-1}\right) + P\left(\sum_{i=0}^{k-1}\begin{bmatrix} \mathbf{0} & \mathbf{0} \\ \mathbf{0} & J_2^i \end{bmatrix} P^{-1}\delta_{k-i-1}\right). \tag{2.28}$$

Notice $\|J_1\| = 1$ and $\|J_2\| = \lambda_{\max} = \max\{\lambda_{m+1}, \ldots, \lambda_n\}$. Hence, from (2.28)

$$\lim_{k\to\infty}\|\mathbf{x}_k\| \leq \|P\|\left(\sum_{i=0}^{m}\left\|\begin{bmatrix} J_1 & \mathbf{0} \\ \mathbf{0} & \mathbf{0} \end{bmatrix}\right\|^i \|P^{-1}\|\cdot\|\delta_{k-i-1}\| + \sum_{i=0}^{k-1}\left\|\begin{bmatrix} \mathbf{0} & \mathbf{0} \\ \mathbf{0} & J_2 \end{bmatrix}\right\|^i \|P^{-1}\|\cdot\|\delta_{k-i-1}\|\right). \tag{2.29}$$

Since $\lambda_{\max} < 1$ for a stable system,

$$\sum_{i=0}^{\infty} \|J_2\|^i = \frac{1}{1-\lambda_{\max}} \quad \& \quad \sum_{i=0}^{m} \|J_1\|^i = m.$$

Using Tustin's approximation

$$\lambda_{\max} = \frac{2+Tp}{2-Tp} \Rightarrow \frac{1}{1-\lambda_{\max}} = \frac{1}{1-\frac{2+Tp}{2-Tp}} = \frac{2-Tp}{-2Tp} \leq O\left(T^{-1}\right) \tag{2.30}$$

where $p \geq O(1)$ is the corresponding pole in continuous-time. Assuming $m \in O(1)$, and using the fact $\|P^{-1}\| = \|P\|^{-1}$, it can be derived from (2.29) that

$$\lim_{k\to\infty} \|\mathbf{x}_k\| \leq O(1) \cdot O\left(T^2\right) + O\left(T^{-1}\right) \cdot O\left(T^2\right) = O(T). \tag{2.31}$$

Remark 2.3 Under practical considerations, it is generally advisable to select the pole p large enough such that the system has a fast enough response. With the selection of a small sampling period T, a pole of order $O(T)$ would lead to an undesirably slow response. Thus, it makes sense to select a pole of order $O(1)$ or larger.

Remark 2.4 The SM control in [150] guarantees that the sliding variable σ_k is of order $O\left(T^2\right)$, but cannot guarantee the same order of magnitude of steady-state errors for the system state variables. In the next section, we show that an integral sliding mode design can achieve a more precise state regulation.

2.3.2 Output Tracking

Consider the discrete-time sliding manifold given below, [150, 154],

$$\boldsymbol{\sigma}_k = D_o\left(\mathbf{r}_k - \mathbf{y}_k\right) \tag{2.32}$$

where D_o is a constant matrix of rank m and $\mathbf{r}_k \in \Re^m$. The objective is to force the output $\mathbf{y}_k$ to track the reference $\mathbf{r}_k$. The property for this class of sampled-data SM control is given by the following lemma:

Lemma 2.3 *For* $\boldsymbol{\sigma}_k = D_o\left(\mathbf{r}_k - \mathbf{y}_k\right)$ *and control based on a disturbance estimate*

$$\hat{\mathbf{d}}_k = \mathbf{x}_k - \Phi\mathbf{x}_{k-1} - \Gamma\mathbf{u}_{k-1}$$

the closed-loop system has the following properties

$$\boldsymbol{\sigma}_{k+1} \in O\left(T^2\right)$$

$$\mathbf{r}_{k+1} - \mathbf{y}_{k+1} \in O\left(T^2\right)$$

Proof Similar to the regulation problem the discrete-time equivalent control is defined by solving $\boldsymbol{\sigma}_{k+1} = 0$, [150]. This leads to

$$\mathbf{u}_k^{\mathrm{eq}} = \left(D_{\mathrm{o}} C \Gamma\right)^{-1} D_{\mathrm{o}} \left(\mathbf{r}_{k+1} - C\Phi \mathbf{x}_k - C\mathbf{d}_k\right) \tag{2.33}$$

with D_{o} selected such that $D_{\mathrm{o}} C \Gamma$ is invertible. As in the regulation case, the control cannot be implemented in the same form as in (2.33) because of the lack of knowledge of $\mathbf{d}_k$ which requires a priori knowledge of the disturbance $\mathbf{f}(t)$. Thus, the delayed disturbance $\mathbf{d}_{k-1}$ will be used

$$\hat{\mathbf{d}}_k = \mathbf{d}_{k-1} = \mathbf{x}_k - \Phi \mathbf{x}_{k-1} - \Gamma \mathbf{u}_{k-1} \tag{2.34}$$

Thus, the control becomes

$$\mathbf{u}_k = \left(D_{\mathrm{o}} C \Gamma\right)^{-1} D_{\mathrm{o}} \left(\mathbf{r}_{k+1} - C\Phi \mathbf{x}_k - C\mathbf{d}_{k-1}\right) \tag{2.35}$$

The closed-loop system under the control given by (2.35) is

$$\mathbf{x}_{k+1} = \left[\Phi - \Gamma \left(D_{\mathrm{o}} C \Gamma\right)^{-1} D_{\mathrm{o}} C \Phi\right] \mathbf{x}_k + \Gamma \left(D_{\mathrm{o}} C \Gamma\right)^{-1} D_{\mathrm{o}} \mathbf{r}_{k+1} + \mathbf{d}_k - \Gamma \left(D_{\mathrm{o}} C \Gamma\right)^{-1} D_{\mathrm{o}} C \mathbf{d}_{k-1}. \tag{2.36}$$

Note that $\left(D_{\mathrm{o}} C \Gamma\right)^{-1} D_{\mathrm{o}} = (C\Gamma)^{-1}$. Simplifying (2.36) further gives

$$\mathbf{x}_{k+1} = \left[\Phi - \Gamma \left(C \Gamma\right)^{-1} C \Phi\right] \mathbf{x}_k + \Gamma \left(C \Gamma\right)^{-1} \mathbf{r}_{k+1} + \mathbf{d}_k - \Gamma \left(C \Gamma\right)^{-1} C \mathbf{d}_{k-1}. \tag{2.37}$$

where the eigenvalues of the matrix $\left[\Phi - \Gamma\left[(C\Gamma)^{-1} C\Phi\right]\right.$ are the transmission zeros of the system, [167]. Postmultiplication of (2.37) with C results in,

$$\mathbf{y}_{k+1} = C x_{k+1} = \mathbf{r}_{k+1} + C\left(\mathbf{d}_k - \mathbf{d}_{k-1}\right) = \mathbf{r}_{k+1} + O\left(T^2\right). \tag{2.38}$$

Substitution of (2.38) into the forward expression of (2.32) results in

$$\boldsymbol{\sigma}_{k+1} = D_{\mathrm{o}} \left(\mathbf{r}_{k+1} - \mathbf{y}_{k+1}\right) \in O\left(T^2\right). \tag{2.39}$$

The above result shows that with the control given by (2.35) the output is stable and that the tracking error converges to a bound of order $O\left(T^2\right)$. However, the stability of the whole system is guaranteed only if the transmission zeros are stable. Looking back at (2.37), it is very simple to show that $\left[\Phi - \Gamma\left(C\Gamma\right)^{-1}C\Phi\right]$ has m eigenvalues in the origin. Note, that those m deadbeat eigenvalues correspond to the output deadbeat response. If the matrices are partitioned as shown

$$\Phi = \begin{bmatrix} \Phi_{11} & \Phi_{12} \\ \Phi_{21} & \Phi_{22} \end{bmatrix}, \quad \Gamma = \begin{bmatrix} \Gamma_1 \\ \Gamma_2 \end{bmatrix} \quad \& \quad C = \left[C_1 \vdots C_2 \right]$$

where $(\Phi_{11}, C_1, \Gamma_1) \in \Re^{m\times m}$, $(\Phi_{12}, C_2) \in \Re^{m\times n-m}$, $(\Phi_{21}, \Gamma_2) \in \Re^{n-m\times m}$ and $\Phi_{22} \in \Re^{n-m\times n-m}$. The eigenvalues of $\left[\Phi - \Gamma\left(C\Gamma\right)^{-1}C\Phi\right]$ can be found from

$$\det\left[\lambda I_n - \left(\Phi - \Gamma\left(C\Gamma\right)^{-1}C\Phi\right)\right] =$$
$$\det\begin{bmatrix} \lambda I_m - \Phi_{11} + \Gamma_1\left(C\Gamma\right)^{-1}C\begin{bmatrix}\Phi_{11}\\ \Phi_{21}\end{bmatrix} & -\Phi_{12} + \Gamma_1\left(C\Gamma\right)^{-1}C\begin{bmatrix}\Phi_{12}\\ \Phi_{22}\end{bmatrix} \\ -\Phi_{21} + \Gamma_2\left(C\Gamma\right)^{-1}C\begin{bmatrix}\Phi_{11}\\ \Phi_{21}\end{bmatrix} & \lambda I_{n-m} - \Phi_{22} + \Gamma_2\left(C\Gamma\right)^{-1}C\begin{bmatrix}\Phi_{12}\\ \Phi_{22}\end{bmatrix} \end{bmatrix} = 0$$

If the top row is premultiplied with C_1 and the bottom row premultiplied by C_2 and the results summed and used as the new top row, the following is obtained

$$\det\begin{bmatrix} \lambda C_1 & \lambda C_2 \\ -\Phi_{21} + \Gamma_2\left(C\Gamma\right)^{-1}C\begin{bmatrix}\Phi_{11}\\ \Phi_{21}\end{bmatrix} & \lambda I_{n-m} - \Phi_{22} + \Gamma_2\left(C\Gamma\right)^{-1}C\begin{bmatrix}\Phi_{12}\\ \Phi_{22}\end{bmatrix} \end{bmatrix} = 0 \tag{2.40}$$

This can be further simplified to

$$\lambda^m \det\begin{bmatrix} C_1 & C_2 \\ -\Phi_{21} + \Gamma_2\left(C\Gamma\right)^{-1}C\begin{bmatrix}\Phi_{11}\\ \Phi_{21}\end{bmatrix} & \lambda I_{n-m} - \Phi_{22} + \Gamma_2\left(C\Gamma\right)^{-1}C\begin{bmatrix}\Phi_{12}\\ \Phi_{22}\end{bmatrix} \end{bmatrix} = 0 \tag{2.41}$$

The above result shows that there are m eigenvalues at the origin.

Remark 2.5 The conventional method guarantees that the sliding surface is of order $O\left(T^2\right)$ and deadbeat tracking error of order $O\left(T^2\right)$. However, deadbeat response is not practical as it requires large control effort and the addition of input saturation would sacrifice global stability.

2.4 Discrete-Time Integral Sliding Mode Control

2.4.1 State Regulation with ISM

Consider the new discrete-time integral sliding manifold defined below

$$\begin{aligned}\boldsymbol{\sigma}_k &= D\mathbf{x}_k - D\mathbf{x}_0 + \boldsymbol{\varepsilon}_k \\ \boldsymbol{\varepsilon}_k &= \boldsymbol{\varepsilon}_{k-1} + E\mathbf{x}_{k-1}\end{aligned} \tag{2.42}$$

where $\boldsymbol{\sigma}_k \in \Re^m$, $\boldsymbol{\varepsilon}_k \in \Re^m$, and matrices D and E are constant and of rank m. The term $D\mathbf{x}_0$ is used to eliminate the reaching phase. (2.42) is the discrete-time counterpart of the following sliding manifold, [40]

$$\boldsymbol{\sigma}(t) = D\mathbf{x}(t) - D\mathbf{x}(0) + \int_0^t E\mathbf{x}(\tau)d\tau = 0. \tag{2.43}$$

Theorem 2.1 *Based on assumption that the pair (Φ, Γ) in (2.2) is controllable, there exists a matrix K such that the eigenvalues of $\Phi - \Gamma K$ are distinct and within the unit circle. Choose the control law*

$$\mathbf{u}_k = (D\Gamma)^{-1} D\mathbf{x}_0 - (D\Gamma)^{-1}\left((D\Phi + E)\,\mathbf{x}_k + D\hat{\mathbf{d}}_k + \boldsymbol{\varepsilon}_k\right) \tag{2.44}$$

where $D\Gamma$ is invertible,

$$E = -D\,(\Phi - I - \Gamma K) \tag{2.45}$$

and $\hat{\mathbf{d}}_k$ is the disturbance compensation (2.67). Then the closed-loop dynamics is

$$\mathbf{x}_{k+1} = (\Phi - \Gamma K)\,\mathbf{x}_k + \boldsymbol{\zeta}_k \tag{2.46}$$

with $\boldsymbol{\zeta}_k \in \Re^n$ is $O\left(T^3\right)$, and

$$\lim_{k\to\infty} \|\mathbf{x}_k\| \le O\left(T^2\right).$$

Proof Consider a forward expression of (2.42)

$$\begin{aligned}\boldsymbol{\sigma}_{k+1} &= D\mathbf{x}_{k+1} - D\mathbf{x}_0 + \boldsymbol{\varepsilon}_{k+1} \\ \boldsymbol{\varepsilon}_{k+1} &= \boldsymbol{\varepsilon}_k + E\mathbf{x}_k\end{aligned} \tag{2.47}$$

Substituting $\boldsymbol{\varepsilon}_{k+1}$ and (2.2) into the expression of the sliding manifold in (2.47) leads to

$$\boldsymbol{\sigma}_{k+1} = (D\Phi + E)\,\mathbf{x}_k + D\,(\Gamma \mathbf{u}_k + \mathbf{d}_k) + \boldsymbol{\varepsilon}_k - D\mathbf{x}_0. \tag{2.48}$$

The equivalent control is found by solving for $\boldsymbol{\sigma}_{k+1} = 0$

$$\mathbf{u}_k^{\mathrm{eq}} = (D\Gamma)^{-1}\,D\mathbf{x}_0 - (D\Gamma)^{-1}\,((D\Phi + E)\,\mathbf{x}_k + D\mathbf{d}_k + \boldsymbol{\varepsilon}_k)\,. \tag{2.49}$$

Similar to the classical case with control given by (2.33), implementation of (2.49) would require a priori knowledge of the disturbance $\mathbf{d}_k$. By replacing the disturbance in (2.49) with its estimate $\hat{\mathbf{d}}_k$, which is defined in (2.67), the practical control law is

$$\mathbf{u}_k = (D\Gamma)^{-1}\,D\mathbf{x}_0 - (D\Gamma)^{-1}\left((D\Phi + E)\,\mathbf{x}_k + D\hat{\mathbf{d}}_k + \boldsymbol{\varepsilon}_k\right) \tag{2.50}$$

Substitution of $\mathbf{u}_k$ defined by (2.50) into (2.2) leads to the closed-loop equation in the sliding mode

$$\begin{aligned}\mathbf{x}_{k+1} = &\left[\Phi - \Gamma\,(D\Gamma)^{-1}\,(D\Phi + E)\right]\mathbf{x}_k - \Gamma\,(D\Gamma)^{-1}\,\boldsymbol{\varepsilon}_k + \Gamma\,(D\Gamma)^{-1}\,D\mathbf{x}_0\\ &+ \mathbf{d}_k - \Gamma\,(D\Gamma)^{-1}\,D\hat{\mathbf{d}}_k.\end{aligned} \tag{2.51}$$

Let us derive the sliding dynamics. Rewriting (2.47)

$$\boldsymbol{\sigma}_{k+1} = D\mathbf{x}_{k+1} + E\mathbf{x}_k - D\mathbf{x}_0 + \boldsymbol{\varepsilon}_k. \tag{2.52}$$

Substitution of (2.51) into (2.52) leads to

$$\boldsymbol{\sigma}_{k+1} = D\mathbf{d}_k - D\hat{\mathbf{d}}_k = D\mathbf{d}_k - D\mathbf{d}_{k-1} \in O\left(T^2\right), \tag{2.53}$$

that is, the introduction of ISM control leads to the same sliding dynamics as in [150]. Next, solving $\boldsymbol{\varepsilon}_k$ in (2.42) in terms of $\mathbf{x}_k$ and $\boldsymbol{\sigma}_k$

$$\boldsymbol{\varepsilon}_k = \boldsymbol{\sigma}_k - D\mathbf{x}_k + D\mathbf{x}_0, \tag{2.54}$$

and substituting it into (2.51), the closed-loop dynamics becomes

$$\mathbf{x}_{k+1} = \left[\Phi - \Gamma\,(D\Gamma)^{-1}\,(D\,(\Phi - I) + E)\right]\mathbf{x}_k - \Gamma\,(D\Gamma)^{-1}\,\boldsymbol{\sigma}_k + \mathbf{d}_k - \Gamma\,(D\Gamma)^{-1}\,D\hat{\mathbf{d}}_k. \tag{2.55}$$

In (2.55), $\boldsymbol{\sigma}_k$ can be substituted by $\boldsymbol{\sigma}_k = D\mathbf{d}_{k-1} - D\mathbf{d}_{k-2}$ as can be inferred from (2.53). Also, under the condition (2.45), $D\,(\Phi - I) + E = D\Gamma K$. Therefore, $\Phi - \Gamma\,(D\Gamma)^{-1}\,(D\,(\Phi - I) + E) = \Phi - \Gamma K$. Since it is assumed that the pair (Φ, Γ) is controllable, there exists a matrix K such that eigenvalues of $\Phi - \Gamma K$ can be placed anywhere inside the unit circle. Note that, the selection of matrix D is arbitrary as long as it guarantees the invertibility of $D\Gamma$ while matrix E, computed using (2.45), guarantees the desired closed-loop performance. Thus, we have

$$\mathbf{x}_{k+1} = (\Phi - \Gamma K)\,\mathbf{x}_k + \mathbf{d}_k - \Gamma\,(D\Gamma)^{-1}\,D\mathbf{d}_{k-1} - \Gamma\,(D\Gamma)^{-1}\,D\,(\mathbf{d}_{k-1} - \mathbf{d}_{k-2})\,. \tag{2.56}$$

Note that in (2.56), the disturbance estimate $\hat{\mathbf{d}}_k$ has been replaced by $\mathbf{d}_{k-1}$. Further simplification of (2.56) leads to

$$\mathbf{x}_{k+1} = (\Phi - \Gamma K)\,\mathbf{x}_k + \boldsymbol{\zeta}_k \tag{2.57}$$

where

$$\boldsymbol{\zeta}_k = \mathbf{d}_k - 2\Gamma\,(D\Gamma)^{-1}\,D\mathbf{d}_{k-1} + \Gamma\,(D\Gamma)^{-1}\,D\mathbf{d}_{k-2}. \tag{2.58}$$

The magnitude of $\boldsymbol{\zeta}_k$ can be evaluated as below. Adding and subtracting $2\mathbf{d}_{k-1}$ and $\mathbf{d}_{k-2}$ from the right hand side of (2.58) yield

$$\boldsymbol{\zeta}_k = (\mathbf{d}_k - 2\mathbf{d}_{k-1} + \mathbf{d}_{k-2}) + \left(I - \Gamma\,(D\Gamma)^{-1}\,D\right)(2\mathbf{d}_{k-1} - \mathbf{d}_{k-2})\,. \tag{2.59}$$

In *Lemma* 2.1, it has been shown that $(\mathbf{d}_k - 2\mathbf{d}_{k-1} + \mathbf{d}_{k-2}) \in O\left(T^3\right)$. On the other hand, from (2.4) we have

$$\begin{aligned}&\left(I - \Gamma\,(D\Gamma)^{-1}\,D\right)(2\mathbf{d}_{k-1} - \mathbf{d}_{k-2})\\ &= \left(I - \Gamma\,(D\Gamma)^{-1}\,D\right)\left(\Gamma\,(2\mathbf{f}_{k-1} - \mathbf{f}_{k-2}) + \frac{T}{2}\Gamma\,(2\mathbf{v}_{k-1} - \mathbf{v}_{k-2}) + O\left(T^3\right)\right)\end{aligned}$$

Note that $\left(I - \Gamma\,(D\Gamma)^{-1}\,D\right)\Gamma = 0$, thus

$$\left(I - \Gamma\,(D\Gamma)^{-1}\,D\right)\left(\Gamma\,(2\mathbf{f}_{k-1} - \mathbf{f}_{k-2}) + \frac{T}{2}\Gamma\,(2\mathbf{v}_{k-1} - \mathbf{v}_{k-2})\right) = 0.$$

Furthermore, $\|\left(I - \Gamma\,(D\Gamma)^{-1}\,D\right)\| \le 1$, thus $\left(I - \Gamma\,(D\Gamma)^{-1}\,D\right) O\left(T^3\right)$ remains $O\left(T^3\right)$. This concludes that

$$\boldsymbol{\zeta}_k \in O\left(T^3\right).$$

Comparing (2.57) with (2.22), the difference is that $\boldsymbol{\delta}_k \in O\left(T^2\right)$ whereas $\boldsymbol{\zeta}_k \in O\left(T^3\right)$. Further, by doing a similarity decomposition for dynamics of (2.57), only the J_2 matrix of dimension n exists. Thus the derivation procedure shown in (2.22)–(2.31) holds for (2.57), and the solution is

$$\mathbf{x}_k = (\Phi - \Gamma K)^k \mathbf{x}_0 + \sum_{i=0}^{k-1} (\Phi - \Gamma K)^i \boldsymbol{\zeta}_{k-i-1}. \tag{2.60}$$

Assuming distinct eigenvalues of $\Phi - \Gamma K$ and following the procedure that resulted in (2.31), it can be shown that

$$\lim_{k\to\infty} \left\| \sum_{i=0}^{k-1} (\Phi - \Gamma K)^i \boldsymbol{\zeta}_{k-i-1} \right\| \in O\left(T^2\right). \tag{2.61}$$

Finally, it is concluded that

$$\lim_{k\to\infty} \|\mathbf{x}_k\| \leq O\left(T^2\right). \tag{2.62}$$

Remark 2.6 From the foregoing derivations, it can be seen that the state errors are always one order higher than the disturbance term $\boldsymbol{\zeta}_k$ in the worst case due to convolution as shown by (2.61). After incorporating the integral sliding manifold, the off-set from the disturbance can be better compensated, in the sequel leading to a smaller steady state error boundary.

Remark 2.7 It is evident from the above analysis that, for the class of systems considered in this work and in [150, 154], the equivalent control based SM control with disturbance observer guarantees the motion of the states within an $O\left(T^2\right)$ bound, which is smaller than $O(T)$ for T sufficiently small, and is lower than what can be achieved by SM control using switching control, [40, 155]. In such circumstance, without the loss of precision we can relax the necessity of incorporating a switching term, in the sequel avoid exciting chattering.

2.4.2 Output-Tracking ISM Control: State Feedback Approach

In this section we discuss the state-feedback-based output ISM control. We first present the controller design using an appropriate integral sliding surface and a delay based disturbance estimation. Next the stability condition of the closed-loop system and the error dynamics under output ISM control are derived. The ultimate tracking error bound is analyzed.

2.4.2.1 Controller Design

Consider the discrete-time integral sliding surface defined below,

$$\begin{aligned} \boldsymbol{\sigma}_k &= \mathbf{e}_k - \mathbf{e}_0 + \boldsymbol{\varepsilon}_k \\ \boldsymbol{\varepsilon}_k &= \boldsymbol{\varepsilon}_{k-1} + E\mathbf{e}_{k-1} \end{aligned} \tag{2.63}$$

where $\mathbf{e}_k = \mathbf{r}_k - \mathbf{y}_k$ is the tracking error, $\boldsymbol{\sigma}_k, \boldsymbol{\varepsilon}_k \in \Re^m$ are the sliding function and integral vectors, and $E \in \Re^{m\times m}$ is an integral gain matrix.

By virtue of the concept of equivalent control, a SM control law can be derived by letting $\boldsymbol{\sigma}_{k+1} = 0$. From (2.63), $-\mathbf{e}_0 + \boldsymbol{\varepsilon}_k = \boldsymbol{\sigma}_k - \mathbf{e}_k$, we have

$$\begin{aligned}\boldsymbol{\sigma}_{k+1} &= \mathbf{e}_{k+1} - \mathbf{e}_0 + \boldsymbol{\varepsilon}_{k+1} = \mathbf{e}_{k+1} - \mathbf{e}_0 + \boldsymbol{\varepsilon}_k + E\mathbf{e}_k \\ &= \mathbf{e}_{k+1} - (I_m - E)\,\mathbf{e}_k + \boldsymbol{\sigma}_k.\end{aligned} \tag{2.64}$$

From the system dynamics (2.2), the output error $\mathbf{e}_{k+1}$ is

$$\mathbf{e}_{k+1} = \mathbf{r}_{k+1} - [C\Phi\mathbf{x}_k + C\Gamma\mathbf{u}_k + C\mathbf{d}_k]$$

and

$$\begin{aligned}\boldsymbol{\sigma}_{k+1} &= \mathbf{r}_{k+1} - [C\Phi\mathbf{x}_k + C\Gamma\mathbf{u}_k + C\mathbf{d}_k] - (I_m - E)\,\mathbf{e} + \boldsymbol{\sigma}_k \\ &= \mathbf{a}_k - C\Gamma\mathbf{u}_k - C\mathbf{d}_k.\end{aligned} \tag{2.65}$$

where $\mathbf{a}_k = \mathbf{r}_{k+1} - \Lambda\mathbf{e}_k - C\Phi\mathbf{x}_k + \boldsymbol{\sigma}_k$, and $\Lambda = I_m - E$. Assuming $\boldsymbol{\sigma}_{k+1} = 0$, we can derive the equivalent control

$$\mathbf{u}_k^{eq} = (C\Gamma)^{-1}(\mathbf{a}_k - C\mathbf{d}_k). \tag{2.66}$$

Note that the control (2.66) is based on the current value of the disturbance $\mathbf{d}_k$ which is unknown and therefore cannot be implemented in the current form. To overcome this, the disturbance estimate will be used. When the system states are accessible, a delay based disturbance estimate can be easily derived from the system (2.2)

$$\hat{\mathbf{d}}_k = \mathbf{d}_{k-1} = \mathbf{x}_k - \Phi\mathbf{x}_{k-1} - \Gamma\mathbf{u}_{k-1}. \tag{2.67}$$

Note that $\mathbf{d}_{k-1}$ is the exogenous disturbance and bounded, therefore $\hat{\mathbf{d}}_k$ is bounded for all k. Using the disturbance estimation (2.67), the actual ISM control law is given by

$$\mathbf{u}_k = (C\Gamma)^{-1}\left(\mathbf{a}_k - C\hat{\mathbf{d}}_k\right). \tag{2.68}$$

2.4.2.2 Stability Analysis

Since the integral switching surface (2.63) consists of outputs only, it is necessary to examine the closed-loop stability in state space when the ISM control (2.68) and disturbance estimation (2.67) are used.

Expressing $\mathbf{e}_k = \mathbf{r}_k - C\mathbf{x}_k$, the ISM control law (2.68) can be rewritten as

$$\begin{aligned}\mathbf{u}_k &= (C\Gamma)^{-1}\left(\mathbf{r}_{k+1} - \Lambda\mathbf{e}_k - C\Phi\mathbf{x}_k + \boldsymbol{\sigma}_k - C\hat{\mathbf{d}}_k\right)\\ &= -(C\Gamma)^{-1}(C\Phi - \Lambda C)\,\mathbf{x}_k - (C\Gamma)^{-1}C\hat{\mathbf{d}}_k\\ &\quad + (C\Gamma)^{-1}(\mathbf{r}_{k+1} - \Lambda\mathbf{r}_k) + (C\Gamma)^{-1}\boldsymbol{\sigma}_k.\end{aligned} \tag{2.69}$$

Substituting the above control law (2.69) into the system (2.2) yields the closed-loop state dynamics

$$\begin{aligned}\mathbf{x}_{k+1} &= \left[\Phi - \Gamma(C\Gamma)^{-1}(C\Phi - \Lambda C)\right]\mathbf{x}_k + \mathbf{d}_k - \Gamma(C\Gamma)^{-1}C\hat{\mathbf{d}}_k\\ &\quad + \Gamma(C\Gamma)^{-1}(\mathbf{r}_{k+1} - \Lambda\mathbf{r}_k) + \Gamma(C\Gamma)^{-1}\boldsymbol{\sigma}_k.\end{aligned} \tag{2.70}$$

It can be seen from the dynamics (2.70) that the stability of $\mathbf{x}_k$ is determined by the matrix $\left[\Phi - \Gamma(C\Gamma)^{-1}(C\Phi - \Lambda C)\right]$ and the boundedness of $\boldsymbol{\sigma}_k$.

Lemma 2.4 *The eignvalues of* $\left[\Phi - \Gamma(C\Gamma)^{-1}(C\Phi - \Lambda C)\right]$ *are the eigenvalues of* Λ *and the non-zero eigenvalues of* $\left[\Phi - \Gamma(C\Gamma)^{-1}C\Phi\right]$.

Proof Consider the matrices Φ, Γ and C which can be partitioned as shown

$$\Phi = \begin{bmatrix}\Phi_{11} & \Phi_{12}\\ \Phi_{21} & \Phi_{22}\end{bmatrix},\quad \Gamma = \begin{bmatrix}\Gamma_1\\ \Gamma_2\end{bmatrix}\quad \&\quad C = \left[C_1 \,\vdots\, C_2\right]$$

where $(\Phi_{11}, C_1, \Gamma_1) \in \Re^{m\times m}$, $(\Phi_{12}, C_2) \in \Re^{m\times n-m}$, $(\Phi_{21}, \Gamma_2) \in \Re^{n-m\times m}$ and $\Phi_{22} \in \Re^{n-m\times n-m}$. The eigenvalues of $\left[\Phi - \Gamma(C\Gamma)^{-1}(C\Phi - \Lambda C)\right]$ are found from

$$\det\left[\lambda I_n - \Phi + \Gamma(C\Gamma)^{-1}(C\Phi - \Lambda C)\right] = 0 \tag{2.71}$$

$$\det\begin{bmatrix}\lambda I - \Phi_{11} + \Gamma_1(C\Gamma)^{-1}\left(C\begin{bmatrix}\Phi_{11}\\ \Phi_{21}\end{bmatrix} - \Lambda C_1\right) & -\Phi_{12} + \Gamma_1(C\Gamma)^{-1}\left(C\begin{bmatrix}\Phi_{12}\\ \Phi_{22}\end{bmatrix} - \Lambda C_2\right)\\ -\Phi_{21} + \Gamma_2(C\Gamma)^{-1}\left(C\begin{bmatrix}\Phi_{11}\\ \Phi_{21}\end{bmatrix} - \Lambda C_1\right) & \lambda I - \Phi_{22} + \Gamma_2(C\Gamma)^{-1}\left(C\begin{bmatrix}\Phi_{12}\\ \Phi_{22}\end{bmatrix} - \Lambda C_2\right)\end{bmatrix} = 0 \tag{2.72}$$

If the top row is premultiplied with C_1 and the bottom row is premultiplied with C_2 and the the results summed and used as the new top row, using the fact that $C_1\Gamma_1 + C_2\Gamma_2 = C\Gamma$ the following is obtained

$$\det\begin{bmatrix} (\lambda I_m - \Lambda)C_1 & (\lambda I_m - \Lambda)C_2 \\ -\Phi_{21} + \Gamma_2(C\Gamma)^{-1}\left(C\begin{bmatrix}\Phi_{11}\\ \Phi_{21}\end{bmatrix} - \Lambda C_1\right) & \lambda I - \Phi_{22} + \Gamma_2(C\Gamma)^{-1}\left(C\begin{bmatrix}\Phi_{12}\\ \Phi_{22}\end{bmatrix} - \Lambda C_2\right)\end{bmatrix} = 0 \tag{2.73}$$

factoring the term $(\lambda I_m - \Lambda)$ and premultipying the top row with $\Gamma_2(C\Gamma)^{-1}\Lambda$ and adding to the bottom row leads to

$$\det(\lambda I_m - \Lambda)\det\begin{bmatrix} C_1 & C_2 \\ -\Phi_{21} + \Gamma_2(C\Gamma)^{-1}C\begin{bmatrix}\Phi_{11}\\ \Phi_{21}\end{bmatrix} & \lambda I_{n-m} - \Phi_{22} + \Gamma_2(C\Gamma)^{-1}C\begin{bmatrix}\Phi_{12}\\ \Phi_{22}\end{bmatrix}\end{bmatrix} = 0 \tag{2.74}$$

Thus, we can conclude that m eigenvalues of $\left[\Phi - \Gamma(C\Gamma)^{-1}(C\Phi - \Lambda C)\right]$ are the eigenvlaues of Λ.

Now, consider

$$\det\begin{bmatrix} C_1 & C_2 \\ -\Phi_{21} + \Gamma_2(C\Gamma)^{-1}C\begin{bmatrix}\Phi_{11}\\ \Phi_{21}\end{bmatrix} & \lambda I_{n-m} - \Phi_{22} + \Gamma_2(C\Gamma)^{-1}C\begin{bmatrix}\Phi_{12}\\ \Phi_{22}\end{bmatrix}\end{bmatrix} = 0 \tag{2.75}$$

Using the following relations

$$C_2\Phi_{21} - C_2\Gamma_2(C\Gamma)^{-1}C\begin{bmatrix}\Phi_{11}\\ \Phi_{21}\end{bmatrix} = -C_1\Phi_{11} + C_1\Gamma_1(C\Gamma)^{-1}C\begin{bmatrix}\Phi_{11}\\ \Phi_{21}\end{bmatrix} \tag{2.76}$$

$$-C_2\Phi_{22} + C_2\Gamma_2(C\Gamma)^{-1}C\begin{bmatrix}\Phi_{12}\\ \Phi_{22}\end{bmatrix} = -C_1\Phi_{12} + C_1\Gamma_1(C\Gamma)^{-1}C\begin{bmatrix}\Phi_{12}\\ \Phi_{22}\end{bmatrix} \tag{2.77}$$

Multiplying (2.75) with $\lambda^{-m}\lambda^{m}$ we get

$$\lambda^{-m}\det\begin{bmatrix} \lambda C_1 & \lambda C_2 \\ -\Phi_{21} + \Gamma_2(C\Gamma)^{-1}C\begin{bmatrix}\Phi_{11}\\ \Phi_{21}\end{bmatrix} & \lambda I_{n-m} - \Phi_{22} + \Gamma_2(C\Gamma)^{-1}C\begin{bmatrix}\Phi_{12}\\ \Phi_{22}\end{bmatrix}\end{bmatrix} = 0 \tag{2.78}$$

Premultiplying the bottom row with C_2 and subtracting from the top row and using the result as the new top row we get

$$\lambda^{-m}\det\begin{bmatrix} \lambda C_1 + C_2\Phi_{21} - C_2\Gamma_2(C\Gamma)^{-1}C\begin{bmatrix}\Phi_{11}\\ \Phi_{21}\end{bmatrix} & C_2\Phi_{22} - C_2\Gamma_2(C\Gamma)^{-1}C\begin{bmatrix}\Phi_{12}\\ \Phi_{22}\end{bmatrix} \\ -\Phi_{21} + \Gamma_2(C\Gamma)^{-1}C\begin{bmatrix}\Phi_{11}\\ \Phi_{21}\end{bmatrix} & \lambda I_{n-m} - \Phi_{22} + \Gamma_2(C\Gamma)^{-1}C\begin{bmatrix}\Phi_{12}\\ \Phi_{22}\end{bmatrix}\end{bmatrix} = 0 \tag{2.79}$$

using relations (2.76) and (2.77) we finally get

$$\lambda^{-m}\det\begin{bmatrix} \lambda C_1 - C_1\Phi_{11} + C_1\Gamma_1(C\Gamma)^{-1}C\begin{bmatrix}\Phi_{11}\\ \Phi_{21}\end{bmatrix} & -C_1\Phi_{12} + C_1\Gamma_1(C\Gamma)^{-1}C\begin{bmatrix}\Phi_{12}\\ \Phi_{22}\end{bmatrix} \\ -\Phi_{21} + \Gamma_2(C\Gamma)^{-1}C\begin{bmatrix}\Phi_{11}\\ \Phi_{21}\end{bmatrix} & \lambda I_{n-m} - \Phi_{22} + \Gamma_2(C\Gamma)^{-1}C\begin{bmatrix}\Phi_{12}\\ \Phi_{22}\end{bmatrix}\end{bmatrix} = 0 \tag{2.80}$$

We can factor out the matrix C_1 from the top row to get

$$\lambda^{-m}\det(C_1)\det\begin{bmatrix} \lambda I_m - \Phi_{11} + \Gamma_1(C\Gamma)^{-1}C\begin{bmatrix}\Phi_{11}\\ \Phi_{21}\end{bmatrix} & -\Phi_{12} + \Gamma_1(C\Gamma)^{-1}C\begin{bmatrix}\Phi_{12}\\ \Phi_{22}\end{bmatrix} \\ -\Phi_{21} + \Gamma_2(C\Gamma)^{-1}C\begin{bmatrix}\Phi_{11}\\ \Phi_{21}\end{bmatrix} & \lambda I_{n-m} - \Phi_{22} + \Gamma_2(C\Gamma)^{-1}C\begin{bmatrix}\Phi_{12}\\ \Phi_{22}\end{bmatrix}\end{bmatrix} = 0 \tag{2.81}$$

which finally simplifies to

$$\lambda^{-m}\det(C_1)\det\left[\Phi - \Gamma(C\Gamma)C\Phi\right] = 0 \tag{2.82}$$

It is well known that $[\Phi - \Gamma(C\Gamma)C\Phi]$ has atleast m zero eigenvlaues which would be cancelled out by λ^{-m} and, thus, we finally conclude that the eigenvalues of $\left[\Phi - \Gamma(C\Gamma)^{-1}(C\Phi - \Lambda C)\right]$ are the eigenvalues of Λ and the non-zero eigenvlaues of $[\Phi - \Gamma(C\Gamma)C\Phi]$.

According to *Lemma* 2.1, the matrix $\left[\Phi - \Gamma\,(C\Gamma)^{-1}\,(C\Phi - \Lambda C)\right]$ has m poles to be placed at desired locations while the remaining $n - m$ poles are the open-loop zeros of the system (Φ, Γ, C). Since, the system (2.2) is assumed to be minimum-phase, the $n-m$ poles are stable. Therefore, stability of the closed-loop state dynamics is guaranteed. Note that if Λ is a zero matrix then m poles are zero and the performance will be the same as the conventional deadbeat sliding-mode controller design.

Since we use the estimate of the disturbance, $\boldsymbol{\sigma}_k \neq 0$. To show the boundedness and facilitate later analysis on the tracking performance, we derive the relationship between the switching surface and the disturbance estimate, as well as the relationship between the output tracking error and the disturbance estimate.

Theorem 2.2 *Assume that the system (2.2) is minimum-phase and the eigenvalues of the matrix Λ are within the unit circle. Then by the control law (2.68) we have*

$$\boldsymbol{\sigma}_{k+1} = C\left(\hat{\mathbf{d}}_k - \mathbf{d}_k\right) \tag{2.83}$$

and the error dynamics

$$\mathbf{e}_{k+1} = \Lambda\mathbf{e}_k + \boldsymbol{\delta}_k \tag{2.84}$$

where $\boldsymbol{\delta}_k = C\left(\hat{\mathbf{d}}_k - \mathbf{d}_k + \mathbf{d}_{k-1} - \hat{\mathbf{d}}_{k-1}\right)$.

Proof In order to verify the first part of *Theorem* 2.2, rewrite (2.65) as

$$\begin{aligned}\boldsymbol{\sigma}_{k+1} &= \mathbf{a}_k - C\Gamma\mathbf{u}_k - C\mathbf{d}_k = \mathbf{a}_k - C\Gamma\mathbf{u}_k^{\text{eq}} - C\mathbf{d}_k + C\Gamma\left(\mathbf{u}_k^{\text{eq}} - \mathbf{u}_k\right) \\ &= C\Gamma\left(\mathbf{u}_k^{\text{eq}} - \mathbf{u}_k\right),\end{aligned}$$

where we use the property of equivalent control $\boldsymbol{\sigma}_{k+1} = \mathbf{a}_k - C\Gamma\mathbf{u}_k^{\text{eq}} - C\mathbf{d}_k = 0$. Comparing two control laws (2.66) and (2.68), we obtain

$$\boldsymbol{\sigma}_{k+1} = C\left(\hat{\mathbf{d}}_k - \mathbf{d}_k\right).$$

Note that the switching surface $\boldsymbol{\sigma}_{k+1}$ is no longer zero as desired but a function of the difference $\mathbf{d}_k - \hat{\mathbf{d}}_k$. This, however, is acceptable since the difference is $\mathbf{d}_k - \hat{\mathbf{d}}_k = \mathbf{d}_k - \mathbf{d}_{k-1}$ by the delay based disturbance estimation; thus, according to *Lemma* 2.1 the difference is $O\left(T^2\right)$ which is quite small in practical applications.

To derive the second part of *Theorem* 2.2 regarding the error dynamics, rewriting (2.64) as

$$\mathbf{e}_{k+1} = \Lambda\mathbf{e}_k + \boldsymbol{\sigma}_{k+1} - \boldsymbol{\sigma}_k$$

and substituting the relationship (2.83), lead to

$$\begin{aligned}\mathbf{e}_{k+1} &= \Lambda\mathbf{e}_k + C\left(\hat{\mathbf{d}}_k - \mathbf{d}_k\right) - C\left(\hat{\mathbf{d}}_{k-1} - \mathbf{d}_{k-1}\right) \\ &= \Lambda\mathbf{e}_k + C\left(\hat{\mathbf{d}}_k - \mathbf{d}_k + \mathbf{d}_{k-1} - \hat{\mathbf{d}}_{k-1}\right) = \Lambda\mathbf{e}_k + \boldsymbol{\delta}_k. \end{aligned} \tag{2.85}$$

Since $\hat{\mathbf{d}}_k = \mathbf{d}_{k-1}$, $\boldsymbol{\delta}_k$ is bounded, from *Lemma* 2.1 we can conclude the boundedness of $\mathbf{e}_k$.

2.4.2.3 Tracking Error Bound

The tracking performance of the ISM control can be evaluated in terms of the error dynamics (2.84).

Theorem 2.3 *Using the delay based disturbance estimation (2.67), the ultimate tracking error bound with ISM control is given by*

$$\|\mathbf{e}_k\| = O\left(T^2\right)$$

where $\|\cdot\|$ *represents the Euclidean norm.*

Proof In order to calculate the tracking error bound we must find the bound of $\boldsymbol{\delta}_k$. From 2.2

$$\boldsymbol{\delta}_k = C\left(\hat{\mathbf{d}}_k - \mathbf{d}_k + \mathbf{d}_{k-1} - \hat{\mathbf{d}}_{k-1}\right).$$

Substituting $\hat{\mathbf{d}}_k = \mathbf{d}_{k-1}$ from (2.67)

$$\boldsymbol{\delta}_k = C\left(\mathbf{d}_{k-1} - \mathbf{d}_k + \mathbf{d}_{k-1} - \mathbf{d}_{k-2}\right)$$

which simplifies to

$$\boldsymbol{\delta}_k = -C\left(\mathbf{d}_k - 2\mathbf{d}_{k-1} + \mathbf{d}_{k-2}\right). \tag{2.86}$$

According to *Lemma* 2.1, $\mathbf{d}_k - 2\mathbf{d}_{k-1} + \mathbf{d}_{k-2} = O\left(T^3\right)$, therefore $\boldsymbol{\delta}_k = O\left(T^3\right)$. Also from *Lemma* 2.2 the ultimate error bound on $\|\mathbf{e}_k\|$ in the expression $\mathbf{e}_{k+1} = \Lambda\mathbf{e}_k + \boldsymbol{\delta}_k$ will be one order higher than the bound on $\boldsymbol{\delta}_k$ due to convolution. Since the bound on $\boldsymbol{\delta}_k$ is $O\left(T^3\right)$, the ultimate bound on $\|\mathbf{e}_k\|$ is $O\left(T^2\right)$.

Remark 2.8 In practical control a disturbance could be piece-wise smooth. The delay based estimation (2.67) can quickly capture the varying disturbance after one sampling period. Assume that the disturbance $\mathbf{f}(t)$ undergoes an abrupt change at the time interval $[(k-1)T, kT]$, then *Lemma* 2.1 does not hold for the time instant k because $\mathbf{v}(t)$ becomes extremely big. Nevertheless, *Lemma* 2.1 will be satisfied immediately after the time instant k if the disturbance becomes smooth again. From this point of view, the delay based estimation has a very small time-delay or equivalently a large bandwidth.

Remark 2.9 Although the state-feedback approach may seem to be not very practical for a number of output tracking tasks, this section serves as a precursor to the output-feedback-based and state-observer-based approaches to be explored in subsequent sections.

2.4.3 Output Tracking ISM: Output Feedback Approach

In this section we derive ISM control that only uses the output tracking error. The new design will require a reference model and a dynamic disturbance observer due to the lack of the state information. The reference model will be constructed such that its output is the reference trajectory $\mathbf{r}_k$.

2.4.3.1 Controller Design

In order to proceed we will first define a reference model

$$\begin{aligned}\mathbf{x}_{m,k+1} &= (\Phi - K_1)\,\mathbf{x}_{m,k} + K_2\mathbf{r}_{k+1}\\ \mathbf{y}_{m,k} &= C\mathbf{x}_{m,k}\end{aligned} \tag{2.87}$$

where $\mathbf{x}_{m,k} \in \Re^n$ is the state vector, $\mathbf{y}_{m,k} \in \Re^m$ is the output vector, and $\mathbf{r}_k \in \Re^m$ is a bounded reference trajectory. K_1 is selected such that $(\Phi - K_1)$ is stable. The selection criteria for the matrices K_1 and K_2 will be discussed in detail in Sects. 2.4.3.2 and 2.4.3.5.

Now consider a new sliding surface

$$\begin{aligned}\boldsymbol{\sigma}_k &= D\left(\mathbf{x}_{m,k} - \mathbf{x}_k\right) + \boldsymbol{\varepsilon}_k\\ \boldsymbol{\varepsilon}_k &= \boldsymbol{\varepsilon}_{k-1} + ED\left(\mathbf{x}_{m,k-1} - \mathbf{x}_{k-1}\right)\end{aligned} \tag{2.88}$$

where $D = C\Phi^{-1}$, $\boldsymbol{\sigma}_k, \boldsymbol{\varepsilon}_k \in \Re^m$ are the switching function and integral vectors, $E \in \Re^{m\times m}$ is an integral gain matrix. Note that $D\mathbf{x}_k = C\Phi^{-1}\left(\Phi\mathbf{x}_{k-1} + \Gamma\mathbf{u}_{k-1} + \mathbf{d}_{k-1}\right) = \mathbf{y}_{k-1} + D\left(\Gamma\mathbf{u}_{k-1} + \mathbf{d}_{k-1}\right)$ is independent of the states, such a simplification was first proposed in [101].

The equivalent control law can be derived from $\boldsymbol{\sigma}_{k+1} = 0$. From (2.88) $\boldsymbol{\varepsilon}_k = \boldsymbol{\sigma}_k - D\left(\mathbf{x}_{m,k} - \mathbf{x}_k\right)$, we have

$$\begin{aligned}\boldsymbol{\sigma}_{k+1} &= D\left(\mathbf{x}_{m,k+1} - \mathbf{x}_{k+1}\right) + \boldsymbol{\varepsilon}_{k+1}\\ &= D\left(\mathbf{x}_{m,k+1} - \mathbf{x}_{k+1}\right) + \boldsymbol{\varepsilon}_k + ED\left(\mathbf{x}_{m,k} - \mathbf{x}_k\right)\\ &= D\left(\mathbf{x}_{m,k+1} - \mathbf{x}_{k+1}\right) + \boldsymbol{\sigma}_k - D\left(\mathbf{x}_{m,k} - \mathbf{x}_k\right) + ED\left(\mathbf{x}_{m,k} - \mathbf{x}_k\right)\\ &= D\mathbf{x}_{m,k+1} - D\mathbf{x}_{k+1} + \boldsymbol{\sigma}_k - \Lambda D\left(\mathbf{x}_{m,k} - \mathbf{x}_k\right)\end{aligned} \tag{2.89}$$

where $\Lambda = I - E$. Substituting the system dynamics (2.2) into (2.89) yields

$$\begin{aligned}\boldsymbol{\sigma}_{k+1} &= D\mathbf{x}_{m,k+1} - D\left(\Phi\mathbf{x}_k + \Gamma\mathbf{u}_k + \mathbf{d}_k\right) + \boldsymbol{\sigma}_k - \Lambda D\left(\mathbf{x}_{m,k} - \mathbf{x}_k\right)\\ &= \mathbf{a}_k - D\Gamma\mathbf{u}_k - D\mathbf{d}_k\end{aligned} \tag{2.90}$$

where $\mathbf{a}_k = -\left(D\Phi - \Lambda D\right)\mathbf{x}_k + \left(D\mathbf{x}_{m,k+1} - \Lambda D\mathbf{x}_{m,k}\right) + \boldsymbol{\sigma}_k$.

Letting $\boldsymbol{\sigma}_{k+1} = 0$, solving for the equivalent control $\mathbf{u}_k^{eq}$, we have

$$\begin{aligned}\mathbf{u}_k^{eq} &= (D\Gamma)^{-1}\left(\mathbf{a}_k - D\mathbf{d}_k\right) \qquad (2.91)\\ &= -(D\Gamma)^{-1}\left(D\Phi - \Lambda D\right)\mathbf{x}_k + (D\Gamma)^{-1}\left(D\mathbf{x}_{m,k+1} - \Lambda D\mathbf{x}_{m,k}\right) - (D\Gamma)^{-1}D\mathbf{d}_k.\end{aligned}$$

Control law (2.91) is not implementable as it requires a priori knowledge of the disturbance. Thus, the estimation of the disturbance should be used

$$\mathbf{u}_k = (D\Gamma)^{-1}\left(\mathbf{a}_k - D\hat{\mathbf{d}}_k\right) \tag{2.92}$$

where $\hat{\mathbf{d}}_k$ is the disturbance estimation.

However, note that the disturbance estimate used in the state feedback controller designed in Sect. 2.4.2 requires full state information which is not available in this case. Therefore, an observer that is based on output feedback is proposed and will be detailed in Sect. 2.4.3.2.

2.4.3.2 Disturbance Observer Design

Note that according to *Lemma* 2.1, the disturbance can be written as

$$\mathbf{d}_k = \Gamma\mathbf{f}_k + \frac{1}{2}\Gamma\mathbf{v}_k T + O\left(T^3\right) = \Gamma\boldsymbol{\eta}_k + O\left(T^3\right) \tag{2.93}$$

where $\boldsymbol{\eta}_k = \mathbf{f}_k + \frac{1}{2}\mathbf{v}_k T$. If $\boldsymbol{\eta}_k$ can be estimated, then the estimation error of $\mathbf{d}_k$ would be $O\left(T^3\right)$ which is acceptable in practical applications.

Define the observer

$$\begin{aligned}\mathbf{x}_{\mathrm{d},k} &= \Phi\mathbf{x}_{\mathrm{d},k-1} + \Gamma\mathbf{u}_{k-1} + \Gamma\hat{\boldsymbol{\eta}}_{k-1}\\ \mathbf{y}_{\mathrm{d},k-1} &= C\mathbf{x}_{\mathrm{d},k-1}\end{aligned} \tag{2.94}$$

where $\mathbf{x}_{\mathrm{d},k-1} \in \Re^n$ is the observer state vector, $\mathbf{y}_{\mathrm{d},k-1} \in \Re^m$ is the observer output vector, $\hat{\boldsymbol{\eta}}_{k-1} \in \Re^m$ is the disturbance estimate and will act as the 'control input' to the observer, therefore we can write $\hat{\mathbf{d}}_{k-1} = \Gamma\hat{\boldsymbol{\eta}}_{k-1}$. Since the disturbance estimate will be used in the final control signal, it must not be overly large. Therefore, it is wise to avoid a deadbeat design. For this reason we design the disturbance observer based on an integral sliding surface

$$\begin{aligned}\boldsymbol{\sigma}_{\mathrm{d},k} &= \mathbf{e}_{\mathrm{d},k} - \mathbf{e}_{\mathrm{d},0} + \boldsymbol{\varepsilon}_{\mathrm{d},k}\\ \boldsymbol{\varepsilon}_{\mathrm{d},k} &= \boldsymbol{\varepsilon}_{\mathrm{d},k-1} + E_{\mathrm{d}}\mathbf{e}_{\mathrm{d},k-1}\end{aligned} \tag{2.95}$$

where $\mathbf{e}_{\mathrm{d},k} = \mathbf{y}_k - \mathbf{y}_{\mathrm{d},k}$ is the output estimation error, $\boldsymbol{\sigma}_{\mathrm{d},k}, \boldsymbol{\varepsilon}_{\mathrm{d},k} \in \Re^m$ are the sliding function and integral vectors, and E_{d} is an integral gain matrix.

Note that the sliding surface (2.95) is analogous to (2.63), that is, the set $\left(\mathbf{y}_k, \mathbf{x}_{\mathrm{d},k}, \mathbf{u}_k + \hat{\boldsymbol{\eta}}_k, \mathbf{y}_{\mathrm{d},k}, \boldsymbol{\sigma}_{\mathrm{d},k}\right)$ has duality with the set $(\mathbf{r}_k, \mathbf{x}_k, \mathbf{u}_k, \mathbf{y}_k, \boldsymbol{\sigma}_k)$, except for an one-step delay in the observer dynamics (2.94). Therefore, let $\boldsymbol{\sigma}_{\mathrm{d},k} = 0$ we can derive the virtual equivalent control $\mathbf{u}_{k-1} + \hat{\boldsymbol{\eta}}_{k-1}$, thus, analogous to (2.69),

$$\hat{\boldsymbol{\eta}}_{k-1} = (C\Gamma)^{-1}\left[\mathbf{y}_k - \Lambda_{\mathrm{d}}\mathbf{e}_{\mathrm{d},k-1} - C\Phi\mathbf{x}_{\mathrm{d},k-1} + \boldsymbol{\sigma}_{\mathrm{d},k-1}\right] - \mathbf{u}_{k-1} \tag{2.96}$$

where $\Lambda_{\mathrm{d}} = I_m - E_{\mathrm{d}}$.

In practice, the quantity $\mathbf{y}_{k+1}$ is not available at the time instant k when computing $\hat{\boldsymbol{\eta}}_k$. Therefore we can only compute $\hat{\boldsymbol{\eta}}_{k-1}$, and in the control law (2.92) we use the delayed estimate $\hat{\mathbf{d}}_k = \Gamma\hat{\boldsymbol{\eta}}_{k-1}$.

The stability and convergence properties of the observer (2.94) and the disturbance estimation (2.96) are analyzed in the following theorem:

Theorem 2.4 *The observer output $\mathbf{y}_{\mathrm{d},k}$ converges asymptotically to the true outputs $\mathbf{y}_k$, and the disturbance estimate $\hat{\mathbf{d}}_k$ converges to the actual disturbance $\mathbf{d}_{k-1}$ with the precision order $O\left(T^2\right)$.*

Proof Substituting (2.96) into (2.94), and using $\mathbf{e}_{\mathrm{d},k-1} = C\left(\mathbf{y}_{k-1} - \mathbf{y}_{\mathrm{d},k-1}\right)$, it is obtained that

$$\begin{aligned}\mathbf{x}_{\mathrm{d},k} = &\left[\Phi - \Gamma\left(C\Gamma\right)^{-1}\left(C\Phi - \Lambda_{\mathrm{d}}C\right)\right]\mathbf{x}_{\mathrm{d},k-1} + \Gamma\left(C\Gamma\right)^{-1}\left[\mathbf{y}_k - \Lambda_{\mathrm{d}}\mathbf{y}_{k-1}\right]\\ &+ \Gamma\left(C\Gamma\right)^{-1}\boldsymbol{\sigma}_{\mathrm{d},k-1}.\end{aligned} \tag{2.97}$$

Since the control and estimate $\mathbf{u}_{k-1} + \hat{\boldsymbol{\eta}}_{k-1}$ are chosen such that $\boldsymbol{\sigma}_{\mathrm{d},k} = 0$ for any $k > 0$, (2.97) renders to

$$\mathbf{x}_{\mathrm{d},k} = \left[\Phi - \Gamma\left(C\Gamma\right)^{-1}\left(C\Phi - \Lambda_{\mathrm{d}}C\right)\right]\mathbf{x}_{\mathrm{d},k-1} + \Gamma\left(C\Gamma\right)^{-1}\left[\mathbf{y}_k - \Lambda_{\mathrm{d}}\mathbf{y}_{k-1}\right]. \tag{2.98}$$

The second term on the right hand side of (2.98) can be expressed as

$$\begin{aligned}\Gamma\left(C\Gamma\right)^{-1}\left[\mathbf{y}_k - \Lambda_{\mathrm{d}}\mathbf{y}_{k-1}\right] = &\;\Gamma\left(C\Gamma\right)^{-1}\left(C\Phi - \Lambda_{\mathrm{d}}C\right)\mathbf{x}_{k-1} + \Gamma\mathbf{u}_{k-1}\\ &+ \Gamma\left(C\Gamma\right)^{-1}C\mathbf{d}_{k-1}\end{aligned}$$

by using the relations $\mathbf{y}_k = C\Phi\mathbf{x}_{k-1} + C\Gamma\mathbf{u}_{k-1} + C\mathbf{d}_{k-1}$ and $\mathbf{y}_{k-1} = C\mathbf{x}_{k-1}$. Therefore (2.98) can be rewritten as

$$\mathbf{x}_{\mathrm{d},k} = \Phi\mathbf{x}_{\mathrm{d},k-1} + \Gamma\left(C\Gamma\right)^{-1}\left(C\Phi - \Lambda_{\mathrm{d}}C\right)\Delta\mathbf{x}_{\mathrm{d},k-1} + \Gamma\mathbf{u}_k + \Gamma\left(C\Gamma\right)^{-1}C\mathbf{d}_{k-1} \tag{2.99}$$

where $\Delta\mathbf{x}_{\mathrm{d},k-1} = \mathbf{x}_{k-1} - \mathbf{x}_{\mathrm{d},k-1}$.

Further subtracting (2.99) from the system (2.2) we obtain

$$\Delta\mathbf{x}_{\mathrm{d},k} = \left[\Phi - \Gamma\left(C\Gamma\right)^{-1}\left(C\Phi - \Lambda_{\mathrm{d}}C\right)\right]\Delta\mathbf{x}_{\mathrm{d},k-1} + \left[I - \Gamma\left(C\Gamma\right)^{-1}C\right]\mathbf{d}_{k-1} \tag{2.100}$$

where $\left[I - \Gamma (C\Gamma)^{-1} C\right] \mathbf{d}_{k-1}$ is $O\left(T^3\right)$ because

$$\left[I - \Gamma (C\Gamma)^{-1} C\right]\left[\Gamma \eta_{k-1} + O\left(T^3\right)\right] = \left[I - \Gamma (C\Gamma)^{-1} C\right] O\left(T^3\right) = O\left(T^3\right).$$

Applying *Lemma* 2.1, $\Delta \mathbf{x}_{\mathrm{d},k-1} = O\left(T^2\right)$.

From (2.100) we can see that the stability of the disturbance observer depends only on the matrix $\left[\Phi - \Gamma (C\Gamma)^{-1} (C\Phi - \Lambda_{\mathrm{d}} C)\right]$ and is guaranteed by the selection of the matrix Λ_{d} and the fact that system (Φ, Γ, C) is minimum phase. It should also be noted that the residue term $\left[I - \Gamma (C\Gamma)^{-1} C\right] \mathbf{d}_{k-1}$ in the state space is orthogonal to the output space, as $C\left[I - \Gamma (C\Gamma)^{-1} C\right] \mathbf{d}_{k-1} = 0$. Therefore premultliplication of (2.100) with C yields the output tracking error dynamics

$$\mathbf{e}_{\mathrm{d},k} = \Lambda_{\mathrm{d}} \mathbf{e}_{\mathrm{d},k-1} \tag{2.101}$$

which is asymptotically stable through choosing a stable matrix Λ_{d}.

Finally we discuss the convergence property of the estimate $\hat{\mathbf{d}}_{k-1}$. Subtracting (2.94) from (2.2) with one-step delay, we obtain

$$\Delta \mathbf{x}_{\mathrm{d},k} = \Phi \Delta \mathbf{x}_{\mathrm{d},k-1} + \Gamma \left(\eta_{k-1} - \hat{\eta}_{k-1}\right) + O\left(T^3\right). \tag{2.102}$$

Premultiplying (2.102) with C, and substituting (2.101) that describes $C \Delta \mathbf{x}_{\mathrm{d},k}$, yield

$$\hat{\eta}_{k-1} = \eta_{k-1} + (C\Gamma)^{-1} (C\Phi - \Lambda_{\mathrm{d}} C) \Delta \mathbf{x}_{\mathrm{d},k-1} + (C\Gamma)^{-1} O\left(T^3\right). \tag{2.103}$$

The second term on the right hand side of (2.103) is $O(T)$ because $\Delta \mathbf{x}_{\mathrm{d},k-1} = O\left(T^2\right)$ but $(C\Gamma)^{-1} = O\left(T^{-1}\right)$. As a result, from (2.103) we can conclude that $\hat{\eta}_{k-1}$ approaches η_{k-1} with the precision $O(T)$. In terms of the relationship

$$\mathbf{d}_{k-1} - \hat{\mathbf{d}}_k = \Gamma \left(\eta_{k-1} - \hat{\eta}_{k-1}\right) + O\left(T^3\right)$$

and $\Gamma = O(T)$, we conclude $\hat{\mathbf{d}}_k$ converges to $\mathbf{d}_{k-1}$ with the precision of $O\left(T^2\right)$.

Remark 2.10 At the time k, we can guarantee the convergence of $\hat{\eta}_{k-1}$ to η_{k-1} with the precision $O(T)$. In other words, we can guarantee the convergence of the disturbance estimate at the time k, $\hat{\mathbf{d}}_k$, to the actual disturbance at time $k-1$, $\mathbf{d}_{k-1}$, with the precision $O\left(T^2\right)$. This result is consistent with the state-based estimation presented in Sect. 2.3.1 in which $\hat{\mathbf{d}}_k$ is made equal to $\mathbf{d}_{k-1}$. Comparing differences between the state-based and output-based disturbance estimation, the former has only one-step delay with perfect precision, where as the latter is asymptotic with $O\left(T^2\right)$ precision.

2.4.3.3 Stability Analysis

To analyze the stability of the closed-loop system, substitute $\mathbf{u}_k$ in (2.92) into the system (2.2) leading to the closed-loop equation in the sliding mode

$$\mathbf{x}_{k+1} = \left[\Phi - \Gamma (D\Gamma)^{-1} (D\Phi - \Lambda_{\mathrm{d}} D)\right] \mathbf{x}_k + \mathbf{d}_k - \Gamma (D\Gamma)^{-1} D\hat{\mathbf{d}}_k$$
$$+ \Gamma (D\Gamma)^{-1} \left[D\mathbf{x}_{\mathrm{m},k+1} - \Lambda_{\mathrm{d}} D\mathbf{x}_{\mathrm{m},k} + \boldsymbol{\sigma}_k\right]. \tag{2.104}$$

The stability of the above sliding equation is summarized in the following theorem:

Theorem 2.5 *Using the control law (2.92) the sliding mode is*

$$\boldsymbol{\sigma}_{k+1} = D\left(\hat{\mathbf{d}}_k - \mathbf{d}_k\right).$$

Further, the state tracking error $\Delta\mathbf{x}_k = \mathbf{x}_{\mathrm{m},k} - \mathbf{x}_k$ is bounded if system (Φ, Γ, D) is minimum-phase and the eigenvalues of the matrix Λ_{d} are within the unit circle.

Proof In order to verify the first part of *Theorem* 2.5, rewrite the dynamics of the slidng mode (2.90)

$$\boldsymbol{\sigma}_{k+1} = \mathbf{a}_k - D\Gamma\mathbf{u}_k - D\mathbf{d}_k = \mathbf{a}_k - D\Gamma\mathbf{u}_k^{\mathrm{eq}} - D\mathbf{d}_k + D\Gamma\left(\mathbf{u}_k^{\mathrm{eq}} - \mathbf{u}_k\right)$$
$$= D\Gamma\left(\mathbf{u}_k^{\mathrm{eq}} - \mathbf{u}_k\right)$$

where we use the property of equivalent control $\boldsymbol{\sigma}_{k+1} = \mathbf{a}_k - D\Gamma\mathbf{u}_k^{\mathrm{eq}} - D\mathbf{d}_k = 0$. Comparing two control laws (2.91) and (2.92), we obtain

$$\boldsymbol{\sigma}_{k+1} = D\left(\hat{\mathbf{d}}_k - \mathbf{d}_k\right).$$

Note that if there is no disturbance or we have perfect estimation of the disturbance, then $\boldsymbol{\sigma}_{k+1} = 0$ as desired. From the results of *Theorem* 2.3 and *Lemma* 2.1

$$\hat{\mathbf{d}}_k - \mathbf{d}_k = \hat{\mathbf{d}}_k - \mathbf{d}_{k-1} - (\mathbf{d}_k - \mathbf{d}_{k-1}) = O\left(T^2\right)$$

as $k \to \infty$. Thus $\boldsymbol{\sigma}_{k+1} \to O\left(T^2\right)$ which is acceptable in practice.

To prove the boundedness of the state tracking error $\Delta\mathbf{x}_k$, first derive the state error dynamics. Subtracting both sides of (2.104) from the reference model (2.87), and substituting $\boldsymbol{\sigma}_k = D\left(\hat{\mathbf{d}}_{k-1} - \mathbf{d}_{k-1}\right)$, yields

$$\Delta\mathbf{x}_{k+1} = \left[\Phi - \Gamma (D\Gamma)^{-1} (D\Phi - \Lambda_{\mathrm{d}} D)\right] \Delta\mathbf{x}_k$$
$$+ \left[I - \Gamma (D\Gamma)^{-1} D\right] \left(K_2\mathbf{r}_{k+1} - K_1\mathbf{x}_{\mathrm{m},k}\right) - \boldsymbol{\zeta}_k \tag{2.105}$$

where

$$\boldsymbol{\zeta}_k = \mathbf{d}_k - \Gamma (D\Gamma)^{-1} D \left(\hat{\mathbf{d}}_k - \hat{\mathbf{d}}_{k-1} + \mathbf{d}_{k-1}\right). \tag{2.106}$$

The stability of (2.105) is dependent on $\left[\Phi - \Gamma (D\Gamma)^{-1} (D\Phi - \Lambda_{\mathrm{d}} D)\right]$. From *Lemma* 2.1 the closed-loop poles of (2.105) are the eigenvalues of Λ_{d} and the open-loop zeros of the system (Φ, Γ, D). Thus, m poles of the closed-loop system can be selected by the proper choice of the matrix Λ_{d} while the remaining poles are stable only if the system (Φ, Γ, D) is minimum-phase. Note that both $\mathbf{r}_{k+1}$ and $\mathbf{x}_{\mathrm{m},k}$ are reference signals and are bounded. Therefore we need only to show the boundedness of $\boldsymbol{\zeta}_k$ which is

$$\boldsymbol{\zeta}_k = \left(I - \Gamma (D\Gamma)^{-1} D\right) \mathbf{d}_k + \Gamma (D\Gamma)^{-1} D \left(\mathbf{d}_k - \hat{\mathbf{d}}_k\right) - \Gamma (D\Gamma)^{-1} D \left(\mathbf{d}_{k-1} - \hat{\mathbf{d}}_{k-1}\right). \tag{2.107}$$

From *Theorem* 2.3, the second and third terms on the right hand side of (2.107) are $O\left(T^2\right)$. From *Lemma* 2.1, the first term on the right hand side of (2.107) can be written as

$$\left(I - \Gamma (D\Gamma)^{-1} D\right) \mathbf{d}_k = \left(I - \Gamma (D\Gamma)^{-1} D\right) \left(\Gamma \boldsymbol{\eta}_k + O\left(T^3\right)\right) = O\left(T^3\right).$$

Therefore $\boldsymbol{\zeta}_k = O\left(T^2\right)$ which is bounded.

We have established the stability condition for the closed-loop system, but, have yet to establish the ultimate tracking error bound. From (2.105) it can be seen that the tracking error bound is dependent on the disturbance estimate $\hat{\mathbf{d}}_k$ as well as the selection of K_1 and K_2. Up to this point, not much was discussed in terms of the selection of the reference model (2.87). As it can be seen from (2.105) the selection of the reference model can effect the overall tracking error bound. Since we consider an arbitrary reference $\mathbf{r}_k$, the reference model must be selected such that its output is the reference signal $\mathbf{r}_k$. To achieve this requirement, we explore two possible selections of the reference model.

2.4.3.4 Error Bound for a Minimum-Phase (Φ, Γ, C)

For this case the reference model requires that the matrices $K_1 = \Gamma (C\Gamma)^{-1} C\Phi$ and $K_2 = \Gamma (C\Gamma)^{-1}$, therefore, the reference model (2.87) can be written as

$$\begin{aligned} \mathbf{x}_{\mathrm{m},k+1} &= \left[\Phi - \Gamma (C\Gamma)^{-1} C\Phi\right] \mathbf{x}_{\mathrm{m},k} + \Gamma (C\Gamma)^{-1} \mathbf{r}_{k+1} \\ \mathbf{y}_{\mathrm{m},k} &= C\mathbf{x}_{\mathrm{m},k} = \mathbf{r}_k. \end{aligned} \tag{2.108}$$

It can be seen from (2.108) that it is stable if the matrix $\left[\Phi - \Gamma (C\Gamma)^{-1} C\Phi\right]$ is stable, i.e., the system (Φ, Γ, C) is minimum-phase. Substituting the selected matrices K_1 and K_2 into (2.105) and using the fact that $\left[I - \Gamma (D\Gamma)^{-1} D\right]\Gamma = 0$, we obtain

$$\Delta \mathbf{x}_{k+1} = \left[\Phi - \Gamma (D\Gamma)^{-1} (D\Phi - \Lambda_{\mathrm{d}} D)\right] \Delta \mathbf{x}_k - \boldsymbol{\zeta}_k. \tag{2.109}$$

where $\boldsymbol{\zeta}_k = O\left(T^2\right)$ according to *Theorem* 2.5. As for the output tracking error bound, from *Lemma* 2.1, the ultimate error bound on $\|\Delta \mathbf{x}_k\|$ will be one order higher than the bound on $\boldsymbol{\zeta}_k$. Thus, the ultimate bound on the output tracking error is

$$\|\mathbf{e}_k\| \leq \|C\| \cdot \|\Delta \mathbf{x}_k\| = O(T). \tag{2.110}$$

2.4.3.5 Error Bound for a Minimum-Phase $(\boldsymbol{\Phi}, \boldsymbol{\Gamma}, \boldsymbol{D})$

In the case that it is only possible to satisfy (Φ, Γ, D) to be minimum-phase, a different reference model needs to be selected. For this new reference model, select the matrices $K_1 = \Gamma (D\Gamma)^{-1} D\Phi$ and $K_2 = \Gamma (D\Gamma)^{-1}$. Then the reference model (2.87) can be written as

$$\begin{aligned} \mathbf{x}_{\mathrm{m},k+1} &= \left[\Phi - \Gamma (D\Gamma)^{-1} D\Phi\right] \mathbf{x}_{\mathrm{m},k} + \Gamma (D\Gamma)^{-1} \mathbf{r}_{k+1} \\ \mathbf{y}_{\mathrm{m},k} &= D\mathbf{x}_{\mathrm{m},k} = \mathbf{r}_k. \end{aligned} \tag{2.111}$$

The matrix $\left[\Phi - \Gamma (D\Gamma)^{-1} C\Phi\right]$ is stable only if (Φ, Γ, D) is minimum-phase. Substituting the selected matrices K_1 and K_2 into (2.105), and using the property $\left[I - \Gamma (D\Gamma)^{-1} D\right]\Gamma = 0$, we have

$$\Delta \mathbf{x}_{k+1} = \left[\Phi - \Gamma (D\Gamma)^{-1} (D\Phi - \Lambda_{\mathrm{d}} D)\right] \Delta \mathbf{x}_k - \boldsymbol{\zeta}_k. \tag{2.112}$$

We can see from (2.112) that the tracking error bound is only dependent on the disturbance estimation $\boldsymbol{\zeta}_k$.

On the other hand, the disturbance observer requires (Φ, Γ, C) to be minimum-phase, hence is not implementable in this case. Without the disturbance estimator, using *Lemma* 2.1, (2.106) becomes

$$\begin{aligned} \boldsymbol{\zeta}_k &= \mathbf{d}_k - \Gamma (D\Gamma)^{-1} \mathbf{d}_{k-1} \\ &= \mathbf{d}_k - \mathbf{d}_{k-1} + \left[I - \Gamma (D\Gamma)^{-1} D\right] \left(\Gamma \eta_{k-1} + O\left(T^3\right)\right) \\ &= O\left(T^2\right) + O\left(T^3\right) = O\left(T^2\right). \end{aligned} \tag{2.113}$$

As a result, the closed-loop system is

$$\Delta \mathbf{x}_{k+1} = \left[\Phi - \Gamma (D\Gamma)^{-1} (D\Phi - \Lambda_{\mathrm{d}} D)\right] \Delta \mathbf{x}_k + O\left(T^2\right). \quad (2.114)$$

By *Lemma* 2.1, the ultimate bound on $\|\Delta \mathbf{x}_k\| = O(T)$, and therefore, the ultimate bound on the tracking error is

$$\|\mathbf{e}_k\| \leq \|D\| \cdot \|\Delta \mathbf{x}_k\| = O(T). \quad (2.115)$$

While this approach gives a similar precision in output tracking performance, it only requires (Φ, Γ, D) to be minimum-phase and can be used in the cases (Φ, Γ, C) is not minimum-phase.

2.4.4 Output Tracking ISM: State Observer Approach

In this section we explore the observer-based approach for the unknown states when only output measurement is available. By virtue of state estimation, it is required that (Φ, Γ, C) to be minimum-phase, thus, the observer based disturbance estimation approach in Sect. 2.4.3.2 is applicable. From the derivation procedure in Sect. 2.4.3.2, we can see that the error dynamics (2.101) of the disturbance observer (2.94) is independent of the control inputs $\mathbf{u}_k$. Therefore the same disturbance observer (2.94)–(2.96) can be incorporated in the state-observer approach directly without any modification. In this section we focus on the design and analysis of the control law and state observer.

2.4.4.1 Controller Structure and Closed-Loop System

With the state estimation, the ISM control can be constructed according to the preceding state-feedback based ISM control design (2.69) by substituting $C\Phi\mathbf{x}_k$ with $C\Phi\mathbf{x}_k - C\Phi\tilde{\mathbf{x}}_k$ where $\tilde{\mathbf{x}}_k = \mathbf{x}_k - \hat{\mathbf{x}}_k$ is the state estimation error,

$$\begin{aligned} \mathbf{u}_k = & -(C\Gamma)^{-1} (C\Phi - \Lambda_{\mathrm{d}} C) \mathbf{x}_k - (C\Gamma)^{-1} C\hat{\mathbf{d}}_k + (C\Gamma)^{-1} (\mathbf{r}_{k+1} - \Lambda_{\mathrm{d}} \mathbf{r}_k) \\ & + (C\Gamma)^{-1} \boldsymbol{\sigma}_k + (C\Gamma)^{-1} C\Phi\tilde{\mathbf{x}}_k. \end{aligned} \quad (2.116)$$

In comparison to the control law (2.69), the control law (2.116) has an additional term $(C\Gamma)^{-1} C\Phi\tilde{\mathbf{x}}_k$ due to the state estimation error. Substituting $\mathbf{u}_k$ in (2.116) into (2.2) yields the closed-loop dynamics

$$\begin{aligned} \mathbf{x}_{k+1} = & \left[\Phi - \Gamma (C\Gamma)^{-1} (C\Phi - \Lambda_{\mathrm{d}} C)\right] \mathbf{x}_k + \mathbf{d}_k - \Gamma (C\Gamma)^{-1} C\hat{\mathbf{d}}_k \\ & + \Gamma (C\Gamma)^{-1} (\mathbf{r}_{k+1} - \Lambda_{\mathrm{d}} \mathbf{r}_k) + \Gamma (C\Gamma)^{-1} \boldsymbol{\sigma}_k \\ & + \Gamma (C\Gamma)^{-1} C\Phi\tilde{\mathbf{x}}_k \end{aligned} \quad (2.117)$$

which, comparing with the state-feedback (2.70), is almost the same except for an additional term $\Gamma (C\Gamma)^{-1} C\Phi\tilde{\mathbf{x}}_k$. Hence, following the proof of *Theorem* 2.2, the properties of the closed-loop system (2.117) can be derived.

Theorem 2.6 *Assume that the system (2.2) is minimum-phase and the eigenvalues of the matrix Λ_d are within the unit circle. Then by the control law (2.116) we have*

$$\boldsymbol{\sigma}_{k+1} = C\left(\hat{\mathbf{d}}_k - \mathbf{d}_k\right) - C\Phi\tilde{\mathbf{x}}_k \tag{2.118}$$

and the error dynamics

$$\mathbf{e}_{k+1} = \Lambda_\mathrm{d}\mathbf{e}_k + \boldsymbol{\delta}_k \tag{2.119}$$

where $\boldsymbol{\delta}_k = C\left(\hat{\mathbf{d}}_k - \mathbf{d}_k + \mathbf{d}_{k-1} - \hat{\mathbf{d}}_{k-1}\right) - C\Phi\left(\tilde{\mathbf{x}}_k - \tilde{\mathbf{x}}_{k-1}\right)$.

Proof In order to prove (2.118), notice that (2.64) can be written as

$$\boldsymbol{\sigma}_{k+1} = \mathbf{r}_{k+1} - C\mathbf{x}_{k+1} - \Lambda_\mathrm{d}\mathbf{r}_k + \Lambda_\mathrm{d}C\mathbf{x}_k + \boldsymbol{\sigma}_k. \tag{2.120}$$

Substituting the closed-loop dynamics (2.117) into (2.120) and simplifying we obtain

$$\boldsymbol{\sigma}_{k+1} = C\left(\hat{\mathbf{d}}_k - \mathbf{d}_k\right) - C\Phi\tilde{\mathbf{x}}_k$$

which proves the first part of the theorem.

In order to prove the second part, premultiply (2.117) with C and simplify to obtain the following result

$$\mathbf{y}_{k+1} = \Lambda_\mathrm{d}\mathbf{y}_k + C\mathbf{d}_k - C\hat{\mathbf{d}}_k + \mathbf{r}_{k+1} - \Lambda_\mathrm{d}\mathbf{r}_k + C\Phi\tilde{\mathbf{x}}_k + \boldsymbol{\sigma}_k \tag{2.121}$$

From the result (2.118) it can be obtained that $\boldsymbol{\sigma}_k = C\left(\hat{\mathbf{d}}_{k-1} - \mathbf{d}_{k-1}\right) - C\Phi\tilde{\mathbf{x}}_{k-1}$. Substituting in (2.121) and using the fact that $\mathbf{e}_k = \mathbf{r}_k - \mathbf{y}_k$ we obtain

$$\begin{aligned}\mathbf{e}_{k+1} &= \Lambda_\mathrm{d}\mathbf{e}_k + C\left(\hat{\mathbf{d}}_k - \mathbf{d}_k + \mathbf{d}_{k-1} - \hat{\mathbf{d}}_{k-1}\right) - C\Phi\left(\tilde{\mathbf{x}}_k - \tilde{\mathbf{x}}_{k-1}\right)\\ &= \Lambda_\mathrm{d}\mathbf{e}_k + \boldsymbol{\delta}_k \end{aligned} \tag{2.122}$$

where $\boldsymbol{\delta}_k = C\left(\hat{\mathbf{d}}_k - \mathbf{d}_k + \mathbf{d}_{k-1} - \hat{\mathbf{d}}_{k-1}\right) - C\Phi\left(\tilde{\mathbf{x}}_k - \tilde{\mathbf{x}}_{k-1}\right)$.

As in the state-feedback approach, the output tracking error depends on the proper selection of the eigenvalues of Λ_d, as well as the disturbance estimation and state estimation precision. The influence of the disturbance estimation has been discussed in *Theorem* 2.4. The effect of $\tilde{\mathbf{x}}_k$ on the tracking error bound will be evaluated, we will discuss the state observer in the next section.

2.4.4.2 State Observer

State estimation will be accomplished with the following state-observer

$$\hat{\mathbf{x}}_{k+1} = \Phi \hat{\mathbf{x}}_k + \Gamma \mathbf{u}_k + L\left(\mathbf{y}_k - \hat{\mathbf{y}}_k\right) + \hat{\mathbf{d}}_k \tag{2.123}$$

where $\hat{\mathbf{x}}_k$, $\hat{\mathbf{y}}_k$ are the state and output estimates and L is a design matrix. Observer (2.123) is well-known, however, the term $\hat{\mathbf{d}}_k$ has been added to compensate for the disturbance. It is necessary to investigate the effect of the disturbance estimation on the state estimation. Subtracting (2.123) from (2.2) we get

$$\tilde{\mathbf{x}}_{k+1} = [\Phi - LC]\,\tilde{\mathbf{x}}_k + \mathbf{d}_k - \hat{\mathbf{d}}_k. \tag{2.124}$$

It can be seen that the state estimation is independent of the control inputs. Under the assumption that (Φ, Γ, C) is controllable and observable, we can choose L such that $\Phi - LC$ is asymptotically stable. From *Theorem* 2.4, $\mathbf{d}_k - \hat{\mathbf{d}}_k = O\left(T^2\right)$, thus, from *Lemma* 2.1 the ultimate bound of $\tilde{\mathbf{x}}_k$ is $O\left(T\right)$. Later we will show that, for systems of relative degree greater than 1, by virtue of the integral action in the ISM control, the state estimation error will be reduced to $O\left(T^2\right)$ in the closed-loop system.

2.4.4.3 Tracking Error Bound

In order to calculate the tracking error bound we need first calculate the bound of ζ_k in *Theorem* 2.5. From the error dynamics of the state estimation (2.124), the solution trajectory is

$$\tilde{\mathbf{x}}_k = [\Phi - LC]^k\,\tilde{\mathbf{x}}_0 + \sum_{i=0}^{k-1}\left([\Phi - LC]^{k-1-i}\left(\mathbf{d}_i - \hat{\mathbf{d}}_i\right)\right). \tag{2.125}$$

The difference $\tilde{\mathbf{x}}_k - \tilde{\mathbf{x}}_{k-1}$ can be calculated as

$$\begin{aligned}\tilde{\mathbf{x}}_k - \tilde{\mathbf{x}}_{k-1} &= [(\Phi - LC) - I_n]\,(\Phi - LC)^{k-1}\,\tilde{\mathbf{x}}_0 + \sum_{i=0}^{k-1}\left([\Phi - LC]^{k-1-i}\left(\mathbf{d}_i - \hat{\mathbf{d}}_i\right)\right)\\ &-\sum_{i=0}^{k-2}\left([\Phi - LC]^{k-1-i}\left(\mathbf{d}_i - \hat{\mathbf{d}}_i\right)\right)\end{aligned} \tag{2.126}$$

which can be simplified to

$$\tilde{\mathbf{x}}_k - \tilde{\mathbf{x}}_{k-1} = [(\Phi - LC) - I_n]\,(\Phi - LC)^{k-1}\,\tilde{\mathbf{x}}_0 + \left(\mathbf{d}_k - \hat{\mathbf{d}}_k\right). \tag{2.127}$$

Since $(\Phi - LC)$ is asymptotically stable, the ultimate bound is

$$\tilde{\mathbf{x}}_k - \tilde{\mathbf{x}}_{k-1} = \mathbf{d}_k - \hat{\mathbf{d}}_k, \tag{2.128}$$

and $\boldsymbol{\delta}_k$ can be expressed ultimately as

$$\begin{aligned}\boldsymbol{\delta}_k &= C\left(\hat{\mathbf{d}}_k - \mathbf{d}_k + \mathbf{d}_{k-1} - \hat{\mathbf{d}}_{k-1}\right) - C\Phi\left(\tilde{\mathbf{x}}_k - \tilde{\mathbf{x}}_{k-1}\right)\\ &= C\left(\hat{\mathbf{d}}_k - \mathbf{d}_k + \mathbf{d}_{k-1} - \hat{\mathbf{d}}_{k-1}\right) - C\Phi\left(\mathbf{d}_k - \hat{\mathbf{d}}_k\right).\end{aligned} \tag{2.129}$$

From *Theorem* 2.4, the disturbance estimation error $\mathbf{d}_k - \hat{\mathbf{d}}_k$ is $O\left(T^2\right)$. Therefore we have

$$\boldsymbol{\delta}_k = C\cdot\left(O\left(T^2\right) + O\left(T^2\right)\right) - C\Phi\cdot O\left(T^2\right) = O\left(T^2\right).$$

The ultimate bound on $\boldsymbol{\sigma}_k$ is $O\left(T^2\right)$ according to (2.118), and, from (2.122) the ultimate bound on $\|\mathbf{e}_k\|$ is $O\left(T\right)$.

Remark 2.11 Note that the guaranteed tracking precision is $O\left(T\right)$ because the control problem becomes highly challenging in the presence of state estimation and disturbance estimation errors, and meanwhile aiming at arbitrary reference tracking.

In many motion control tasks the system relative degree is 2, for instance from the torque or force to position tracking in motion control. Now we derive an interesting property by the following corollary.

Corollary 2.1 *For a continuous system of relative degree greater than 1, the ultimate bound of* $\|\mathbf{e}_k\|$ *is* $O\left(T^2\right)$.

Proof From (2.84) in *Theorem* 2.5 and *Lemma* 2.1, $\|\mathbf{e}_k\|$ is $O\left(T^2\right)$ if $\boldsymbol{\delta}_k = O\left(T^3\right)$. When the system relative degree is 2, $CB = 0$, and

$$C\Gamma = C\left(BT + \frac{1}{2!}ABT^2 + \frac{1}{3!}A^2BT^3 + \cdots\right) = \frac{1}{2!}CABT^2 + \frac{1}{3!}CA^2BT^3 + \cdots = O\left(T^2\right). \tag{2.130}$$

Similarly

$$C\Phi\Gamma = C\left(I + AT + \frac{1}{2!}A^2T^2 + \cdots\right)\Gamma = C\left(I + O\left(T\right)\right)\Gamma = C\Gamma + O\left(T^2\right) = O\left(T^2\right). \tag{2.131}$$

Now rewrite

$$\begin{aligned}\boldsymbol{\delta}_k &= C\left(\hat{\mathbf{d}}_k - \mathbf{d}_k + \mathbf{d}_{k-1} - \hat{\mathbf{d}}_{k-1}\right) - C\Phi\left(\mathbf{d}_k - \hat{\mathbf{d}}_k\right) \\ &= C\Gamma\left(\hat{\boldsymbol{\eta}}_k - \boldsymbol{\eta}_k + \boldsymbol{\eta}_{k-1} - \hat{\boldsymbol{\eta}}_{k-1}\right) - C\Phi\Gamma\left(\boldsymbol{\eta}_k - \hat{\boldsymbol{\eta}}_k\right) + O\left(T^3\right). \quad (2.132)\end{aligned}$$

Note that the ultimate bound of $\boldsymbol{\eta}_k - \hat{\boldsymbol{\eta}}_k$, derived in *Theorem* 2.3, is $O(T)$. Thus we conclude from (2.132)

$$\boldsymbol{\delta}_k = O\left(T^2\right)\cdot\left(O(T) + O(T)\right) - O\left(T^2\right)\cdot O(T) + O\left(T^3\right) = O\left(T^3\right)$$

and consequently $\|\mathbf{e}_k\|$ is ultimately $O\left(T^2\right)$.

Remark 2.12 From the result we see that even though the state estimation error is $O(T)$ we can still obtain $O\left(T^2\right)$ output tracking by virtue of the integral action in the controller design for relative degree greater than 1.

2.4.5 Systems with a Piece-Wise Smooth Disturbance

In practice, the disturbance of a motion system, $\mathbf{f}(\mathbf{x}, t)$, may become discontinuous at certain circumstances. For example, due to the static friction force, a discontinuity occurs when the motion speed drops to zero. It is thus vital to examine the system performance around the time the discontinuity occurs.

Suppose the the discontinuity of $\mathbf{f}(\mathbf{x}, t)$ occurs at the jth sampling instant. The immediate consequence of the discontinuity in $\mathbf{f}(\mathbf{x}, t)$ is the loss of the property $\mathbf{d}_j - \mathbf{d}_{j-1} = O(T)$ instead of $O\left(T^2\right)$, and the loss of the property $\mathbf{d}_j - 2\mathbf{d}_{j-1} + \mathbf{d}_{j-2} = O(T)$ instead of $O\left(T^3\right)$.

Since we focus on tracking tasks in this work, it can be reasonably assumed that the discontinuity occurs only occasionally. As such, the property $\mathbf{d}_j - \mathbf{d}_{j-1} = O\left(T^2\right)$ will be restored one step after the occurrence of the discontinuity, and the property: $\mathbf{d}_j - 2\mathbf{d}_{j-1} + \mathbf{d}_{j-2} = O\left(T^3\right)$ will be restored two step after the occurrence of the discontinuity. The discontinuity presents an impulse-like impact to the system behavior at the instant $k = j$. It is worth to investigate the applicability of *Lemma* 2.1 for this case. Write

$$\mathbf{e}_k = \Lambda_{\mathrm{d}}^k\mathbf{e}_0 + \sum_{i=0}^{k-1}\Lambda_{\mathrm{d}}^i\boldsymbol{\delta}_{k-i-1}.$$

For simplicity assume $\Lambda_{\mathrm{d}}^k\mathbf{e}_0$ can be ignored and $\mathbf{e}_k = O\left(T^2\right)$ at $k = j$. Then $\boldsymbol{\delta}_j$ will give $\mathbf{e}_k$ an offset with the magnitude $O(T)$, which can be viewed as a new initial error at the time instant j and will disappear exponentially with the rate Λ_{d}^k.

Note that Λ_d is a design matrix, thus can be chosen to be sufficiently small such that the impact from $\boldsymbol{\delta}_j$, which is $O(T)$, can be quickly attenuated. As a result, the analysis of the preceding sections still holds for discontinuous disturbances.

In the worst case the non-smooth disturbance presents for a long interval, we can consider a nonlinear switching control action. Denote $\mathbf{u}_k$ the ISM control designed in preceding sections, a new nonlinear control law is

$$\mathbf{u}_k^\mathrm{n} = \mathbf{u}_k + \mu\,\mathrm{sat}\,(\boldsymbol{\sigma}_k) \tag{2.133}$$

where sat $(\boldsymbol{\sigma}_k)$ is a vector with each of the elements given by

$$\mathrm{sat}\,(x) = \begin{cases} 1 & \forall x \in (\varepsilon, \infty) \\ \frac{x}{\varepsilon} & \forall x \in [-\varepsilon, \varepsilon] \\ -1 & \forall x \in (-\infty, -\varepsilon) \end{cases}$$

and ε is the required bound on the sliding function $\boldsymbol{\sigma}_k$.

Remark 2.13 Let $\varepsilon \to 0$, it is known that sat $(\cdot)$ renders to a signum function and improves the control system bandwidth, hence supress the discontinuous disturbance. In digital implementation, however, the actual bandwidth, being limited by the sampling frequency, is $\pi/2T$ where T is the sampling period.

2.4.6 Illustrative Example

2.4.6.1 State Regulation

Consider the system (2.1) with the following parameters

$$A = \begin{bmatrix} 1 & -2 & 3 \\ -4 & 5 & -6 \\ 7 & -8 & 9 \end{bmatrix}, \quad B = \begin{bmatrix} 1 & -2 \\ -3 & 4 \\ 5 & 6 \end{bmatrix},$$
$$C = \begin{bmatrix} 0 & 1 & 2 \\ 4 & -1 & 2 \end{bmatrix} \quad \& \quad \mathbf{f}\,(t) = \begin{bmatrix} 0.3\sin(4\pi t) \\ 0.3\cos(4\pi t) \end{bmatrix}$$

The initial states are $\mathbf{x}_0 = [1 \quad 1 \quad -1]^T$. The system will be simulated for a sample interval of 1 ms. For the classical SM control, the D matrix is chosen such that the non-zero pole of the sliding dynamics is $p = -5$ in continuous-time, or $z = 0.9950$ in discrete-time. Hence, the poles of the system with SM are $[0 \quad 0.9950 \quad 0]^T$ respectively. Accordingly the D matrix is

$$D = \begin{bmatrix} 0.2621 & -0.3108 & -0.0385 \\ 3.4268 & 2.4432 & 1.1787 \end{bmatrix}.$$

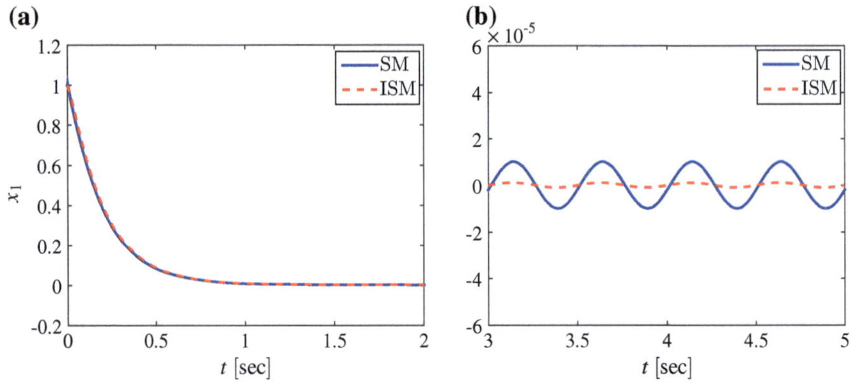

Fig. 2.1 System state x_1 **a** Transient performance **b** Steady-State performance

Using the same D matrix given above, the system with ISM is designed such that the dominant (non-zero) pole remains the same, but, the remaining poles are not deadbeat. The poles are selected as $z = [0.9048 \quad 0.9950 \quad 0.8958]^T$ respectively.

Using the pole placement command of Matlab, the gain matrices can be obtained

$$K = \begin{bmatrix} 66.6705 & 9.4041 & 15.8872 \\ 18.2422 & 21.3569 & 8.5793 \end{bmatrix}.$$

According to (2.45)

$$E = \begin{bmatrix} 0.0297 & -0.0313 & -0.0034 \\ 0.3147 & 0.2366 & 0.1115 \end{bmatrix}.$$

The delayed disturbance compensation is used. Figure 2.1a shows that the system state $x_1(t)$ is asymptotically stable for both discrete-time SM control and ISM control, which show almost the same behavior globally. On the other hand, the difference in the steady-state response between discrete-time SM control and ISM control can be seen from Fig.2.1b. The control inputs are shown in Fig. 2.2. Note that the control inputs of the SM control has much larger values at the initial phase in comparison with ISM control, due to the existence of deadbeat poles. Another reason for the lower value of the control inputs in the ISM control is the elimination of the reaching phase by compensating for the non-zero intial condition in (2.50).

To demonstrate the quality of both designs, the open-loop transfer function matrices, $G_{i,j}^{\mathrm{OL}}$, for the systems with SM and ISM are computed and Bode plots of some elements are shown in Fig. 2.3. In addition, the sensitivity function of state x_1 with respect to the disturbance components $f_1(t)$ and $f_2(t)$ is plotted in Fig. 2.4. It can be seen from Figs. 2.3 and 2.4 that ISM control greatly reduces the effect of the disturbance as compared to SM control. Moreover ISM presents a larger open-loop gain at the lower frequency band by virtue of the integral action in the sliding manifold, which ensures a more accurate closed-loop response to possible reference inputs.

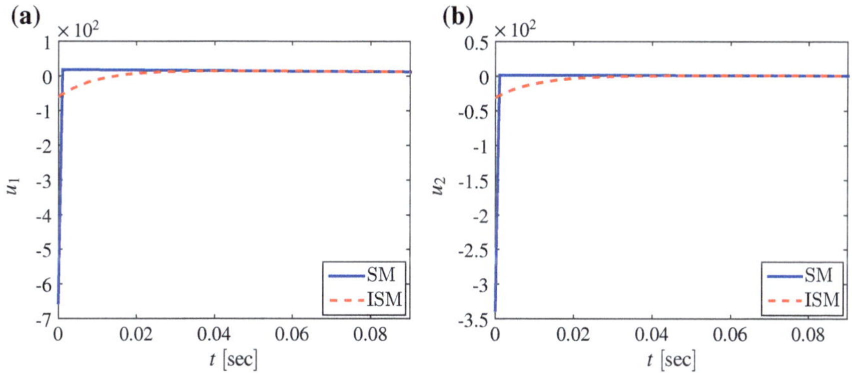

Fig. 2.2 **a** System control input u_1 **b** System control input u_2

Fig. 2.3 Bode plot of some of the elements of the open-loop transfer matrix

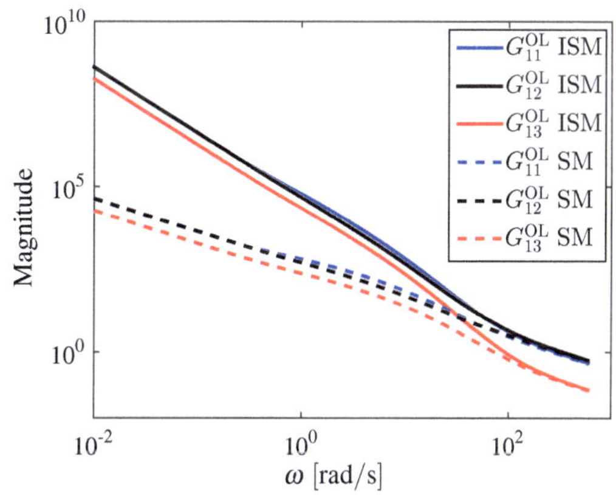

Fig. 2.4 Sensitivity function of x_1 with respect to f_1 and f_2

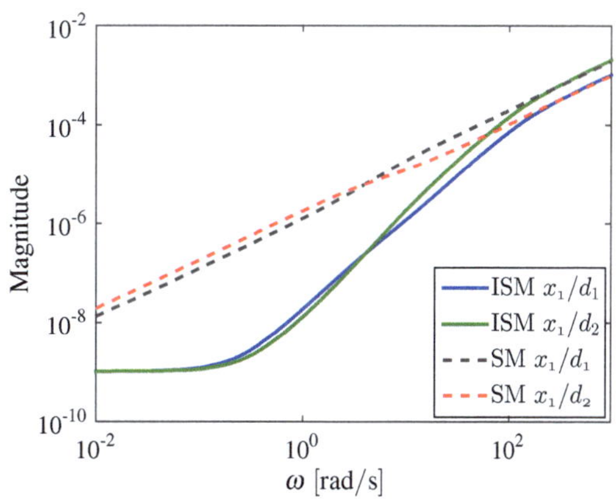

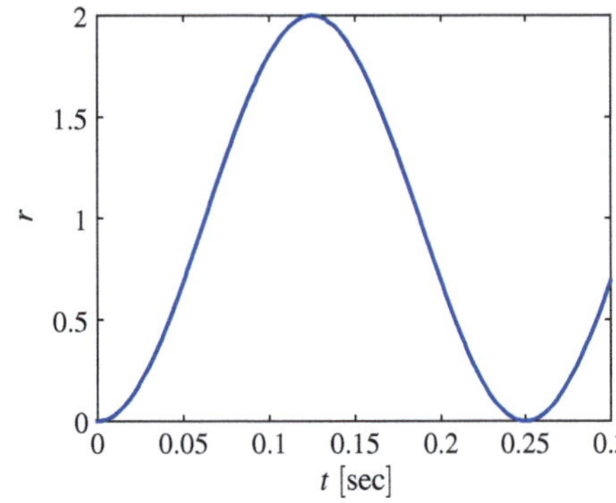

Fig. 2.5 The output reference trajectory

2.4.6.2 State Feedback Approach

Consider the following second order system

$$A = \begin{bmatrix} 10 & 1 \\ -10 & -10 \end{bmatrix}, \quad B = \begin{bmatrix} 4 \\ 4.2 \end{bmatrix} \quad \& \quad C = \begin{bmatrix} 1 & 0 \end{bmatrix}.$$

The sampled-data system obtained with a sampling period $T = 1\,\text{ms}$ is

$$\Phi = \begin{bmatrix} 1.01 & -0.001 \\ -0.01 & 0.99 \end{bmatrix}, \quad \Gamma = \begin{bmatrix} 0.0040 \\ 0.0042 \end{bmatrix} \quad \& \quad C = \begin{bmatrix} 1 & 0 \end{bmatrix}.$$

The zero of (Φ, Γ, C) is $z = 0.989$ and, therefore, the system is minimum-phase. Let the desired pole, the remaining pole of the closed-loop system to be designed, be $z = 0.75$, then the design parameter is given by $E = 0.25$. The system is simulated with an output reference $r_k = 1 + \sin\left(8\pi kT - \frac{\pi}{2}\right)$, shown in Fig. 2.5. The disturbance acting on the system will be non-smooth when speed crosses zero and given by

$$f(\mathbf{x}, t) = \begin{cases} 10 & \forall x_2 \in (-\infty, -\varepsilon_f) \\ 0 & \forall x_2 \in [-\varepsilon_f, \varepsilon_f] \\ -10 & \forall x_2 \in (\varepsilon_f, \infty) \end{cases} \tag{2.134}$$

where the coefficient ε_f is sufficiently small. The system is simulated using control law (2.68). The controller performance is compared with that of a PI controller having a proportional gain of $k_p = 240$ and integral gain of $k_i = 8$. In Fig. 2.6 the tracking error is 4×10^{-6} which corresponds to $O\left(T^2\right)$ and is almost invisible as compared with the PI controller performance. Note worthy is the fact that the control signal for the PI control is much larger initially as compared to the ISM control. A smaller control signal is more desirable in practice as it would not create a heavy burden on actuators (Fig. 2.7).

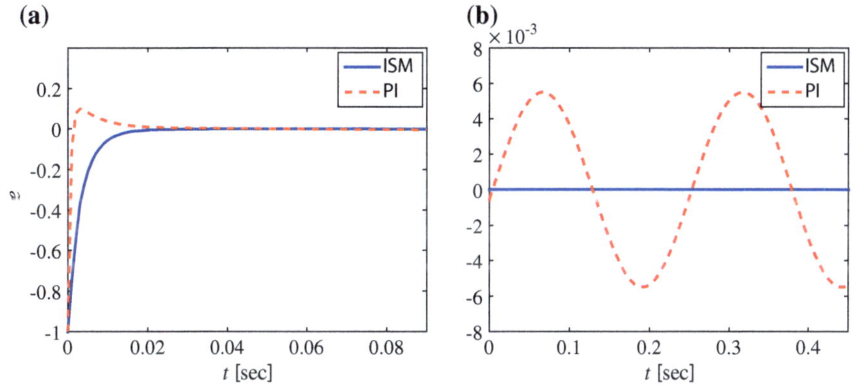

Fig. 2.6 Tracking error of ISM control and PI controllers **a** Transient performance **b** Steady-State performance

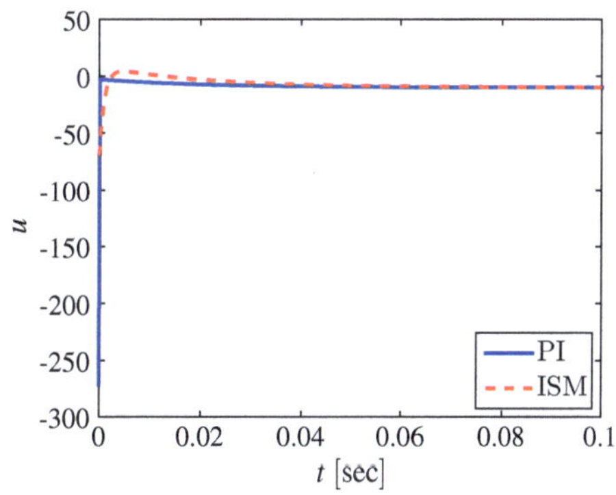

Fig. 2.7 Control inputs of ISM control with state feedback and PI

2.4.6.3 Output Feedback Approach

Consider the system

$$A = \begin{bmatrix} -60 & -10 \\ 10 & -10 \end{bmatrix}, \quad B = \begin{bmatrix} 4 \\ 4.2 \end{bmatrix} \quad \& \quad C = \begin{bmatrix} 1 & 0 \end{bmatrix}.$$

After sampling the system at $T = 1\,\text{ms}$, the discretized system is

$$\Phi = \begin{bmatrix} 0.9417 & -0.0097 \\ 0.0097 & 0.9900 \end{bmatrix}, \quad \Gamma = \begin{bmatrix} 0.0039 \\ 0.0042 \end{bmatrix} \quad \& \quad C = \begin{bmatrix} 1 & 0 \end{bmatrix}.$$

For this system the zero of (Φ, Γ, C) is $z = 1.001$ whereas, using $D = C\Phi^{-1}$, the zero of (Φ, Γ, D) is $z = 0.998$. Therefore, the output-feedback approach with the reference model in Sect. 2.4.3.5 is the only option. Using the same disturbance $f(t)$ and reference trajectory r_k, the system is simulated. The controller performance is

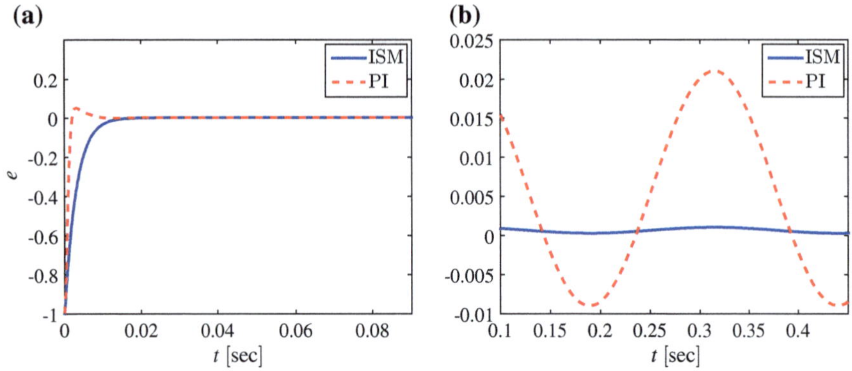

Fig. 2.8 Tracking error of ISM control and PI controllers **a** Transient performance **b** Steady-State performance

Fig. 2.9 Control inputs of ISM control and PI output feedback

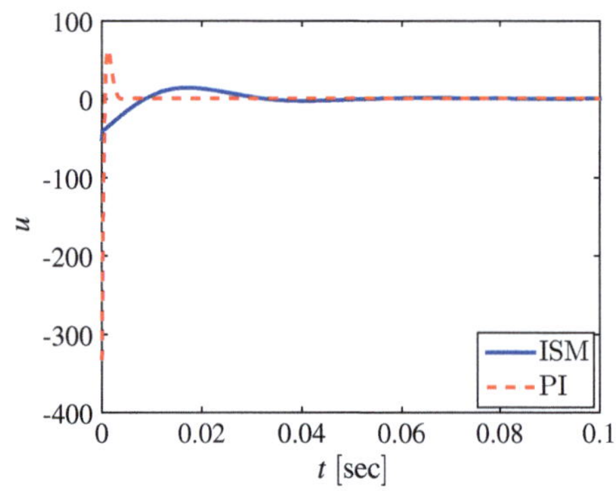

compared with that of a PI controller having a proportional gain of $k_\text{p} = 200$ and integral gain of $k_\text{i} = 30$. As it can be seen from Fig. 2.8, the performance is quite good and better than that of a PI controller. The tracking error for the ISM control is about 17×10^{-6} which corresponds to $O\left(T^2\right)$ at steady state. Note that even though the worst case scenario of $O\left(T\right)$ was predicted for this approach it was possible to achieve $O\left(T^2\right)$ at steady state. Also, similar to the state feedback approach the control signal of the ISM control controller is much smaller than that of the PI controller at the onset of motion (Fig. 2.9).

2.4.6.4 State Observer Approach

For this approach we will go back to the system (2.135), and estimate x_2 using the observer (2.123). The observer has a gain of $L = [1.19 \quad 342.23]$ and is designed such that two poles are at $z = 0.4$ allowing a fast enough convergence. From Fig. 2.10 the estimation of error $\tilde{x}_{2,k}$ is plotted. As we can see the estimation is quite good

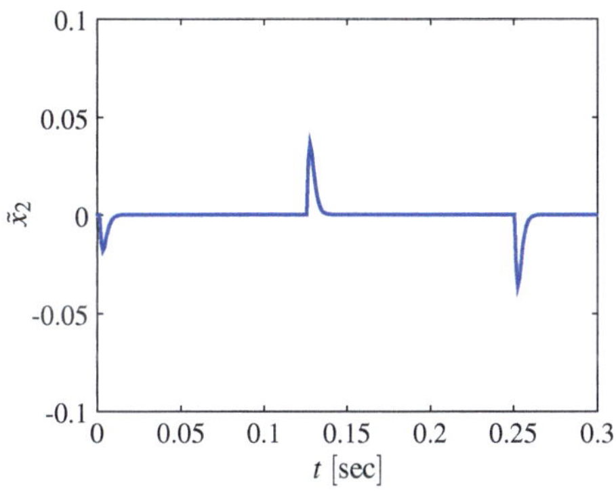

Fig. 2.10 Observer state estimation error $\tilde{x}_2$

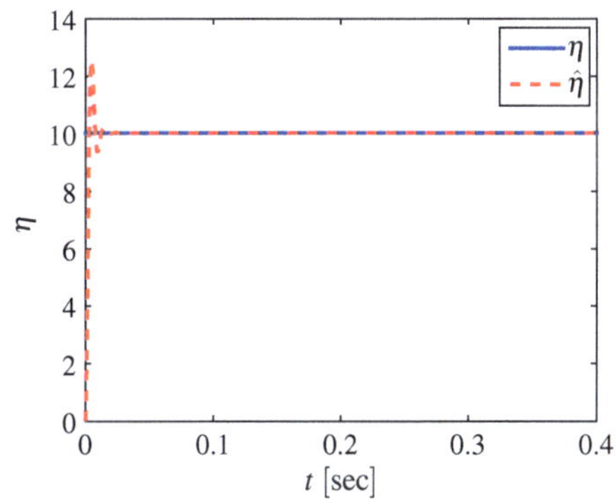

Fig. 2.11 Disturbance η and estimate $\hat{\eta}$

and deviating only when the discontinuities occur but attenuates very quickly. The disturbance estimation is seen in Fig. 2.11 and the estimation converges quickly to the actual disturbance. From Fig. 2.12 we can see the tracking error performance. The tracking error is about 6×10^{-6} which matches the theoretical results of $O\left(T^2\right)$ bound. Again like the previous two approaches, the control signal of the ISM control is smaller than that of the PI controller at the initial phase (Fig. 2.13).

Finally we need to show the effects of a more frequently occurring discontinuous disturbance and how adding a nonlinear term, (2.133), would improve the performance. The disturbance is shown in Fig. 2.15. As it can be seen from Fig. 2.14 the rapid disturbance degrades the performance of the ISM control and PI controllers, however, addition of a switching term with $\mu = 10$ and $\varepsilon = 0.01$ improves the tracking performance. Note also from Fig. 2.15 that the disturbance estimate $\hat{\eta}_k$ converges quickly to the disturbance.

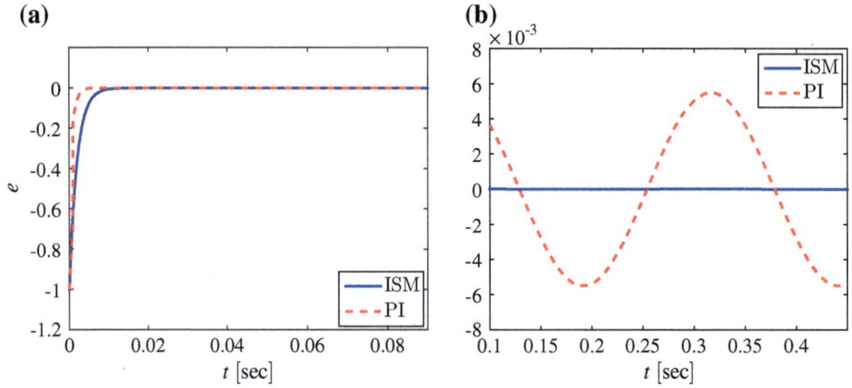

Fig. 2.12 Tracking error of ISM control and PI controllers **a** Transient performance **b** Steady-State performance

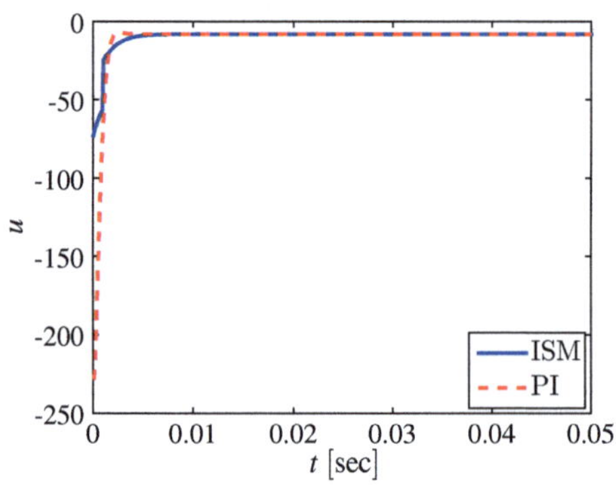

Fig. 2.13 Control inputs of ISM control with state observer and PI

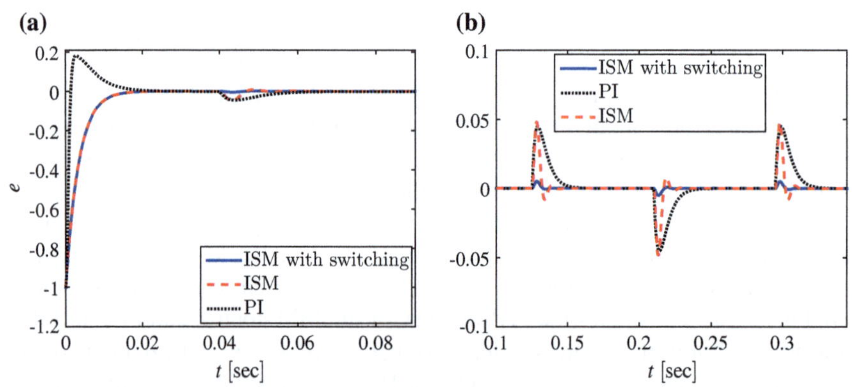

Fig. 2.14 Tracking errors of ISM control, PI and ISM control with switching under a more frequent discontinuous disturbance **a** Transient performance **b** Steady-State performance

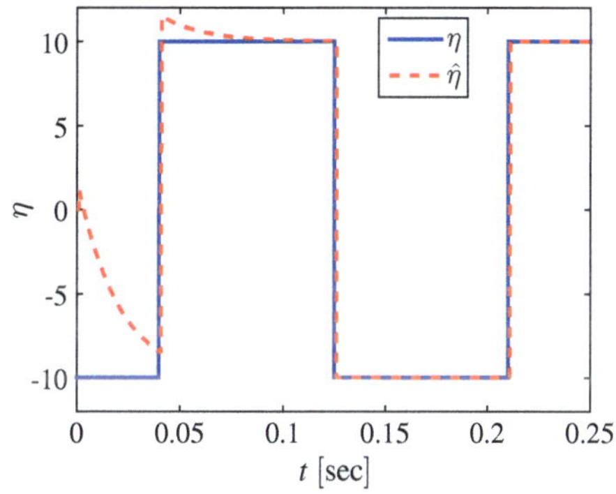

Fig. 2.15 Disturbance η and estimate $\hat{\eta}$

2.5 Discrete-Time Terminal Sliding Mode Control

In this section we will discuss the design of the tracking controller for the system. The controller will be designed based on an appropriate sliding surface. Further, the stability conditions of the closed-loop system will be analyzed. The relation between TSM control properties and the closed-loop eigenvalue will be explored. Note that the subsequent analysis and derivations will be based on a 2nd order system example, however, it is possible to apply the same principles for a system of higher order as long as it is SISO.

2.5.1 Controller Design and Stability Analysis

Consider a 2nd order SISO version of (2.2) given by

$$\begin{aligned} \mathbf{x}_{k+1} &= \boldsymbol{\Phi}\mathbf{x}_k + \boldsymbol{\gamma}u_k + \mathbf{d}_k \\ y_k &= \mathbf{c}\mathbf{x}_k = x_{1,k}, \quad y_0 = y(0) \end{aligned} \tag{2.135}$$

where $\mathbf{x}_k \in \Re^2$, $u_k \in \Re$, $\mathbf{d}_k \in \Re^2$, $y_k \in \Re$, $\boldsymbol{\Phi} \in \Re^{2\times 2}$, $\boldsymbol{\gamma} \in \Re^2$ and $\mathbf{c} \in \Re^2$ respectively. Further, consider the discrete-time sliding surface below,

$$\sigma_k = \mathbf{s}\mathbf{e}_k + \beta e_{k-1}^p \tag{2.136}$$

where $\mathbf{e}_k^\top = \left[e_{1,k} \,\vdots\, e_{2,k}\right]$, $e_{1,k} = r_{1,k} - x_{1,k}$, $e_{2,k} = r_{2,k} - x_{2,k}$, $r_{1,k}$, $r_{2,k}$ are arbitrary time-varying references, σ_k is the sliding function, and $\mathbf{s}$, β, p are positive design constants. The tracking problem is to force $x_{1,k} \to r_{1,k}$. The selection of $p < 1$ guarantees a steeper slope of the sliding surface as the states approach the origin which is desirable as seen in Fig. 2.16. Also note that p should be selected as

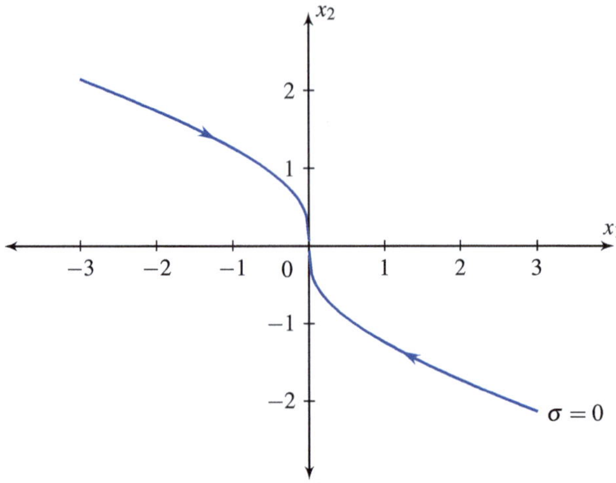

Fig. 2.16 Phase portrait of the sliding surface

a rational number with odd numerator and denominator to gaurantee that the sign of the error remains intact. Let us first derive the discrete-time TSM control law by using the concept of equivalent control and discuss the TSM control properties associated with stability.

Theorem 2.7 *The new TSM control law porposed is*

$$u_k = (\mathbf{s}\boldsymbol{\gamma})^{-1}\left(\mathbf{s}\mathbf{r}_{k+1} - \mathbf{s}\boldsymbol{\Phi}\mathbf{r}_k + \mathbf{s}\boldsymbol{\Phi}\mathbf{e}_k + \beta e_k^p\right) - (\mathbf{s}\boldsymbol{\gamma})^{-1}\mathbf{s}\hat{\mathbf{d}}_k \tag{2.137}$$

where $\mathbf{r}_k^\top = \left[r_{1,k} \,\vdots\, r_{2,k}\right]$ *and* $\hat{\mathbf{d}}_k$ *is the estimate of the disturbance. The controller (2.137) drives the sliding variable to*

$$\sigma_{k+1} = \mathbf{s}\left(\hat{\mathbf{d}}_k - \mathbf{d}_k\right)$$

and results in the closed-loop error dynamics

$$\mathbf{e}_{k+1} = \left[\boldsymbol{\Phi} - \boldsymbol{\gamma}(\mathbf{s}\boldsymbol{\gamma})^{-1}\mathbf{s}\boldsymbol{\Phi}\right]\mathbf{e}_k - \beta(\mathbf{s}\boldsymbol{\gamma})^{-1}e_k^p + \boldsymbol{\delta}_k$$

where $\boldsymbol{\delta}_k = \boldsymbol{\gamma}(\mathbf{s}\boldsymbol{\gamma})^{-1}\mathbf{s}\hat{\mathbf{d}}_k - \mathbf{d}_k$ *is due to the disturbance estimation error and is of* $O\left(T^2\right)$*. Further, the stable range of closed-loop system is nonlinearly dependent on the tracking error* e_k*.*

Proof The control law (2.137) can be derived using the design method based on equivalent control. To proceed, consider a forward expression of (2.136)

$$\sigma_{k+1} = \mathbf{s}\mathbf{e}_{k+1} + \beta e_k^p. \tag{2.138}$$

The objective of a sliding mode controller is to achieve $\sigma_{k+1} = 0$, therefore, we need to derive an explicit expression in terms of the error dynamics. For this we rewrite the system dynamics (2.2) in terms of the error dynamics. It can be shown that the error dynamics is of the form

$$\mathbf{e}_{k+1} = \Phi \mathbf{e}_k - \gamma u_k - \mathbf{d}_k + \mathbf{r}_{k+1} - \Phi \mathbf{r}_k. \tag{2.139}$$

Substitution of (2.139) into (2.138) and equating the resulting expression of σ_{k+1} to zero we obtain the expression for the equivalent control u_k^{eq},

$$u_k^{\mathrm{eq}} = (\mathbf{s}\gamma)^{-1} \left(\mathbf{s}\mathbf{r}_{k+1} - \mathbf{s}\Phi\mathbf{r}_k + \mathbf{s}\Phi\mathbf{e}_k + \beta e_k^p\right) - (\mathbf{s}\gamma)^{-1} \mathbf{s}\mathbf{d}_k. \tag{2.140}$$

Note that the control (2.140) is based the current value of the disturbance $d_{1,k}$ which is unknown and therefore cannot be implemented in this current form. To overcome this, the disturbance will be estimated with the so called delay estimate as follows,

$$\hat{\mathbf{d}}_k = \mathbf{d}_{k-1} = \mathbf{x}_k - \Phi\mathbf{x}_{k-1} - \gamma u_{k-1} \tag{2.141}$$

therefore, the final controller structure is given by

$$u_k = (\mathbf{s}\gamma)^{-1} \left(\mathbf{s}\mathbf{r}_{k+1} - \mathbf{s}\Phi\mathbf{r}_k + \mathbf{s}\Phi\mathbf{e}_k + \beta e_k^p\right) - (\mathbf{s}\gamma)^{-1} \mathbf{s}\hat{\mathbf{d}}_k. \tag{2.142}$$

In order to verify the second part of *Theorem* 2.7 with regard to the closed-loop stability, first derive the closed-loop error dynamics. Substitute u_k in (2.137) into (2.139), we obtain

$$\begin{aligned}\mathbf{e}_{k+1} = & \left[\Phi - \gamma (\mathbf{s}\gamma)^{-1} \mathbf{s}\Phi\right] \mathbf{e}_k - \beta (\mathbf{s}\gamma)^{-1} e_k^p - \mathbf{d}_k \\ & + \gamma (\mathbf{s}\gamma)^{-1} \mathbf{s}\hat{\mathbf{d}}_k + \left(I - \gamma (\mathbf{s}\gamma)^{-1} \mathbf{s}\right) (\mathbf{r}_{k+1} - \Phi\mathbf{r}_k).\end{aligned} \tag{2.143}$$

Next, in order to eliminate the term $\left(I - \gamma (\mathbf{s}\gamma)^{-1} \mathbf{s}\right) (\mathbf{r}_{k+1} - \Phi\mathbf{r}_k)$ from the closed-loop dynamics (2.143), we note that since the objective is to have $\mathbf{x}_k \to \mathbf{r}_k$ then there must exist a control input $u_{\mathrm{m},k}$ such that $\mathbf{r}_{k+1} = \Phi\mathbf{r}_k + \gamma u_{\mathrm{m},k}$. Thus,

$$\left(I - \gamma (\mathbf{s}\gamma)^{-1} \mathbf{s}\right) (\mathbf{r}_{k+1} - \Phi\mathbf{r}_k) = \left(I - \gamma (\mathbf{s}\gamma)^{-1} \mathbf{s}\right) \gamma u_{\mathrm{m},k} = 0. \tag{2.144}$$

and the final closed-loop error dynamics is

$$\mathbf{e}_{k+1} = \left[\Phi - \gamma (\mathbf{s}\gamma)^{-1} \mathbf{s}\Phi\right] \mathbf{e}_k - \beta (\mathbf{s}\gamma)^{-1} e_k^p + \boldsymbol{\delta}_k \tag{2.145}$$

where $\boldsymbol{\delta}_k = \gamma (\mathbf{s}\gamma)^{-1} \mathbf{s}\hat{\mathbf{d}}_k - \mathbf{d}_k$ and is of $O\left(T^2\right)$, [1]. The sliding surface dynamics is obtained by substituting (2.145) into (2.138) to get

$$\sigma_{k+1} = \mathbf{s}\left(\hat{\mathbf{d}}_k - \mathbf{d}_k\right) = \mathbf{s}\left(\mathbf{d}_{k-1} - \mathbf{d}_k\right) = O\left(T^2\right). \tag{2.146}$$

To evaluate the stable range of (2.145), rewrite (2.145) in the form

$$\mathbf{e}_{k+1} = \left[\Phi - \boldsymbol{\gamma}\left(\mathbf{s}\boldsymbol{\gamma}\right)^{-1}\left(\mathbf{s}\Phi + \beta e_k^{p-1} C\right)\right]\mathbf{e}_k + \boldsymbol{\delta}_k \tag{2.147}$$

where $C = \mathrm{diag}(1, 0)$. Denote $\mathbf{l}_k = \left[l_{1,k} \,\vdots\, l_{2,k}\right] = \left(\mathbf{s}\boldsymbol{\gamma}\right)^{-1}\left(\mathbf{s}\Phi + \beta e_k^{p-1} C\right)$ the control gain vector, where $l_{1,k}$ is error dependent. The error dynamics (2.148) can be rewritten as

$$\mathbf{e}_{k+1} = \left[\Phi - \boldsymbol{\gamma}\mathbf{l}_k\right]\mathbf{e}_k + \boldsymbol{\delta}_k. \tag{2.148}$$

From (2.148) we see that there must exist certain range for the first element of the gain vector, $l_{1,k}$, such that the closed-loop system is stable. Let $l_{1,\min} \le l_{1,k} \le l_{1,\max}$ where $l_{1,\min}$ and $l_{1,\max}$ denote the minimum and maximum allowable values for $l_{1,k}$.

From the definition of $\mathbf{l}_k$ we can obtain

$$\beta e_k^{p-1} + s_1\phi_{1,1} + s_2\phi_{2,1} = \mathbf{s}\boldsymbol{\gamma} l_{1,k} \tag{2.149}$$

and

$$s_1\phi_{1,2} + s_2\phi_{2,2} = \mathbf{s}\boldsymbol{\gamma} l_{2,k} \tag{2.150}$$

where $\phi_{i,j}$ are elements of the matrix Φ. From (2.149) we can derive the following inequality

$$\mathbf{s}\boldsymbol{\gamma} l_{1,\min} < \beta e_k^{p-1} + s_1\phi_{1,1} + s_2\phi_{2,1} < \mathbf{s}\boldsymbol{\gamma} l_{1,\max} \tag{2.151}$$

from which we can obtain, when $p < 1$,

$$\begin{aligned} |e_k| &> \left(\frac{\beta}{\mathbf{s}\boldsymbol{\gamma} l_{1,\max} - s_1\phi_{1,1} - s_2\phi_{2,1}}\right)^{\frac{1}{1-p}} \\ |e_k| &< \left(\frac{\beta}{\mathbf{s}\boldsymbol{\gamma} l_{1,\min} - s_1\phi_{1,1} - s_2\phi_{2,1}}\right)^{\frac{1}{1-p}} \end{aligned} \tag{2.152}$$

The first relation gives the minimum-bound of the error and the second relation gives the stable operation range. Note, that by selecting a proper s_1 and s_2 such that the denominator in the second relation is zero for a non-zero β then it is possible to guarantee global stability outside of the minimum-error bound.

2.5.2 TSM Control Tracking Properties

First derive the ultimate tracking error bound of (2.1) when the proposed discrete-time TSM control is applied. Using the discrete-time TSM control law (2.137), in the following we show that the ultimate bound of the tracking error is $O\left(T^2\right)$. From the previous section we obtained a minimum error bound based on the selection of β which if selected small enough would result in a small error bound.

However, due to the existence of a disturbance term $\boldsymbol{\delta}_k$ the ultimate error bound may be large. Note that the solution of the closed-loop system (2.148) is

$$\mathbf{e}_k = \left(\Pi_{i=0}^{k}\left[\Phi - \gamma \mathbf{l}_i\right]\right)\mathbf{e}_0 + \sum_{i=0}^{k-1}\left(\Pi_{j=0}^{i}\left[\Phi - \gamma \mathbf{l}_j\right]\right)\boldsymbol{\delta}_{k-i-1}. \tag{2.153}$$

According to our earlier discussion the ultimate error bound would be an order higher than $\boldsymbol{\delta}_k$, which means that the error bound will be $O\left(T\right)$. This property holds if the gain $\mathbf{l}_k$ constant and the term $\left(\Pi_{j=0}^{i}\left[\Phi - \gamma \mathbf{l}_j\right]\right)\boldsymbol{\delta}_{k-i-1}$ is an infinite series. According to [1], the series will be of the order $O\left(\frac{1}{1-\lambda_{\max}}\right)\cdot O\left(\boldsymbol{\delta}\right)$ where $\lambda_{\max}$ is the dominant eigenvalue. Therefore, if the dominant eigenvalue is designed close to the edge of the unit circle, then $O\left(\frac{1}{1-\lambda_{\max}}\right) = O\left(T^{-1}\right)$. This implies a rather bad rejection of the exogenous disturbance. To enhance disturbance rejection, it is desirable to choose the dominant eigenvalue closer to the origin during steady state motion, then $O\left(\frac{1}{1-\lambda_{\max}}\right) = O\left(1\right)$ and

$$O\left(\frac{1}{1-\lambda_{\max}}\right)\cdot O\left(\boldsymbol{\delta}\right) = O\left(\boldsymbol{\delta}\right) = O\left(T^2\right). \tag{2.154}$$

However, in practical consideration of sampled-data processes during transient motion, an eigenvalue closer to the origin will result in large initial control effort of the order $O\left(\frac{1}{1-\lambda_{\max}}\right) = O\left(T^{-1}\right)$.

A very useful property that is acquired by using the terminal switching surface is that the system gain $l_{1,k}$ will increase as the error approaches zero because of the nonlinear term e_k^{1-p} in $\mathbf{l}_k$. This means that it is possible to move the dominant eigenvule of the closed-loop system from an initial position nearby the unit circle towards the origin, thus avoid the large initial control effort during the transient period, obtain very stable operation at steady state, and quickly attentuate exogenous disturbances. We will explore more about this property in the next section.

Finally, we look at what happens to the tracking error when there is a discontinuity in the disturbance. Consider the disturbance term associated with the closed-loop system (2.148). It can be reasonably assumed that the discontinuity occurs rarely, therefore, if we assume that the discontinuity occurs at the kth sampling point, then $\boldsymbol{\delta}_k = O\left(T\right)$ rather than $O\left(T^2\right)$ as the difference $\mathbf{d}_{k-1} - \mathbf{d}_k$ will no longer be $O\left(T^2\right)$

but of the order of $\mathbf{d}_k$ which is $O(T)$. If the discontinuity occurs at a time instant k', then $\boldsymbol{\delta}_k = O(T)$ at $k = k', k'+1, k'+2$, and return to $\boldsymbol{\delta}_k = O(T^2)$ for subsequent sampling instants. Therefore the solution of (2.148) would lead to the worst case error bound

$$|e_k| = O(T) \tag{2.155}$$

for certain time interval but $O(T^2)$ ultimately.

2.5.3 Determination of Controller Parameters

In order to discuss the controller design, consider the case when the system parameters are

$$\Phi = \begin{bmatrix} 1 & 9.313 \times 10^{-4} \\ 0 & 0.8659 \end{bmatrix}, \quad \gamma = \begin{bmatrix} 2.861 \times 10^{-6} \\ 5.6 \times 10^{-3} \end{bmatrix} \quad \& \quad \mathbf{c} = \begin{bmatrix} 1 & 0 \end{bmatrix}.$$

The control law requires the design of three parameters: the vector **s**, the parameter β, and the power p. As was discussed earlier the parameters p and β determine the dynamics of the eigenvalue and, using (2.148), this can be seen from Figs. 2.17, 2.18, 2.19 and 2.20. We can see from these figures, since e_1 leads to high gain feedback, both closed-loop eigenvalues of the discrete-time TSM control will eventually exceed unity and become unstable when e_1 is sufficiently small. This is consistent with the discussion made in previous section that there exists a minimum-bound of tracking error specified by (2.152). From the eigenvalue figures it is clear that the minimum error bound is determined by a critical value of e_1 where at least one eigenvalue becomes marginal stable. The minimum error bound can be reduced by shifting the curves of eigenvalues leftwards. As shown in Figs. 2.17, 2.18, 2.19 and 2.20, this can be achieved by either reducing β or increasing p. However, a smaller β implies a

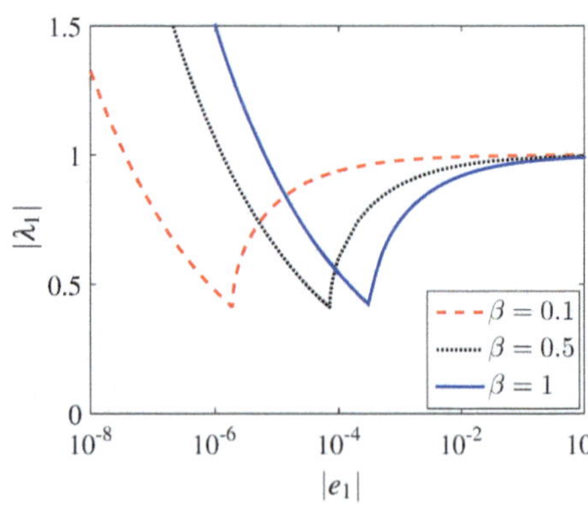

Fig. 2.17 System eigenvalue λ_1 w.r.t e_1 for different choices of β and $p = \frac{5}{9}$

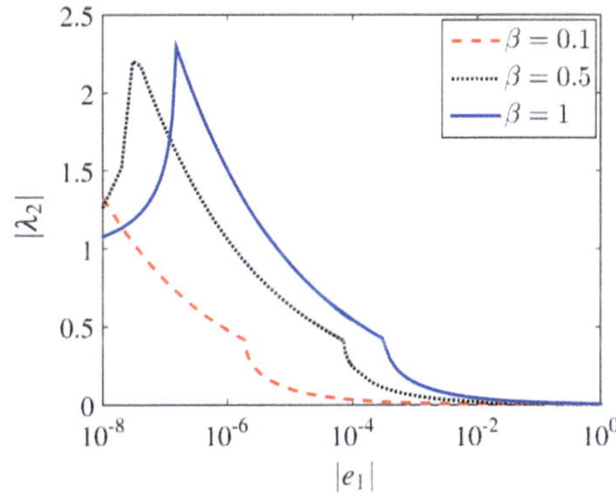

Fig. 2.18 System eigenvalue λ_2 w.r.t e_1 for different choices of β and $p = \frac{5}{9}$

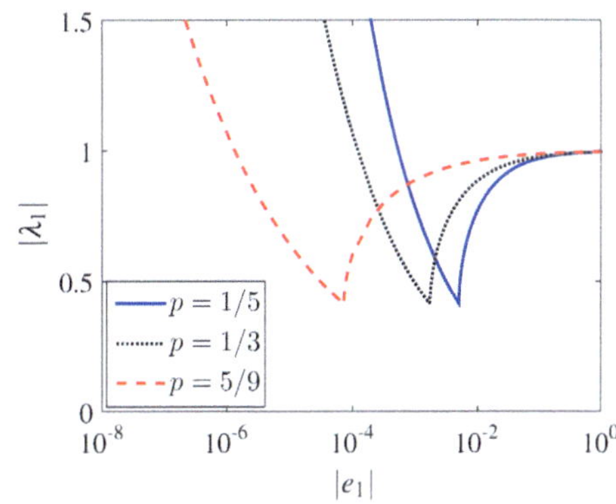

Fig. 2.19 System eigenvalue λ_1 w.r.t e_1 for different choices of p and $\beta = 0.5$

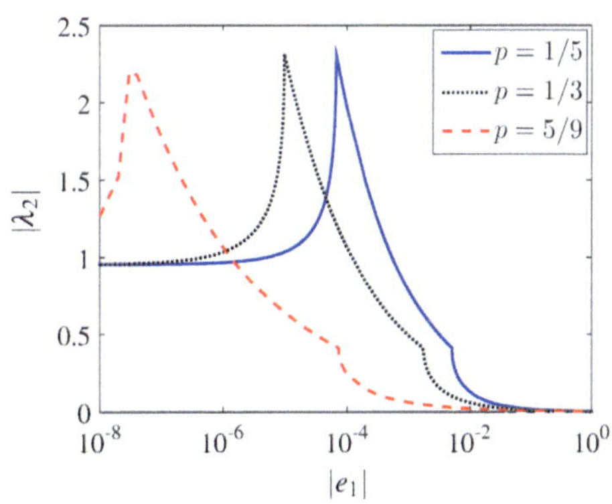

Fig. 2.20 System eigenvalue λ_2 w.r.t e_1 for different choices of p and $\beta = 0.5$

smaller range of stability as shown in (2.152). Therefore in TSM control design, a relatively larger p is preferred.

From Figs. 2.17 and 2.19, the first eigenvalue, λ_1, is always near unity when the error e_1 is large. Hence this eigenvalue does not generate large initial control efforts, but may generate large steady state error in the presence of disturbance if the eigenvalue does not decrease with respect to error. By incorporating TSM, λ_1 will first drop when e_1 decreases, then rise when e_1 further decreases. A smaller β and a larger p will speed up this variation pattern as e_1 decreases.

The variation pattern of the second eigenvalue, λ_2, is opposite to the first eigenvalue as can be seen from Figs. 2.18 and 2.20.

For the system $\mathbf{e}_{k+1} = \left[\Phi - \boldsymbol{\gamma}\mathbf{l}_k\right]\mathbf{e}_k$ the solution is $\mathbf{e}_k = V \cdot \mathrm{diag}\left(\lambda_1^k, 0\right) V^{-1}\mathbf{e}_0$ where the matrix V consists of the eigenvectors of $\left[\Phi - \boldsymbol{\gamma}\mathbf{l}_k\right]$. This leads us to conclude that the control law (2.142) is proportional to V^{-1} which is a function of λ_1 and will take large values as λ_1 moves towards 0. Thus, it is desirable to have λ_1 closer to the edge of the unit circle so that V^{-1} does not take large values. This is evident from Figs. 2.21 and 2.22 where the controller gains, at the initial time step of $k = 1$, $(\mathbf{s}\boldsymbol{\gamma})^{-1}\mathbf{s}\Phi V \cdot \mathrm{diag}\left(\lambda_1, 0\right) V^{-1}$ was plotted w.r.t λ_1. Note, that the nonlinear term βe_k^p has been disregarded as its contribution at the initial time step is negligible w.r.t the linear term.

Based on the above dicussions the design guideline for discrete TSM control is determined. The controller gains $\mathbf{l}_k$ can be determined according to the selection of closed-loop eigenvalues. Note that eigenvalue λ_1 should take a larger value initially and drop when approaching steady state. Thus we can choose λ_1 varying from the initial value 0.995 to the final 0. The other eigenvalue is $\lambda_2 = 0$ when the closed-loop system is in sliding mode. As functions of eigenvalues, the range of the feedback gain vectors can be calculated as $[894 \quad 155] \leq \mathbf{l}_k \leq \left[1.79 \times 10^5 \quad 240\right]$.

Next, from relations (2.150) and (2.152) we can obtain $\mathbf{s}$, β, and p. The selected $\mathbf{s}$ should ensure that the denominator of the second expression in (2.152) is close to zero so that the upper limit of e_k is maximized, and at the same time (2.150) should be satisfied for the given range of l_2. It is not necessary to select $\mathbf{s}$ to make the upper bound on e_1 infinite, as the real system has a maximum displacement limitation of 60×10^{-3} m. Thus, we select the denominator to take a value of 0.01 and select $l_2 = 200$ from the specified range of l_2. Solving the two simultaneous equations (2.150) and the denominator of the second expression in (2.152) being zero for s_1 and s_2 gives $\mathbf{s} = [0.49 \quad 0.100]$.

To determine the parameters β and p, firtst look into the relations between these two parameters and the closed-loop eigenvalues. The behavior of the eigenvalues under these parameters are shown in Figs. 2.23 and 2.24. It is possible to divide the plots into three regions in which the system has different behaviors. The transient response region is the region in which the error is large enough and the dominant eigenvalue, λ_1, remains close to the edge of the unit circle. As a result, the control effort can be kept at appropriate level despite any large initial error. At this region, disturbance rejection is not a main concern as an $O(T)$ or $O\left(T^2\right)$ disturbance would be much smaller than the state errors.

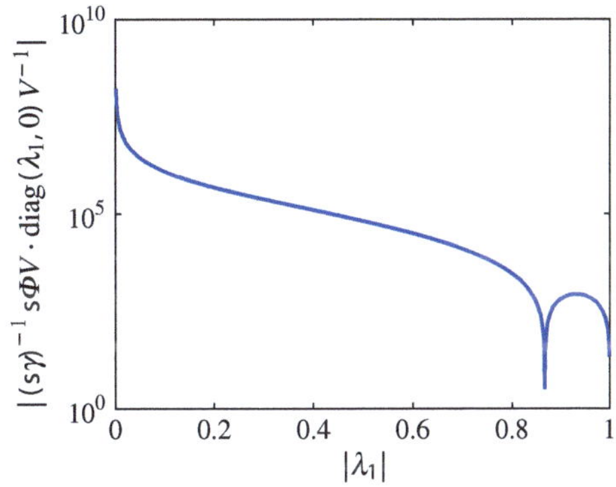

Fig. 2.21 First element of the system gain $(\mathbf{s}\boldsymbol{\gamma})^{-1}\,\mathbf{s}\boldsymbol{\Phi} V \cdot \text{diag}\,(\lambda_1, 0)\, V^{-1}$ w.r.t λ_1

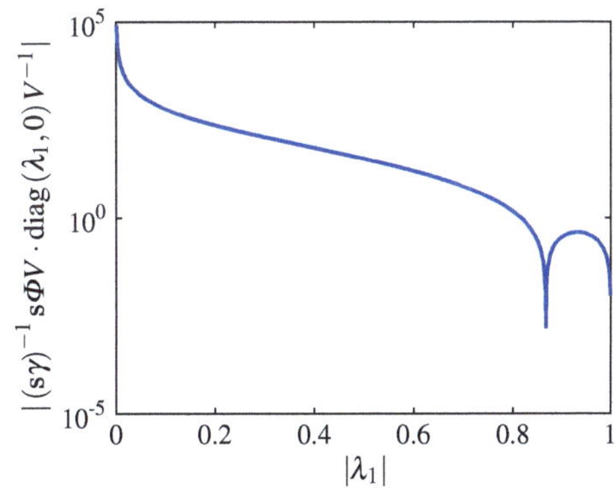

Fig. 2.22 Second element of the system gain $(\mathbf{s}\boldsymbol{\gamma})^{-1}\,\mathbf{s}\boldsymbol{\Phi} V \cdot \text{diag}\,(\lambda_1, 0)\, V^{-1}$ w.r.t λ_1

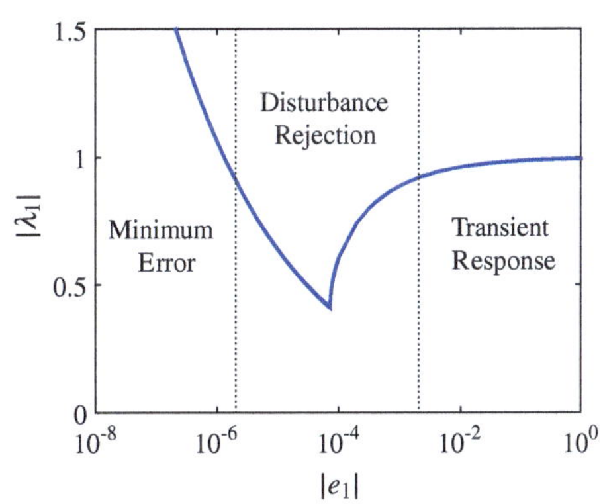

Fig. 2.23 System eigenvalue λ_1 w.r.t e_1

The disturbance rejection region is when the state errors become smaller, reaching $O\,(T)$ or $O\left(T^2\right)$ level. Now, since the eigenvalue also becomes smaller as the controller gain increases, the robustness and disturbance rejection property of the system are enhanced. Finally, the minimum error region is the region in which the eigenvalue goes beyond the unit circle. Therefore, the error will stay around the

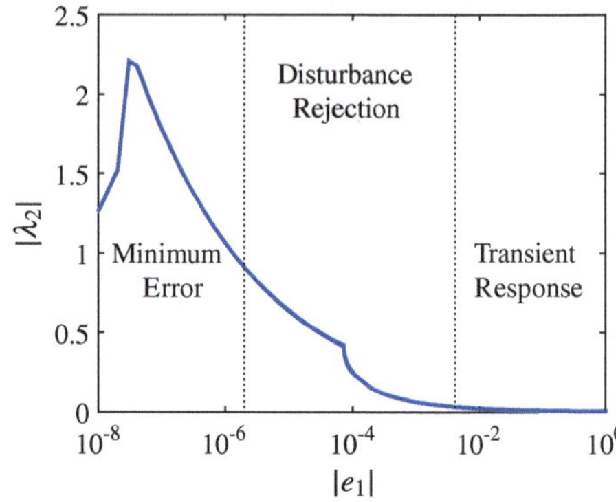

Fig. 2.24 System eigenvalue λ_2 w.r.t e_1

boundary between the disturbance rejection region and the minimum error region, and the boundary determines the minimum error bound.

Based on the above discussions, from Figs. 2.17 and 2.19 it can be see that a larger $\beta = 1$ would lead to a faster response because the dominant eigenvalue, λ_1, drops quickly in the transient response region. However, it may also lead to a larger steady state error because λ_1 rises quickly and produces a rather large minimum error region. When smaller $\beta = 0.1$ is used, we can achieve a much smaller minimum error region, but λ_1 drops slowly in the transient response region. In the real-time implementation, we choose a mid value $\beta = 0.5$ as a tradeoff.

Looking into Figs. 2.18 and 2.20, we can observe that the variation of p will produce the similar trend as β. Namely, when p is close to 0, the transient response region is improved but the minimum error region is larger. When p is approaching 1, on the other hand, the transient response is slower as the dominant eigenvalue λ_1 drops slowly, but the minimum error region is getting smaller. In the real-time implementation, we choose a mid value $p = \frac{5}{9}$ as a tradeoff.

It should also be noted that, in the real-time implementation, the presence of the disturbance will limit the best possible tracking error e_1 to $O\left(T^2\right)$ or $O\left(10^{-6}\right)$ for a sampling period of 1 ms. Therefore, any selection of larger p and smaller β which result in the minimum tracking region below $O\left(T^2\right)$ would lead to little or no improvements.

It is well known that sliding mode control requires a fast switching frequency in order to maintain the sliding motion. The proposed TSM control however does not produce much chattering though with the limited sampling frequency.

Finally, it is also interesting to check the tracking error bound according to the theoretical analysis and experimental result. Through analysis we have shown the tracking error bound is proportional to the size of the sampling period T. Therefore, we can expect a smooth control response and low tracking error when the sampling period is 1ms. In Chap. 7 it will be shown that the magnitude of the tracking error obtained in experiment is consistent with the theoretical error bound of $O\left(T^2\right) = O\left(0.001^2\right) = O\left(10^{-6}\right)$ when $T = 1$ ms.

2.6 Conclusion

This chapter discussed a discrete-time integral sliding (ISM) control design and a discrete-time terminal sliding mode (TSM) control design for sampled-data systems.

It was shown that with the discrete-time integral type sliding manifold, the ISM design retains the deadbeat structure of the discrete-time sliding mode, and at the same time allocates the closed-loop eigenvalues for the full-order multi-input system. The discrete-time ISM control achieves accurate control performance for the sliding mode, state regulation and output tracking, meanwhile eliminates the reaching phase and avoids overlarge control efforts.

In addition it was shown through theoretical investigation that the revised discrete-time TSM controller can achieve very good performance and along with discrete-time ISM control is a superior alternative to classical SM control.

Chapter 3
Discrete-Time Periodic Adaptive Control

Abstract In this study a periodic adaptive control approach is discussed for a class of nonlinear discrete-time systems with time-varying parametric uncertainties which are periodic, and the only prior knowledge is the periodicity. The adaptive controller updates the parameters and the control signal periodically in a pointwise manner over one entire period, in the sequel achieves the asymptotic tracking convergence. The result is further extended to a scenario with mixed time-varying and time-invariant parameters, and a hybrid classical and periodic adaptation law is proposed to handle the scenario more appropriately. Extension of the periodic adaptation to systems with unknown input gain, higher order dynamics, and tracking problems are also discussed.

3.1 Introduction

Adaptive control theory for continuous-time systems is one of the most well established control theories, and numerous results have been reported, e.g., [58, 87, 96, 121, 141]. In the classical adaptive control, the parametric adaptation mechanism essentially consists of a number of integrators, thus the adaptive control system is able to achieve asymptotic tracking convergence in the presence of constant parametric uncertainties. In [160], a method for dealing with a class of time-varying periodic uncertain parameters is introduced that is based on pointwise integration relying on the *a priori* knowledge of the periodicity of the parameters.

Considering the fact that, as a function of time, the classes of time-varying parameters are in essence infinite, it would be extremely difficult to find a general solution to such a broad control problem. A more realistic way is first to classify the time-varying parametric uncertainties into subclasses, and then look for an appropriate adaptive control approach for each subclass. Instead of classifying parameters into slow versus rapid time-varying, in this work we classify parameters into periodic versus nonperiodic ones. When the periodicity of system parameters is known a priori, an adaptive controller with periodic updating can be constructed by means of a pointwise update mechanism.

Periodic variations are encountered in many real systems. These variations can exist in the system parameters, [130, 165], or as a disturbance to the system, [41, 90, 157]. This necessitates the effort in formulating an adaptive control scheme that

K. Abidi and J.-X. Xu, *Advanced Discrete-Time Control*, Studies in Systems, Decision and Control 23, DOI 10.1007/978-981-287-478-8_3

can handle a class of systems with time-varying periodic uncertain parameters or disturbances by taking into account the periodicity of the variations.

In this chapter, we apply the concept of periodic adaptation, originally proposed for continuous-time systems, to discrete-time systems. In particular, it will be shown that the new periodic adaptive controller can work effectively to nullify the influence from the time-varying parametric uncertainties to the control error, in the sequel achieve the asymptotic convergence. Comparing with the continuous-time adaptive control, the discrete-time periodic adaptation is a more natural extension of the classical adaptive control: From the updating in two consecutive instances to the updating in the same instance of two consecutive periods. This is owing to the fact that the value of a periodic parameter will be invariant if shifting the time by one period. This feature necessitates the need to explore a new parametric estimation law and a new convergence property analysis tool. In the periodic adaptive control, the parametric values of the preceding period, instead of the preceding time instance, are used to update the current parameter estimate. Analogously, the boundedness of the parametric estimate and convergence analysis are conducted by deriving the difference between two consecutive periods, that is, the convergence is asymptotic with respect to the number of periods, instead of the time instances. When there exist both time-invariant and time-varying periodic parameters, the classical adaptation and the new periodic adaptation laws can be employed simultaneously, whereas the convergence analysis will be based on the number of periods.

3.2 Discrete-Time Periodic Adaptive Control

In order to clearly demonstrate the idea, first consider a scalar discrete-time system with only one uncertain parameter

$$x_{k+1} = \theta_k \xi_k + u_k, \quad x_0 = x(0) \tag{3.1}$$

where $x_k \in \Re$, $u_k \in \Re$, $\theta_k \in \Re$ and $\xi_k \equiv \xi(x_k) \in \Re$ is a known nonlinear function.

Assumption 3.1 The uncertain parameter θ_k is periodic, i.e. $\theta(k) = \theta(k-N)$ with a known period $N > 1$.

For simplicity we will consider the regulation problem only and leave the extension to tracking tasks to the next section.

3.2.1 Discrete-Time Adaptive Control Revisited

If θ_k is time invariant, i.e. $\theta(k) = \theta(k-1)$, then the standard approach would be to combine a certainty equivalence controller with a least-squares estimator

$$u_k = -\hat{\theta}_k \xi_k \tag{3.2}$$
$$\hat{\theta}_k = \hat{\theta}_{k-1} + P_k \xi_{k-1} x_k, \tag{3.3}$$
$$P_k = P_{k-1} - \frac{P_{k-1}^2 \xi_{k-1}^2}{1 + P_{k-1}\xi_{k-1}^2}, \quad P_0 > 0 \tag{3.4}$$

where P_k is a scalar for this case. Defining $\tilde{\theta} = \theta - \hat{\theta}$, and substituting the adaptive law (3.2) into the dynamical relation (3.1), the closed-loop system can be expressed as

$$\begin{aligned} x_{k+1} &= \tilde{\theta}_k \xi_k, \\ \tilde{\theta}_{k+1} &= \tilde{\theta}_k - P_k \xi_k x_{k+1} \\ P_{k+1} &= P_k - \frac{P_k^2 \xi_k^2}{1 + P_k \xi_k^2}. \end{aligned} \tag{3.5}$$

The least-squares estimator has several desirable properties [69], for instance the boundedness and positivity of P_k which can be found by rewriting (3.4) as

$$P_{k+1}^{-1} = P_k^{-1} + \xi_k^2 \tag{3.6}$$

implying $P_k^{-1} \geq P_0^{-1} > 0$ or $P_0 \geq P_k > 0$ for all $k \geq 0$. Now, consider a non-negative function $V_k = P_k^{-1}\tilde{\theta}_k^2$, its difference over one step is

$$\begin{aligned} V_{k+1} - V_k &= P_{k+1}^{-1}\tilde{\theta}_{k+1}^2 - P_k^{-1}\tilde{\theta}_k^2 \\ &= P_{k+1}^{-1}\left(\tilde{\theta}_k - P_{k+1}\xi_k x_{k+1}\right)^2 - P_k^{-1}\tilde{\theta}_k^2 \\ &= -\frac{x_{k+1}^2}{1 + P_k \xi_k^2} \leq 0. \end{aligned} \tag{3.7}$$

From (3.7), the term $x_{k+1}/(1 + P_k \xi_k^2)^{1/2}$ converges to zero as $k \to \infty$. Further, if the nonlinear function $\xi(x)$ is sector-bounded

$$|\xi(x)| \leq c_1 + c_2 |x| \tag{3.8}$$

with c_1 and c_2 arbitrary positive constants, then it is possible to establish the condition [91],

$$|\xi(x_k)| \leq c_1^0 + c_2^0 \max_{j \in [0,k]} |x_{j+1}| \tag{3.9}$$

where c_1^0 and c_2^0 are constants, then the *Key Technical Lemma* ensures that $x_k \to 0$ as $k \to \infty$.

3.2.2 Periodic Adaptation

Can the concept of adaptation control still be applied to periodic but arbitrarily time-varying θ_k? Note that, the dynamics $\tilde{\theta}_{k+1}$ in the closed-loop system (3.5) is derived by subtracting the parameter adaptation law (3.3) from the time invariance relationship $\theta_k = \theta_{k-1}$, which however does not hold for a time-varying θ_k. On the other hand, for the periodic parameter we have $\theta_k = \theta_{k-N}$. Note that $N = 1$ renders a periodic θ_k to a constant. Hence a periodic parameter, with the periodicity $N > 1$, can be viewed as a 'constant' with respect to the interval N. As such, we can modify the standard adaptation law (3.3), originally designed to update the parameter estimate between two consecutive instances, namely from $k-1$ to k, into a new periodic adaptation law that updates the parameter estimate after a fixed interval N, namely from $k-N$ to k. In the following we will verify and demonstrate this simple idea.

Revise the adaptive control mechanism (3.2)–(3.4) into the following periodic one

$$u_k = -\hat{\theta}_k \xi_k \tag{3.10}$$

$$\hat{\theta}_k = \begin{cases} \hat{\theta}_{k-N} + P_k \xi_{k-N} x_{k-N+1} & \forall k \in [N, \infty) \\ \hat{\theta}_0 & \forall k = [0, N) \end{cases} \tag{3.11}$$

$$P_k = \begin{cases} P_{k-N} - \dfrac{P_{k-N}^2 \xi_{k-N}^2}{1 + P_{k-N} \xi_{k-N}^2} & \forall k \in [N, \infty) \\ P_0 > 0 & \forall k \in [0, N) \end{cases} \tag{3.12}$$

where $k = k_0 + nN$ and n is the total number of periods in the interval $[0, k)$. From (3.12) we can derive a result similar to (3.6)

$$P_k^{-1} = P_{k-N}^{-1} + \xi_{k-N}^2 \tag{3.13}$$

which implies that $P_k^{-1} \geq P_0^{-1} > 0$ and $P_0 \geq P_k > 0$ for all $k \geq N$.

Remark 3.1 Note that the adaptation process starts only after the first cycle is completed or $k \geq N$. The estimate $\hat{\theta}_k$ for $k < N$ is set to $\hat{\theta}_0$, which can be chosen according to some prior knowledge, or simply zero if no prior knowledge is available. Similarly, we can choose P_0 to be a sufficiently large constant over the interval $[0, N)$.

3.2.3 Convergence Analysis

Define the parameter estimation error $\tilde{\theta}_k = \theta_k - \hat{\theta}_k$. Substituting the adaptive control (3.10) into the dynamical relation (3.1), and subtracting the adaptive law (3.11) from $\theta_k = \theta_{k-N}$, the closed-loop system, for any $k \geq N$, can be expressed as

$$\begin{aligned} x_{k+1} &= \tilde{\theta}_k \xi_k \\ \tilde{\theta}_k &= \tilde{\theta}_{k-N} - P_k \xi_{k-N} x_{k-N+1} \\ P_k &= P_{k-N} - \frac{P_{k-N}^2 \xi_{k-N}^2}{1 + P_{k-N} \xi_{k-N}^2}. \end{aligned} \tag{3.14}$$

The convergence property of the periodic adaptive control system (3.14) is summarized in the following theorem.

Theorem 3.1 *For the closed-loop system (3.14), the parameter estimation error $\tilde{\theta}$ is bounded and the regulation error x_k converges to zero asymptotically.*

Proof Similar to the time-invariant case, select a nonnegative function $V_k = P_k^{-1}\tilde{\theta}_k^2$, its difference with respect to the interval N for any $k \geq N$ is

$$\begin{aligned} \Delta V_k &= V_k - V_{k-N} \\ &= P_k^{-1}\tilde{\theta}_k^2 - P_{k-N}^{-1}\tilde{\theta}_{k-N}^2 \\ &= P_k^{-1}\left(\tilde{\theta}_{k-N} - P_k \xi_{k-N} x_{k-N+1}\right)^2 - P_{k-N}^{-1}\tilde{\theta}_{k-N}^2 \\ &= (P_k^{-1} - P_{k-N}^{-1})\tilde{\theta}_{k-N}^2 - 2\tilde{\theta}_{k-N}\xi_{k-N}x_{k-N+1} + P_k \xi_{k-N}^2 x_{k-N+1}^2 \\ &= -\frac{x_{k-N+1}^2}{1 + P_{k-N}\xi_{k-N}^2} \leq 0. \end{aligned} \tag{3.15}$$

Thus V_k is nonincreasing, implying that $\tilde{\theta}_k$ is bounded. Applying (3.15) repeatedly for any $k \in [pN, (p+1)N]$, and noticing $k_0 = k - pN$, we have

$$V_k = V_{k_0} + \sum_{i=1}^{p} \Delta V_{k_0+iN} \tag{3.16}$$

Since $k_0 \in [0, N)$, and

$$p = \frac{k - k_0}{N} \to \infty$$

when $k \to \infty$, according to (3.15)

$$\lim_{p\to\infty} V_k < \max_{k_0 \in [0,N)} V_{k_0} - \lim_{p\to\infty} \sum_{i=1}^{p} \frac{x_{k_0+(i-1)N+1}^2}{1 + P_{k_0+(i-1)N}\xi_{k_0+(i-1)N}^2}. \tag{3.17}$$

Consider that V_k is nonnegative, V_{k_0} is finite in the interval $[0, N)$, thus according to the convergence theorem of the sum of series, we have

$$\lim_{k\to\infty} \frac{x_{k-N+1}^2}{1 + P_{k-N}\xi_{k-N}^2} = 0. \tag{3.18}$$

Using (3.18) and the sector condition (3.9), the *Key Technical Lemma* guarantees that ξ_k is bounded and consequently the regulation error $x_k \to 0$ as $k \to \infty$.

Remark 3.2 Since difference equations do not have a finite escape time, the finiteness of V_{k_0} is obvious. From the viewpoint of achieving asymptotic convergence, the initial phase control performance in $[0, N)$ is not as crucial as that of the continuous-time periodic adaptive control. One can choose any controller, not necessarily the same as (3.10), for the initial phase $[0, N)$, so long as a better performance can be obtained.

3.3 Extension to More General Cases

In this section we consider four extensions to multiple periodic parameters, mixed periodic and time-invariant parameters, the trajectory tracking problem, and the higher order systems, respectively.

3.3.1 Extension to Multiple Parameters

For simplicity, we will still consider a scalar system

$$x_{k+1} = \left(\boldsymbol{\theta}_k^0\right)^{\top} \boldsymbol{\xi}_k^0 + b_k u_k, \quad x(0) = x_0 \tag{3.19}$$

where $\boldsymbol{\theta}_k^0 = \left[\theta_{1,k}^0 \,\vdots\, \theta_{2,k}^0 \,\vdots\, \cdots \,\vdots\, \theta_{m,k}^0\right]^{\top} \in \Re^m$ and $\boldsymbol{\xi}_k^0 = \left[\xi_{1,k}^0 \,\vdots\, \xi_{2,k}^0 \,\vdots\, \cdots \,\vdots\, \xi_{m,k}^0\right]^{\top} \in \Re^m$ are the vectors of uncertain periodic parameters and known functions respectively. Further, the input gain is $b_k \in C[0, \infty) \in \Re$ is also considered to be periodic time-varying and uncertain. The prior information with regards to b_k is that the control direction is known and invariant, that is, b_k is either positive or negative and nonsingular for all k. Without loss of generality, assume that $b_k \geq b_{\min}$ where $b_{\min} > 0$ is a known lower bound. Note that each uncertain parameter, $\theta_{i,k}^0$ or b_k, may have its own period N_i or N_b. The periodic adaptive control will still be applicable if there exists a common period N, such that each N_i and N_b can divide N with an integer quotient. This is always true in discrete-time since N_i and N_b are integers, therefore N can be either the least common multiple of N_i and N_b, or simply the product of N_i and N_b when both are prime. Therefore, N is used as the updating period. The presence of the uncertain system input gain makes the controller design more complex. To derive the periodic adaptive control law, define $\hat{b}_k$ to be the estimate of b_k and let $\tilde{b}_k = b_k - \hat{b}_k$, then (3.19) can be rewritten as

$$x_{k+1} = \left(\boldsymbol{\theta}_k^0\right)^{\top} \boldsymbol{\xi}_k^0 + b_k u_k - \hat{b}_k u_k + \hat{b}_k u_k = \left(\boldsymbol{\theta}_k^0\right)^{\top} \boldsymbol{\xi}_k^0 + \hat{b}_k u_k + \tilde{b}_k u_k. \tag{3.20}$$

By observation, we can choose the control law

$$u_k = -\hat{b}_k^{-1}\left(\hat{\boldsymbol{\theta}}_k^0\right)^\top \boldsymbol{\xi}_k^0 \tag{3.21}$$

where $\hat{\boldsymbol{\theta}}_k^0 = \left[\hat{\theta}_{1,k}^0 \,\vdots\, \hat{\theta}_{2,k}^0 \,\vdots\, \cdots \,\vdots\, \hat{\theta}_{\mathrm{m},k}^0\right]^\top \in \Re^m$. Substituting (3.21) into (3.20) yields the closed-loop system

$$x_{k+1} = \tilde{\boldsymbol{\theta}}_k^\top \boldsymbol{\xi}_k \tag{3.22}$$

where $\tilde{\boldsymbol{\theta}}_k^\top = \left[\left(\tilde{\boldsymbol{\theta}}_k^0\right)^\top \,\vdots\, \tilde{b}_k\right] \in \Re^{m+1}$ and $\boldsymbol{\xi}_k^\top = \left[\left(\boldsymbol{\xi}_k^0\right)^\top \,\vdots\, -\hat{b}_k^{-1}\left(\hat{\boldsymbol{\theta}}_k^0\right)^\top \boldsymbol{\xi}_k^0\right] \in \Re^{m+1}$ are the augmented parameter estimation error and the augmented functions vector respectively. Note that the parameter estimation error is given by $\tilde{\boldsymbol{\theta}}_k^0 = \boldsymbol{\theta}_k^0 - \hat{\boldsymbol{\theta}}_k^0$. Note that the computation of $\boldsymbol{\xi}_k$ requires the inverse of the system input gain estimate $\hat{b}_k$ and may cause a singularity in the solution if the estimate of the system input gain is zero. To ensure this never occurs a semi-saturator must be applied on the input gain estimator such that the estimate never goes below the lower bound. For this purpose and based on (3.22), the adaptation law is

$$\hat{\boldsymbol{\theta}}_k = \begin{cases} L[\hat{\boldsymbol{\theta}}_{k-N} + P_k \boldsymbol{\xi}_{k-N} x_{k-N+1}] & \forall k \in [N, \infty) \\ L[\hat{\boldsymbol{\theta}}_0] & \forall k \in [0, N) \end{cases} \tag{3.23}$$

$$P_k = \begin{cases} P_{k-N} - \dfrac{P_{k-N}\boldsymbol{\xi}_{k-N}^\top \boldsymbol{\xi}_{k-N} P_{k-N}}{1 + \boldsymbol{\xi}_{k-N}^\top P_{k-N} \boldsymbol{\xi}_{k-N}} & \forall k \in [N, \infty) \\ P_0 > 0 & \forall k \in [0, N) \end{cases} \tag{3.24}$$

where the covariance P_k is a positive definite matrix of dimension $m+1$ and derived from the relationship $P_k^{-1} = P_{k-N}^{-1} + \boldsymbol{\xi}_{k-N}\boldsymbol{\xi}_{k-N}^\top$ by means of the *Matrix Inversion Lemma*, [96]. Let $\mathbf{a}^\top = \left[\mathbf{a}_1^\top \,\vdots\, a_2\right]$ denote the vector $\hat{\boldsymbol{\theta}}_{k-N} + P_k \boldsymbol{\xi}_{k-N} x_{k-N+1}$, the semi-saturator is defined as

$$L[\mathbf{a}] = \begin{cases} \left[\mathbf{a}_1^\top \,\vdots\, a_2\right]^\top & \forall a_2 \in [b_{\min}, \infty) \\ \left[\mathbf{a}_1^\top \,\vdots\, b_{\min}\right]^\top & \forall a_2 \in (-\infty, b_{\min})\,. \end{cases} \tag{3.25}$$

The validity of the above periodic adoption law is verified by the following theorem.

Theorem 3.2 *Under the periodic adaptation law (3.23) and (3.24), the closed-loop dynamics (3.22) is asymptotically stable.*

Proof The convergence analysis is analogous to the preceding case. Selecting a nonnegative function $V_k = \tilde{\boldsymbol{\theta}}_k^\top P_k^{-1} \tilde{\boldsymbol{\theta}}_k$. Note that $\tilde{\boldsymbol{\theta}}_k = \boldsymbol{\theta}_k - \hat{\boldsymbol{\theta}}_k = \boldsymbol{\theta}_k - L[\mathbf{a}]$, where $\mathbf{a} = \hat{\boldsymbol{\theta}}_{k-N} + P_k \boldsymbol{\xi}_{k-N} x_{k-N+1}$. The difference of V_k with respect to the interval N is

$$\begin{aligned}\Delta V_k &= V_k - V_{k-N}\\ &= \tilde{\boldsymbol{\theta}}_k^\top P_k^{-1}\tilde{\boldsymbol{\theta}}_k - \tilde{\boldsymbol{\theta}}_{k-N}^\top P_{k-N}^{-1}\tilde{\boldsymbol{\theta}}_{k-N}\\ &= (\boldsymbol{\theta}_k - L[\mathbf{a}])^\top P_k^{-1}(\boldsymbol{\theta}_k - L[\mathbf{a}]) - \tilde{\boldsymbol{\theta}}_{k-N}^\top P_{k-N}^{-1}\tilde{\boldsymbol{\theta}}_{k-N} \end{aligned} \tag{3.26}$$

Note that, because the actual input gain is assumed to be positive and so should the minimum bound, $b_{\min}$, the magnitude of the estimation error would be the same or larger if no saturator was implemented since for estimates below $b_{\min}$ the difference $|b_k - b_{\min}| < |b_k - \hat{b}_k|$. Thus, we can conclude that

$$(\boldsymbol{\theta}_k - L[\mathbf{a}])^\top (\boldsymbol{\theta}_k - L[\mathbf{a}]) \le (\boldsymbol{\theta}_k - \mathbf{a})^\top (\boldsymbol{\theta}_k - \mathbf{a}) \tag{3.27}$$

and, furthermore, for a positive-definite matrix P_k^{-1} then the following is also true

$$(\boldsymbol{\theta}_k - L[\mathbf{a}])^\top P_k^{-1}(\boldsymbol{\theta}_k - L[\mathbf{a}]) \le (\boldsymbol{\theta}_k - \mathbf{a})^\top P_k^{-1}(\boldsymbol{\theta}_k - \mathbf{a}) \tag{3.28}$$

Therefore we can simplify (3.26) further

$$\begin{aligned}\Delta V_k &= (\boldsymbol{\theta}_k - L[\mathbf{a}])^\top P_k^{-1}(\boldsymbol{\theta}_k - L[\mathbf{a}]) - \tilde{\boldsymbol{\theta}}_{k-N}^\top P_{k-N}^{-1}\tilde{\boldsymbol{\theta}}_{k-N}\\ &\le (\boldsymbol{\theta}_k - \mathbf{a})^\top P_k^{-1}(\boldsymbol{\theta}_k - \mathbf{a}) - \tilde{\boldsymbol{\theta}}_{k-N}^\top P_{k-N}^{-1}\tilde{\boldsymbol{\theta}}_{k-N}\\ &\le \left(\tilde{\boldsymbol{\theta}}_{k-N} - P_k\boldsymbol{\xi}_{k-N}x_{k-N+1}\right)^\top P_k^{-1}\left(\tilde{\boldsymbol{\theta}}_{k-N} - P_k\boldsymbol{\xi}_{k-N}x_{k-N+1}\right) - \tilde{\boldsymbol{\theta}}_{k-N}^\top P_{k-N}^{-1}\tilde{\boldsymbol{\theta}}_{k-N}\\ &\le \tilde{\boldsymbol{\theta}}_{k-N}^\top(P_k^{-1} - P_{k-N}^{-1})\tilde{\boldsymbol{\theta}}_{k-N} - 2\tilde{\boldsymbol{\theta}}_{k-N}^\top\boldsymbol{\xi}_{k-N}x_{k-N+1} + \boldsymbol{\xi}_{k-N}^\top P_k\boldsymbol{\xi}_{k-N}x_{k-N+1}^2\\ &\le \tilde{\boldsymbol{\theta}}_{k-N}^\top(\boldsymbol{\xi}_{k-N}\boldsymbol{\xi}_{k-N}^\top)\tilde{\boldsymbol{\theta}}_{k-N} - 2\tilde{\boldsymbol{\theta}}_{k-N}^\top\boldsymbol{\xi}_{k-N}x_{k-N+1} + \boldsymbol{\xi}_{k-N}^\top P_k\boldsymbol{\xi}_{k-N}x_{k-N+1}^2\\ &\le -\frac{x_{k-N+1}^2}{1+\boldsymbol{\xi}_{k-N}^\top P_{k-N}\boldsymbol{\xi}_{k-N}} \le 0. \end{aligned} \tag{3.29}$$

Following the same steps that lead to (3.18) in *Theorem* 3.1, we conclude that

$$\lim_{k\to\infty}\frac{x_{k-N+1}^2}{1+\boldsymbol{\xi}_{k-N}^\top P_{k-N}\boldsymbol{\xi}_{k-N}} = 0. \tag{3.30}$$

The result (3.29) shows that $\hat{\boldsymbol{\theta}}_k^0$ and $\hat{b}_k$ are bounded because V_k is non-increasing and, thus the control signal $\|u_k\| \le \|\hat{b}_k^{-1}\| \cdot \|\hat{\boldsymbol{\theta}}_k^0\| \cdot \|\boldsymbol{\xi}_k^0\| \le q\|\boldsymbol{\xi}_k^0\|$ for some constant q. If the nonlinear function is sector-bounded, i.e. $\|\boldsymbol{\xi}_k^0\| \le c_1^0 + c_2^0|x_k|$ for some positive constants c_1^0 and c_2^0, then $\|\boldsymbol{\xi}_k\| \le \|\boldsymbol{\xi}_k^0\| + |u_k| \le c_1 + c_2|x_k|$ for some positive constants c_1 and c_2. Thus, establishing the condition for (3.9) required by the *Key Technical Lemma* that guarantees $x_k \to 0$ as $k \to \infty$.

3.3.2 Extension to Mixed Parameters

Often, we have some prior knowledge about the system parametric uncertainties, for instance we may know that some uncertain parameters are time invariant, whereas the rest are time-varying. This is a nontrivial case, as the more we know, the better we should be able to improve the control performance. It would be far-fetched if we still apply the periodic adaptation to those constant parameters, and the traditional adaptation is more suitable.

Consider the simplest scalar case

$$x_{k+1} = \theta_{1,k}\xi_{1,k} + \theta_2\xi_{2,k} + b_k u, \qquad x_0 = x(0) \tag{3.31}$$

where $\theta_{1,k} \in \Re$, $b_k \in \Re$ are uncertain periodic parameters with common periodicity N, $\theta_2 \in \Re$ is an uncertain constant, $\xi_1 \in \Re$ and $\xi_2 \in \Re$ are known sector-bounded nonlinear functions. Using the same rationale behind controller (3.21) the control law is chosen to be

$$u_k = -\frac{1}{\hat{b}_k}\left(\hat{\theta}_{1,k}\xi_{1,k} + \hat{\theta}_{2,k}\xi_{2,k}\right). \tag{3.32}$$

Define $\boldsymbol{\theta}_k^\top = \left[\theta_{1,k} \mid b_k\right] \in \Re^2$ as the augmented vector of periodic parameters, q_1, q_2 and q_3 as positive scalar constants. The hybrid periodic adaptation law is chosen to be

$$\hat{\boldsymbol{\theta}}_k = \begin{cases} L\left[\hat{\boldsymbol{\theta}}_{k-N} + Q\dfrac{\boldsymbol{\xi}_{k-N}x_{k-N+1}}{1+\bar{\boldsymbol{\xi}}_{k-N}^\top \bar{Q}\bar{\boldsymbol{\xi}}_{k-N}}\right] & \forall k \in [N, \infty) \\ L\left[\hat{\boldsymbol{\theta}}_0\right] & \forall k \in [0, N) \end{cases}$$

$$\hat{\theta}_{2,k} = \hat{\theta}_{2,k-1} + q_3\frac{\xi_{2,k-1}x_k}{1+\bar{\boldsymbol{\xi}}_{k-1}^\top \bar{Q}\bar{\boldsymbol{\xi}}_{k-1}}$$

where $\hat{\boldsymbol{\theta}}_k^\top = \left[\hat{\theta}_{1,k} \mid \hat{b}_k\right] \in \Re^2$, $\boldsymbol{\xi}_k^\top = \left[\xi_{1,k} \mid -\hat{b}_k^{-1}\left(\hat{\theta}_{1,k}\xi_{1,k} + \hat{\theta}_{2,k}\xi_{2,k}\right)\right] \in \Re^2$, $Q = \mathrm{diag}(q_1, q_2) \in \Re^{2\times 2}$, $\bar{Q} = \mathrm{diag}(q_1, q_2, q_3) \in \Re^{3\times 3}$, $\bar{\boldsymbol{\xi}}_k^\top = \left[\boldsymbol{\xi}_k^\top \mid \xi_{2,k}\right] \in \Re^2$, and the value of $\hat{\theta}_{1,0}$ can be chosen to be zero for the initial period $[0, N)$ if no prior information is available. The semi-saturator $L[\cdot]$ which is the same as (3.25) is again used to prevent the occurrence of a singular solution. Substituting the control law (3.32) into (3.31), the closed-loop system is

$$x_{k+1} = \tilde{\boldsymbol{\theta}}_k^\top \boldsymbol{\xi}_k + \tilde{\theta}_{2,k}\xi_{2,k}. \tag{3.33}$$

where $\tilde{\boldsymbol{\theta}}_k = \boldsymbol{\theta}_k - \hat{\boldsymbol{\theta}}_k$ and $\tilde{\theta}_{2,k} = \theta_2 - \hat{\theta}_{2,k}$. Now let us show the asymptotical stability of the closed-loop system with the hybrid adaptive control.

Theorem 3.3 *For the closed-loop system defined by (3.33) the parameter estimation errors* $\tilde{\boldsymbol{\theta}}_k$ *and* $\tilde{\theta}_{2,k}$ *are bounded, and the regulation error* x_k *approaches to zero asymptotically.*

Proof Choose a nonnegative function below

$$V_k = \sum_{i=k-N}^{k-1} \left(\tilde{\boldsymbol{\theta}}_i^\top Q^{-1}\tilde{\boldsymbol{\theta}}_i\right) + \frac{1}{q_3}\tilde{\theta}_{2,k-N}^2, \tag{3.34}$$

its difference with respect to the interval N is

$$\begin{aligned}\Delta V_k &= V_k - V_{k-N}\\ &= \sum_{i=k-N}^{k-1} \left(\tilde{\boldsymbol{\theta}}_i^\top Q^{-1}\tilde{\boldsymbol{\theta}}_i - \tilde{\boldsymbol{\theta}}_{i-N}^\top Q^{-1}\tilde{\boldsymbol{\theta}}_{i-N}\right) + \frac{1}{q_3}\left(\tilde{\theta}_{2,k-N}^2 - \tilde{\theta}_{2,k-2N}^2\right).\end{aligned} \tag{3.35}$$

This can be rewritten as

$$\begin{aligned}\Delta V_k &= \sum_{i=k-N}^{k-1} \left(\tilde{\boldsymbol{\theta}}_i^\top Q^{-1}\tilde{\boldsymbol{\theta}}_i - \tilde{\boldsymbol{\theta}}_{i-N}^\top Q^{-1}\tilde{\boldsymbol{\theta}}_{i-N} + \frac{1}{q_3}\left(\tilde{\theta}_{2,i-N+1}^2 - \tilde{\theta}_{2,i-N}^2\right)\right)\\ &= \sum_{i=k-N}^{k-1} \left(\left[\tilde{\boldsymbol{\theta}}_i^\top \ \ \tilde{\theta}_{2_{i-N+1}}\right]\bar{Q}^{-1}\begin{bmatrix}\tilde{\boldsymbol{\theta}}_i\\ \tilde{\theta}_{2,i-N+1}\end{bmatrix} - \tilde{\bar{\boldsymbol{\theta}}}_{i-N}^\top \bar{Q}^{-1}\tilde{\bar{\boldsymbol{\theta}}}_{i-N}\right)\end{aligned} \tag{3.36}$$

where $\tilde{\bar{\boldsymbol{\theta}}}_k^\top = \left[\tilde{\boldsymbol{\theta}}_k^\top \,\vdots\, \tilde{\theta}_{2,k}\right]$. Let $\mathbf{a}$ denote the vector $\hat{\boldsymbol{\theta}}_{k-N} + Q\dfrac{\boldsymbol{\xi}_{k-N}x_{k-N+1}}{1+\bar{\boldsymbol{\xi}}_{k-N}^\top \bar{Q}\bar{\boldsymbol{\xi}}_{k-N}}$, thus, (3.38) becomes

$$\Delta V_k = \sum_{i=k-N}^{k-1} \left(\left[\boldsymbol{\theta}_i^\top - L[\mathbf{a}]^\top \ \ \tilde{\theta}_{2,i-N+1}\right]\bar{Q}^{-1}\begin{bmatrix}\boldsymbol{\theta}_i - L[\mathbf{a}]\\ \tilde{\theta}_{2,i-N+1}\end{bmatrix} - \tilde{\bar{\boldsymbol{\theta}}}_{i-N}^\top \bar{Q}^{-1}\tilde{\bar{\boldsymbol{\theta}}}_{i-N}\right). \tag{3.37}$$

Following the logic of (3.28) we obtain

$$\Delta V_k \le \sum_{i=k-N}^{k-1} \left(\left[\boldsymbol{\theta}_i^\top - \mathbf{a}^\top \ \ \tilde{\theta}_{2_{i-N+1}}\right]\bar{Q}^{-1}\begin{bmatrix}\boldsymbol{\theta}_i - \mathbf{a}\\ \tilde{\theta}_{2,i-N+1}\end{bmatrix} - \tilde{\bar{\boldsymbol{\theta}}}_{i-N}^\top \bar{Q}^{-1}\tilde{\bar{\boldsymbol{\theta}}}_{i-N}\right). \tag{3.38}$$

Next, shifting the parameter estimate $\hat{\theta}_{2,k}$ back $N-1$ steps and subtracting both sides from θ_2 yields

$$\tilde{\theta}_{2,k-N+1} = \tilde{\theta}_{2,k-N} - q_3\frac{\xi_{2,k-N}x_{k-N+1}}{1+\boldsymbol{\xi}_{k-N}^\top Q\boldsymbol{\xi}_{k-N}}. \tag{3.39}$$

Combining the above expression (3.39) with that of $\boldsymbol{\theta}_k - \mathbf{a}$ results in the following

$$\begin{bmatrix} \boldsymbol{\theta}_k - \mathbf{a} \\ \tilde{\theta}_{2,k-N+1} \end{bmatrix} = \tilde{\bar{\boldsymbol{\theta}}}_{k-N} - \bar{Q}\frac{\bar{\boldsymbol{\xi}}_{k-N} x_{k-N+1}}{1 + \bar{\boldsymbol{\xi}}_{k-N}^{\top} \bar{Q} \bar{\boldsymbol{\xi}}_{k-N}}. \tag{3.40}$$

Substituting (3.40) into (3.38) with calculation leads to

$$\Delta V_k \leq - \sum_{i=k-N}^{k-1} \left(\frac{2}{1 + \bar{\boldsymbol{\xi}}_{i-N}^{\top} \bar{Q} \bar{\boldsymbol{\xi}}_{i-N}} - \frac{\bar{\boldsymbol{\xi}}_{i-N}^{\top} \bar{Q} \bar{\boldsymbol{\xi}}_{i-N}}{\left(1 + \bar{\boldsymbol{\xi}}_{i-N}^{\top} \bar{Q} \bar{\boldsymbol{\xi}}_{i-N}\right)^2} \right) x_{i-N+1}^2 \tag{3.41}$$

$$\leq - \sum_{i=k-N}^{k-1} \left(\frac{x_{i-N+1}^2}{1 + \bar{\boldsymbol{\xi}}_{i-N}^{\top} \bar{Q} \bar{\boldsymbol{\xi}}_{i-N}} \right) \tag{3.42}$$

which implies that V_k is nonincreasing (w.r.t. N) and, thus, $\tilde{\boldsymbol{\theta}}_k$ and $\tilde{\theta}_{2_k}$ are bounded. Similar to previous derivations, applying (3.42) repeatedly for any $k \in [pN, (p+1)N]$, and denoting $k_0 = k - pN$, we have

$$V_k = V_{k_0} + \sum_{i=1}^{p} \Delta V_{k_0 + iN} \tag{3.43}$$

Since $k_0 \in [0, N)$, according to (3.42)

$$\lim_{p\to\infty} V_k < \max_{k_0 \in [0,N)} V_{k_0} - \lim_{p\to\infty} \sum_{j=1}^{p} \left(\sum_{i=k_0(j-1)N}^{k_0+jN-1} \left(\frac{x_{i-N+1}^2}{1 + \bar{\boldsymbol{\xi}}_{i-N}^{\top} \bar{Q} \bar{\boldsymbol{\xi}}_{i-N}} \right) \right). \tag{3.44}$$

Considering the positiveness of V_k and the boundedness of V_{k_0} in the interval $[0, N)$ then, according to the convergence theorem of the sum of series, we have

$$\lim_{k\to\infty} \sum_{i=k-N}^{k-1} \left(\frac{x_{i-N+1}^2}{1 + \bar{\boldsymbol{\xi}}_{i-N}^{\top} \bar{Q} \bar{\boldsymbol{\xi}}_{i-N}} \right) = 0. \tag{3.45}$$

In Sect. 3.3.1, it was shown that if $\|\boldsymbol{\xi}_k\| \leq c_1 + c_2|x_k|$ for some constants c_1 and c_2, then $\|\bar{\boldsymbol{\xi}}_k\| \leq c_1^0 + c_2^0|x_k|$ is also true for some constants c_1^0 and c_2^0. Combining this with (3.45) the *Key Technical Lemma* guarantees that $x_k \to 0$ as $k \to \infty$.

3.3.3 Extension to Tracking Tasks

Consider the scalar system (3.19) with multiple uncertain parameters and the uncertain periodic input gain. It is required that the state, x_k, follow a given reference trajectory $r(k)$. Specifying the tracking error as $e_k = x_k - r_k$, we have

$$e_{k+1} = x_{k+1} - r_{k+1} = \left(\boldsymbol{\theta}_k^0\right)^\top \boldsymbol{\xi}_k^0 + b_k u_k - r_{k+1}. \tag{3.46}$$

Rewrite (3.46) in the form

$$e_{k+1} = \left(\boldsymbol{\theta}_k^0\right)^\top \boldsymbol{\xi}_k^0 + b_k u_k - r_{k+1} - \hat{b}_k u_k + \hat{b}_k u_k = \left(\boldsymbol{\theta}_k^0\right)^\top \boldsymbol{\xi}_k^0 + \hat{b}_k u_k + \tilde{b}_k u_k - r_{k+1}. \tag{3.47}$$

To accommodate the tracking task, the periodic adaptive control (3.21)–(3.24) can be revised as below

$$u_k = \hat{b}_k^{-1}\left(r_{k+1} - \left(\hat{\boldsymbol{\theta}}_k^0\right)^\top \boldsymbol{\xi}_k^0\right) \tag{3.48}$$

$$\hat{\boldsymbol{\theta}}_k = \begin{cases} L\left[\hat{\boldsymbol{\theta}}_{k-N} + P_k \boldsymbol{\xi}_{k-N} e_{k-N+1}\right] & \forall k \in [N, \infty) \\ L\left[\hat{\boldsymbol{\theta}}_0\right] & \forall k \in [0, N) \end{cases} \tag{3.49}$$

$$P_k = \begin{cases} P_{k-N} - \frac{P_{k-N}\boldsymbol{\xi}_{k-N}^\top \boldsymbol{\xi}_{k-N} P_{k-N}}{1+\boldsymbol{\xi}_{k-N}^\top P_{k-N}\boldsymbol{\xi}_{k-N}} & \forall k \in [N, \infty) \\ P_0 > 0 & \forall k \in [0, N) \end{cases} \tag{3.50}$$

where $\hat{\boldsymbol{\theta}}_k^\top = \left[\left(\hat{\boldsymbol{\theta}}_k^0\right)^\top \vdots\, \tilde{b}_k\right]$, and $\boldsymbol{\xi}_k^\top = \left[\left(\boldsymbol{\xi}_k^0\right)^\top \vdots\, \hat{b}_k^{-1}\left(r_{k+1} - \left(\hat{\boldsymbol{\theta}}_k^0\right)^\top \boldsymbol{\xi}_k^0\right)\right]$. The closed-loop system for any $k \geq N$ is given by

$$e_{k+1} = \tilde{\boldsymbol{\theta}}_k^\top \boldsymbol{\xi}_k \tag{3.51}$$

Note that the tracking error dynamics in (3.51) has the same form as (3.22), and the adaption mechanism (3.49) and (3.50) also has the same form as (3.23) and (3.24) with the state x_k replaced by the tracking error e_k. Thus, *Theorem* 3.2 is directly applicable to this case and the asymptotic convergence of the tracking error e_k can easily be verified.

3.3.4 Extension to Higher Order Systems

Finally consider the single input higher order system in canonical form

$$\mathbf{x}_{k+1} = \begin{bmatrix} \mathbf{0} & I_{n-1} \\ 0 & \mathbf{0} \end{bmatrix} \mathbf{x}_k + \begin{bmatrix} \mathbf{0} \\ 1 \end{bmatrix} \left(\boldsymbol{\theta}_k^\top \boldsymbol{\xi}_k + b_k u_k\right) \tag{3.52}$$

where $\mathbf{x}_k \in \Re^n$, $\boldsymbol{\theta}_k^\top = \left[\theta_{1,k} \vdots\, \theta_{2,k} \vdots \cdots \vdots\, \theta_{m,k}\right] \in \Re^m$ are uncertain periodic parameters, $b_k \in \Re$ is the uncertain periodic input gain, $\boldsymbol{\xi}_k^\top = \boldsymbol{\xi}^\top(\mathbf{x}_k) = \left[\xi_{1,k} \vdots\, \xi_{2,k} \vdots \cdots \vdots\, \xi_{m,k}\right]$ is a known vector valued function which is sector bounded,

$\|\boldsymbol{\xi}_k\| \le c_1 + c_2\|\mathbf{x}_k\|$ (c_1 and c_2 being arbitrary positive constants). Similar to the previous case, it is assumed that the uncertain parameters have a common period N.

Assuming that all the states are available, the following control is proposed

$$u_k = -\frac{1}{\hat{b}_k}\hat{\boldsymbol{\theta}}_k^\top \boldsymbol{\xi}_k \tag{3.53}$$

with the following parameter adaptation law

$$\hat{\bar{\boldsymbol{\theta}}}_k = \begin{cases} L\left[\hat{\bar{\boldsymbol{\theta}}}_{k-N} + P_k\bar{\boldsymbol{\xi}}_{k-N}x_n(k-N+1)\right] & \forall k \in [N,\infty) \\ L\left[\hat{\bar{\boldsymbol{\theta}}}_0\right] & \forall k \in [0,N) \end{cases} \tag{3.54}$$

$$P_k = \begin{cases} P_{k-N} - \frac{P_{k-N}\bar{\boldsymbol{\xi}}_{k-N}\bar{\boldsymbol{\xi}}_{k-N}^\top P_{k-N}}{1+\bar{\boldsymbol{\xi}}_{k-N}^\top P_{k-N}\bar{\boldsymbol{\xi}}_{k-N}} & \forall k \in [N,\infty) \\ P_0 > 0 & \forall k \in [0,N) \end{cases} \tag{3.55}$$

where $\hat{\bar{\boldsymbol{\theta}}}_k^\top = \left[\hat{\boldsymbol{\theta}}_k^\top \,\vdots\, \hat{b}_k\right]$, $\bar{\boldsymbol{\xi}}_k^\top = \left[\boldsymbol{\xi}_k^\top \,\vdots\, -\hat{b}_k^{-1}\hat{\boldsymbol{\theta}}_k^\top\boldsymbol{\xi}_k\right]$ and $L[\cdot]$ is the semi-saturator defined by (3.25). The covariance P_k is a positive definite matrix of dimension $m+1$ and derived from the relationship $P_k^{-1} = P_{k-N}^{-1} + \bar{\boldsymbol{\xi}}_{k-N}\bar{\boldsymbol{\xi}}_{k-N}^\top$ by means of the *Matrix Inversion Lemma*. Note that the parameter estimate (3.54) is dependent on x_{n_k} where the subscript n denotes the nth state variable. Substitute the control (3.53) into (3.52) and rewrite the result into two subsystems

$$\mathbf{x}_{\mathrm{a},k+1} = \begin{bmatrix} \mathbf{0} & I_{n-2} \\ 0 & \mathbf{0} \end{bmatrix}\mathbf{x}_{\mathrm{a},k} + \begin{bmatrix} \mathbf{0} \\ 1 \end{bmatrix} x_{n,k} \tag{3.56}$$

$$x_{n,k+1} = \tilde{\boldsymbol{\theta}}_k^\top \boldsymbol{\xi}_k \tag{3.57}$$

where $\mathbf{x}_{\mathrm{a},k}^\top = \left[x_{1,k} \,\vdots\, x_{2,k} \,\vdots\, \cdots \,\vdots\, x_{n-1,k}\right]$. Looking into (3.54), (3.55) and (3.57), it is clear that this problem is transformed to the previous multiple parameter case. Thus the derivations and conclusions in *Theorem* 3.2 hold, as far as the *Key Technical Lemma* is still valid under the sector condition.

In order to establish the sector condition, note that the solution of (3.56) is given by

$$\mathbf{x}_{\mathrm{a},k} = \begin{bmatrix} \mathbf{0} & I_{n-2} \\ 0 & \mathbf{0} \end{bmatrix}^k \mathbf{x}_{\mathrm{a},0} + \sum_{i=0}^{k-1}\begin{bmatrix} \mathbf{0} & I_{n-2} \\ 0 & \mathbf{0} \end{bmatrix}^i \begin{bmatrix} \mathbf{0} \\ 1 \end{bmatrix} x_{n,k-i-1} \tag{3.58}$$

and for $k \ge n-1$ it can be shown that the solution is reduced to

$$\mathbf{x}_{\mathrm{a},k} = \sum_{i=0}^{n-2}\begin{bmatrix} \mathbf{0} & I_{n-2} \\ 0 & \mathbf{0} \end{bmatrix}^i \begin{bmatrix} \mathbf{0} \\ 1 \end{bmatrix} x_{n,k-i-1}. \tag{3.59}$$

Applying the norm on both sides of (3.59) leads to

$$\max_{j\in[0,k]} \|\mathbf{x}_{a,j+1}\| \leq (n-1) \max_{j\in[0,k]} \|x_{n,j}\|. \tag{3.60}$$

The above result is then used to simplify the sector condition $\|\boldsymbol{\xi}_k\| \leq c_1 + c_2\|\mathbf{x}_k\|$ as follows

$$\begin{aligned}\|\boldsymbol{\xi}_k\| &\leq c_1 + c_2\|\mathbf{x}_k\| \leq c_1 + c_2\left(\|\mathbf{x}_{a,k}\| + \|x_{n,k}\|\right)\\ &\leq c_1 + c_2\left((n-1)\max_{j\in[0,k]}\|x_{n,j-1}\| + \max_{j\in[0,k]}\|x_{n,j}\|\right).\end{aligned}$$

Note that $\max_{j\in[0,k]}\|x_{n,j-1}\| \leq \max_{j\in[0,k]}\|x_{n,j}\|$, thus,

$$\begin{aligned}\|\boldsymbol{\xi}_k\| &\leq c_1 + nc_2\max_{j\in[0,k]}\|x_{n,j}\| \leq c_1 + nc_2\left(\|x_{n,0}\| + \max_{j\in[0,k]}\|x_{n,j+1}\|\right)\\ &\leq c_1^0 + c_2^0\max_{j\in[0,k]}\|x_{n,j+1}\|\end{aligned} \tag{3.61}$$

where $c_1^0 = c_1 + nc_2\|x_{n,0}\|$ and $c_2^0 = nc_2$.

Remark 3.3 Extension from the first order to higher order can also be applied to systems with the uncertain input gain, mixed parameters, or tracking problems, as discussed in preceding Sects. 3.3.1–3.3.3.

3.4 Illustrative Example

Consider a system

$$x_{k+1} = \theta_k \sin(x_k + 1) + b_k u_k, \quad x(0) = 1 \tag{3.62}$$

where $\theta_k = \sin\left(\frac{1}{25}\pi k\right)$. We use $|x_i|_{\sup}$ to record the maximum absolute regulation error during the ith period.

First let $b_k = 3 + 0.5\sin\left(\frac{1}{50}\pi k\right)$. The minimum common period is $N = 100$. A typical adaptive controller is used with the Least Squares estimator. Figure 3.1a shows the regulation error over each period. By virtue of the rapid time-varying nature, the tracking error does not converge. Applying the proposed periodic adaptation method, Fig. 3.1b shows the maximum regulation error over each period. We can clearly see the effectiveness, as the regulation error has been reduced to less than 1 % after 50 periods.

Next, let $b_k = 3$ be an uncertain constant. Still using the periodic adaptation law, the result is shown in Fig. 3.2a. Now assume that it is known a priori that b_k is an uncertain constant, the hybrid adaptation law is adopted and the result is shown in Fig. 3.2b. The performance improvement is immediately obvious.

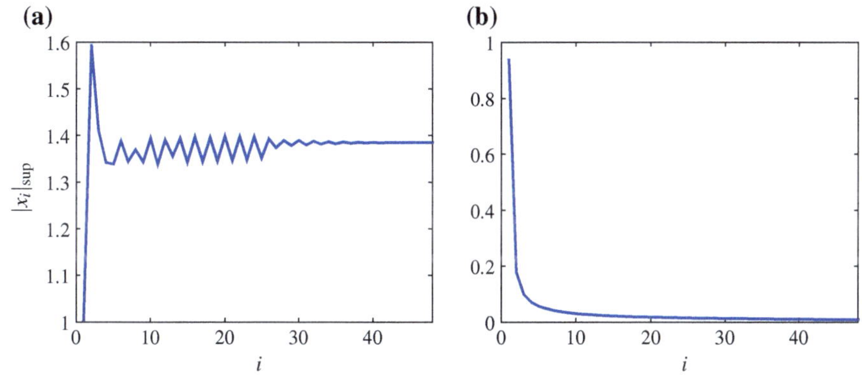

Fig. 3.1 Error convergence using **a** classical adaptation and **b** periodic adaptation

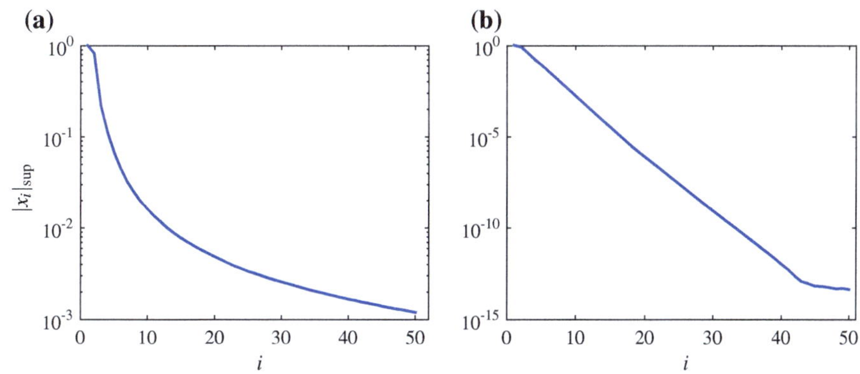

Fig. 3.2 Error convergence with mixed parameters using **a** periodic adaptation and **b** hybrid periodic adaptation

Fig. 3.3 Tracking error convergence using periodic adaptation

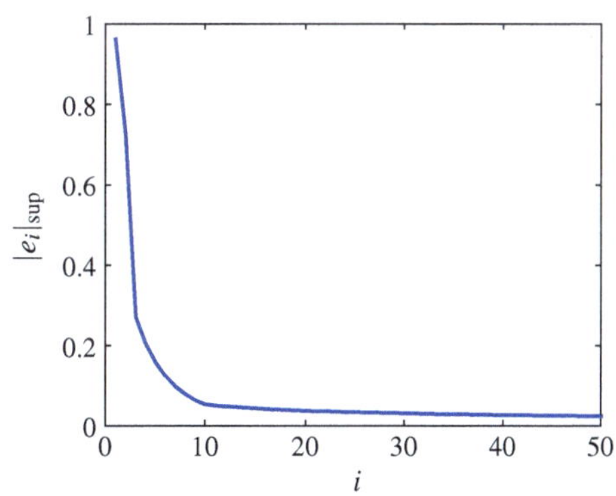

Finally, let $b_k = 3 + 0.5 \sin\left(\frac{1}{50}\pi k\right)$ again, and it is required that x_k track a given reference $r_k = \sin\left(\frac{1}{50}\pi k\right)$. Figure 3.3 shows that the tracking error converges asymptotically.

3.5 Conclusion

In this chapter we presented an adaptive control approach characterized by periodic parameter adaptation, which complements the existing adaptive control characterized by instantaneous adaptation. By virtue of the periodic adaptation, the approach is applicable to system with periodic parameters or periodic disturbances which can be rapidly time-varying. The only prior knowledge needed in the periodic adaptation is the periodicity. A hybrid adaptation scheme is also proposed when more of the parameter knowledge is available. Both regulation and tracking problems were discussed. Extension to higher order processes was also exploited. The validity of the proposed approach is confirmed through theoretical analysis and numerical simulations.

Chapter 4
Discrete-Time Adaptive Posicast Control

Abstract In this study, we discuss the discrete version of the Adaptive Posicast Controller (APC) that deals with parametric uncertainties in systems with input time-delays. The continuous-time APC is based on the Smith Predictor and Finite Spectrum Assignment with time-varying parameters adjusted online. Although the continuous-time APC showed dramatic performance improvements in experimental studies with internal combustion engines, the full benefits could not be realized since the finite integral term in the control law had to be approximated in computer implementation. It is shown in the literature that integral approximation in time-delay compensating controllers degrades the performance if care is not taken. In this study, we discuss a development of the APC in the discrete-time domain, eliminating the need for approximation. Rigorous and complete derivation is provided with a Lyapunov stability proof. The discussed discrete-time APC is developed in State Space to easily accommodate multivariable systems and also allow for the extension to nonlinear systems. In essence, this study presents a unified development of the discrete-time APC for systems that are linear/nonlinear with known input time-delays or linear systems with unknown but upper-bounded time-delays. Performances of the continuous-time and discrete-time APC, as well as conventional Model Reference Adaptive Controller (MRAC) for linear systems with known time-delay are compared in simulation studies. It is shown that discrete-time APC outperforms it's continuous-time counterpart and MRAC. Further simulations studies are also presented to show the performance of the design for nonlinear systems and also for systems with unknown time-delay.

4.1 Introduction

Adaptive Posicast Controller (APC), [169–175], is a model reference adaptive controller for linear time invariant systems with known input time delays. Basic building blocks of this controller are the celebrated Smith Predictor, [16, 71, 106, 124, 125, 131, 138, 147, 148], the finite spectrum assignment controller (FSA), [88, 109, 116, 156], and the adaptive controller developed by Ortega and Lozano [129]. APC has proved to be a powerful candidate for time-delay systems control both in

K. Abidi and J.-X. Xu, *Advanced Discrete-Time Control*, Studies in Systems, Decision and Control 23, DOI 10.1007/978-981-287-478-8_4

simulation and experimental works. Successful experimental implementations include spark ignition engine idle speed control, [171, 172], and fuel-to-air ratio control, [173, 174], while simulation implementation on flight control is presented in [169]. Recently, an extension of APC using combined/composite model reference adaptive control is presented [52]. Although APC has successfully been implemented in various domains with considerable performance improvements, the premise of time-delay compensation using future output prediction, as proven by the theory, had to be approximately realized in these applications. The main reason behind this was that the APC had to be implemented using a microprocessor and therefore all the terms in the control laws had to be digitally approximated. This is a conventional approach in many control implementations and in most of the cases works perfectly well as long as the sampling is fast enough. One exception to this rule is the implementation of the finite spectrum assignment (FSA) controller. It is shown in [156] that, as the sampling frequency increases, the phase margin of the FSA controller decreases. A remedy to this problem is provided in [116]. Since APC is based on FSA controller, fast sampling to achieve good approximation of the continuous control laws may degrade the system performance.

To eliminate the need for approximation and, therefore, to exploit the full benefits of APC, a fully discrete time APC design is provided in this paper. A Lyapunov stability proof is given and the discrete APC is compared with its continuous counterpart in the simulation environment. A comparison with a conventional model reference adaptive controller is also provided. As expected, simulation results verify the advantage of developing the controller in the discrete domain over a continuous-time development followed by a discrete approximation.

There are already many successful methods proposed in the literature to compensate the effect of time-delays in continuous-time control systems. Among them, the very recent ones are presented in [23–26, 89, 97, 111–113, 135]. To see an analysis of robustness of nonlinear predictive laws to delay perturbations and a comprehensive list of delay-compensating controllers see [27]. Also, the book [98] is a very recent important contribution to the field presenting predictive feedback in delay systems with extensions to nonlinear systems, delay-adaptive control and actuator dynamics modeled by PDEs.

Also in the discrete time domain, there are various solutions to model reference adaptive control problem in the literature with natural inclusion of the time delays, [5, 6, 68, 96]. The main contributions of the discrete time APC are that in the controller development, future state estimation, i.e. predictor feedback, is explicit, which helped the extension of the method to the control of nonlinear systems and unknown input time-delay cases, in the discrete-time domain. It is noted that recently, unknown input delay case is solved for the continuous time systems without approximating the delay in [35].

4.2 Problem Formulation

Consider a continuous-time system with state space description given as

$$\begin{aligned}\dot{\mathbf{x}}(t) &= A\mathbf{x}(t) + B_{\rm n}\Lambda\mathbf{u}(t-\tau)\\ \mathbf{y}(t) &= C\mathbf{x}(t)\end{aligned} \tag{4.1}$$

where $\mathbf{x}(t) \in \Re^n$ is the vector of system states, $A \in \Re^{n\times n}$ is a constant unknown matrix, $B_{\rm n} \in \Re^{n\times m}$ is a constant known matrix, $\Lambda \in \Re^{m\times m}$ is a constant unknown positive definite matrix, $\mathbf{u}(t) \in \Re^m$ is the vector of the control inputs, $\tau \geq 0$ is the input time-delay, and $\mathbf{y}(t) \in \Re^m$ is the system output and $C \in \Re^{m\times n}$ is the ouput matrix. For the system (4.1), the following assumptions are made:

Assumption 4.1 Input time-delay τ is known.

Assumption 4.2 System (4.1) is minimum-phase.

Suppose that the reference model, reflecting the desired response characteristics, is given as

$$\dot{\mathbf{x}}_{\rm m}(t) = A_{\rm m}\mathbf{x}_{\rm m} + B_{\rm m}\mathbf{r}(t-\tau) \tag{4.2}$$

where $A_{\rm m} \in \Re^{n\times n}$ is a constant Hurwitz matrix, $B_{\rm m} \in \Re^{n\times m}$ is a constant matrix and $\mathbf{r}(t)$ is the desired reference command. Note that the dynamics given in the reference model (4.2) is the best one can achieve with any kind of control law in terms of handling the time-delay, i.e., the time-delay can not be eliminated from the closed loop dynamics, but only it's effect on the stability can be eliminated. The control problem is finding a bounded control input $\mathbf{u}(t)$ such that $lim_{t\to\infty}\|\mathbf{x}(t)-\mathbf{x}_{\rm m}(t)\| = 0$, while keeping all the system signals bounded.

Remark 4.1 A proper selection of the reference model enables $\mathbf{y}(t) \to \mathbf{r}(t)$ as $t \to \infty$, for a constant reference $\mathbf{r}(t)$. See the details in [169]. For the rest of the chapter, we will assume that the reference model is selected accordingly.

The reference model (4.2) can be obtained by applying a nominal controller to the nominal system which is the system given in (4.1) with no uncertainty, that is, $\Lambda = I_m$ and A does not have any uncertainty leading to a nominal matrix A_n. It is noted that a conventional state feedback or linear quadratic regulator (LQR) controller can not be used to obtain (4.2). This is because the time-delay τ introduces infinitely many eigenvalues to the closed loop system unless a nominal controller which eliminates the effect of the time-delay on the eigenvalues is used. Therefore, a finite spectrum assignment controller [109] is used to design the nominal controller. This point is important also for the theoretically sound augmentation of the nominal controller with the adaptive controller, [169].

4.2.1 Continuous-Time Adaptive Posicast Controller (APC)

APC is a model reference adaptive controller for systems with known input time-delay. Below, the main idea behind the APC is summarized. The reader is referred to [170] for details.

Consider the uncertain system dynamics with input time-delay (4.1) and the reference model (4.2). The reference model can be constructed using a finite spectrum assignment controller, [169]. It can be shown that the following control law

$$\mathbf{u}_{\mathrm{APC}}(t) = \Theta_{\mathrm{x}}^{\top}(t)\mathbf{x}(t) + \int_{-\tau}^{0} \Lambda_{\mathrm{u}}(t,\eta)\mathbf{u}(t+\eta)d\eta + \Theta_{\mathrm{r}}(t)^{\top}\mathbf{r}(t) \tag{4.3}$$

together with the adaptive laws

$$\begin{aligned}
\dot{\Theta}_{\mathrm{x}}(t) &= -\Psi_{\mathrm{x}}\mathbf{x}(t-\tau)\mathbf{e}(t)^{\top}P_{\mathrm{x}}B_{\mathrm{n}} \\
\dot{\Lambda}_{\mathrm{u}}(t,\eta) &= -\Psi_{\Lambda}\mathbf{u}(t-\tau+\eta)\mathbf{e}(t)^{\top}P_{\mathrm{x}}B_{\mathrm{n}} \\
\dot{\Theta}_{\mathrm{r}}(t) &= -\Psi_{\mathrm{r}}\mathbf{r}(t-\tau)\mathbf{e}(t)^{\top}P_{\mathrm{x}}B_{\mathrm{n}}
\end{aligned} \tag{4.4}$$

achieves reference model following. In (4.4), $\Psi_{\mathrm{x}} \in \Re^{m\times n}$, $\Psi_{\Lambda} \in \Re^{m\times m}$ and $\Psi_{\mathrm{r}} \in \Re^{m\times m}$ are adaptation gains and $\mathbf{e}(t) = \mathbf{x}(t) - \mathbf{x}_{\mathrm{m}}(t)$ is the state tracking error. $P_{\mathrm{x}} \in \Re^{n\times n}$ is the symmetric positive definite solution of the Lyapunov equation

$$A_{\mathrm{m}}^{\top}P_{\mathrm{x}} + P_{\mathrm{x}}A_{\mathrm{m}} = -Q_{\mathrm{x}} \tag{4.5}$$

where $Q_{\mathrm{x}} \in \Re^{n\times n}$ is a positive definite matrix. In these derivations it is assumed that there exists a Θ_{x}^{*} such that

$$A - B_{\mathrm{n}}\Lambda\Theta_{\mathrm{x}}^{*\top} = A_{\mathrm{m}}. \tag{4.6}$$

The digital implementation of the controller (4.3) and the adaptation laws given by (4.4) would require discrete-time approximations of the integrals and the derivatives which may cause degradation to the overall controller performance. Thus, it is logical to consider the APC design from a strictly discrete-time perspective in the case of digital implementation.

4.3 Discrete-Time Adaptive Posicast Controller Design

In this section the discrete-time design of the APC will be presented starting with a simple first order system.

4.3.1 *Control of a 1st Order Input Time-Delay System in Discrete-Time*

Consider the following discrete-time system with input time-delay

$$x_{k+1} = ax_k + u_{k-p} \tag{4.7}$$

where the scalar constant parameter a is assumed to be known and p is the known time-delay in number of steps. The goal is to force the system (4.7) to track the reference model

$$x_{\mathrm{m},k+1} = a_{\mathrm{m}} x_{\mathrm{m},k} + b_{\mathrm{m}} r_{k-p} \tag{4.8}$$

where a_{m} is in the unit circle. Extending the work in [109, 147] and assuming there exists a δ such that

$$a - \delta = a_{\mathrm{m}}, \tag{4.9}$$

a controller is chosen as

$$u_k = -\delta x_{k+p} + b_{\mathrm{m}} r_k \tag{4.10}$$

The non-causal controller (4.10) can be converted into a causal one following the same idea in [109]. This is possible by realizing that the system (4.7) can be written as

$$x_{k+p} = a^p x_k + \left(a^{p-1} u_{k-p} + a^{p-2} u_{k-p+1} + \cdots + u_{k-1}\right) \tag{4.11}$$

thus, the control becomes

$$u_k = -\delta a^p x_k - \delta \left(a^{p-1} u_{k-p} + a^{p-2} u_{k-p+1} + \cdots + u_{k-1}\right) + b_{\mathrm{m}} r_k. \tag{4.12}$$

It is possible to compact (4.12) into a vector form as follows

$$u_k = -\delta_{\mathrm{x}} x_k - \boldsymbol{\delta}_{\mathrm{u}}^{\top} \boldsymbol{\xi}_k + b_{\mathrm{m}} r_k \tag{4.13}$$

where $\delta_{\mathrm{x}} = \delta a^p \in \Re$, $\boldsymbol{\delta}_{\mathrm{u}}^{\top} = \delta \left[1 \,\vdots\, a \,\vdots\, a^2 \,\vdots \cdots \vdots\, a^{p-1}\right] \in \Re^p$ is a vector of constant parameters and $\boldsymbol{\xi}_k^{\top} = \left[u_{k-1} \,\vdots\, u_{k-2} \,\vdots \cdots \vdots\, u_{k-p}\right] \in \Re^p$ is a vector of control history. Subtracting (4.8) from (4.7) and using (4.9) it is possible to obtain

$$e_{k+1} = a_{\mathrm{m}} e_k + \delta x_k + u_{k-p} - b_{\mathrm{m}} r_{k-p} \tag{4.14}$$

where $e_k = x_k - x_{\mathrm{m},k}$. Substitution of p time steps delayed (4.11) into (4.14) it is obtained that

$$e_{k+1} = a_{\mathrm{m}} e_k + \delta_{\mathrm{x}} x_{k-p} + \boldsymbol{\delta}_{\mathrm{u}}^{\top} \boldsymbol{\xi}_{k-p} + u_{k-p} - b_{\mathrm{m}} r_{k-p}. \tag{4.15}$$

Using the control law (4.13) leads to a closed-loop system of the form

$$e_{k+1} = a_{\mathrm{m}} e_k \tag{4.16}$$

which is stable.

4.3.2 Adaptive Control of an Input Time-Delay System

Consider now the system with uncertain input gain

$$x_{k+1} = a x_k + b u_{k-p} \tag{4.17}$$

where the scalar parameters a and b are unknown constants and that the nominal value of b is known and represented by b_{n}. Consider that there exists a ϕ and ϕ_{γ} such that

$$a - b_{\mathrm{n}} \phi = a_{\mathrm{m}} \ \& \ b = b_{\mathrm{n}} \phi_{\gamma} \tag{4.18}$$

Proceeding further, using the control law given by

$$u_k = -\phi_{\gamma}^{-1} \left(\phi_{\mathrm{x}} x_k + \boldsymbol{\phi}_{\mathrm{u}}^{\top} \boldsymbol{\xi}_k - \phi_{\mathrm{r}} r_k \right) \tag{4.19}$$

where $\phi_{\mathrm{x}} = \phi a^p \in \Re$, $\boldsymbol{\phi}_{\mathrm{u}}^{\top} = \phi b \left[1 \,\vdots\, a \,\vdots\, a^2 \,\vdots \cdots \vdots\, a^{p-1} \right] \in \Re^p$ and $\phi_{\mathrm{r}} = b_{\mathrm{n}}^{-1} b_{\mathrm{m}}$ the following closed-loop dynamics is obtained

$$x_{k+1} = a x_k - b_{\mathrm{n}} \left(\phi_{\mathrm{x}} x_{k-p} + \boldsymbol{\phi}_{\mathrm{u}}^{\top} \boldsymbol{\xi}_{k-p} - \phi_{\mathrm{r}} r_{k-p} \right) \tag{4.20}$$

and using the fact that

$$x_k = a^p x_{k-p} + b \left(a^{p-1} u_{k-2p} + a^{p-2} u_{k-2p+1} + \cdots + u_{k-p-1} \right) \tag{4.21}$$

and

$$\begin{aligned} \phi x_k &= \phi a^p x_{k-p} + \phi b \left(a^{p-1} u_{k-2p} + a^{p-2} u_{k-2p+1} + \cdots + u_{k-p-1} \right) \\ &= \phi_{\mathrm{x}} x_{k-p} + \boldsymbol{\phi}_{\mathrm{u}}^{\top} \boldsymbol{\xi}_{k-p} \end{aligned} \tag{4.22}$$

then (4.23) simplies to the form

$$x_{k+1} = a_{\mathrm{m}} x_k + b_{\mathrm{m}} r_{k-p} \tag{4.23}$$

which is an exact match with the reference model (4.8). However, since, a and b are uncertain then (4.19) is rewritten as

$$u_k = -\hat{\phi}_{\gamma,k}^{-1}\left(\hat{\phi}_{x,k}x_k + \hat{\boldsymbol{\phi}}_{u,k}^{\top}\boldsymbol{\xi}_k - \phi_r r_k\right) \tag{4.24}$$

where $\hat{\phi}_{\gamma,k}$, $\hat{\phi}_{x,k}$, $\hat{\boldsymbol{\phi}}_{u,k}$ are the estimates of ϕ_γ, ϕ_x, $\boldsymbol{\phi}_u$ respectively. Note that in the computation of the control input u_k, the inverse of $\hat{\phi}_{\gamma,k}$ is required which introduces the danger of division by zero if $\hat{\phi}_{\gamma,k}$ is in the vicinity of zero. Therefore, the adaptive law must be designed such that $\hat{\phi}_{\gamma,k}$ never becomes singular. Proceeding with the derivation of the error dynamics, subtracting (4.8) from (4.17) and using $e_k = x_k - x_{m,k}$ it is obtained that

$$e_{k+1} = a_m e_k + b_n\phi_x x_{k-p} + b_n\boldsymbol{\phi}_u^{\top}\boldsymbol{\xi}_{k-p} + b_n\phi_\gamma u_{k-p} - b_n\phi_r r_{k-p} \tag{4.25}$$

further, adding and subtracting the term $b_n\hat{\phi}_{\gamma,k-p}u_{k-p}$ to the right hand side of (4.25) it is obtained that

$$\begin{aligned} e_{k+1} &= a_m e_k + b_n\phi_x x_{k-p} + b_n\boldsymbol{\phi}_u^{\top}\boldsymbol{\xi}_{k-p} + b_n\phi_\gamma u_{k-p} - b_n\hat{\phi}_{\gamma,k-p}u_{k-p} \\ &\quad + b_n\hat{\phi}_{\gamma,k-p}u_{k-p} - b_n\phi_r r_{k-p}. \end{aligned} \tag{4.26}$$

Let $\tilde{\phi}_{\gamma,k} = \phi_\gamma - \hat{\phi}_{\gamma,k}$ then (4.26) is simplified to the form

$$\begin{aligned} e_{k+1} &= a_m e_k + b_n\phi_x x_{k-p} + b_n\boldsymbol{\phi}_u^{\top}\boldsymbol{\xi}_{k-p} + b_n\tilde{\phi}_{\gamma,k-p}u_{k-p} \\ &\quad + b_n\hat{\phi}_{\gamma,k-p}u_{k-p} - b_n\phi_r r_{k-p}. \end{aligned} \tag{4.27}$$

Substitution of (4.24) into (4.27) and simplifying leads to the closed-loop dynamics

$$e_{k+1} = a_m e_k + b_n\tilde{\phi}_{x,k-p}x_{k-p} + b_n\tilde{\boldsymbol{\phi}}_{u,k-p}^{\top}\boldsymbol{\xi}_k + b_n\tilde{\phi}_{\gamma,k-p}u_{k-p} \tag{4.28}$$

where $\tilde{\phi}_{x,k-p} = \phi_x - \hat{\phi}_{x,k-p}$ and $\tilde{\boldsymbol{\phi}}_{u,k-p} = \boldsymbol{\phi}_u - \hat{\boldsymbol{\phi}}_{u,k-p}$. The closed-loop system (4.28) is rewritten as follows

$$e_{k+1} = a_m e_k + b_n\left(\tilde{\boldsymbol{\psi}}_{k-p}^{*}\right)^{\top}\boldsymbol{\zeta}_{k-p}^{*} \tag{4.29}$$

where $\tilde{\boldsymbol{\psi}}_{k-p}^{*} = \left[\tilde{\phi}_{x,k-p} \,\vdots\, \tilde{\boldsymbol{\phi}}_{u,k-p}^{\top} \,\vdots\, \tilde{\phi}_{\gamma,k-p}\right]^{\top} \in \Re^{2+p}$ is the augmented parameter estimation error and $\boldsymbol{\zeta}_{k-p}^{*} = \left[x_{k-p} \,\vdots\, \boldsymbol{\xi}_{k-p}^{\top} \,\vdots\, u_{k-p}\right]^{\top} \in \Re^{2+p}$. Let $z_{k+1} = b_n^{-1}(e_{k+1} - a_m e_k)$ then (4.28) can be rewritten as

$$z_{k+1} = \left(\tilde{\boldsymbol{\psi}}_{k-p}^{*}\right)^{\top}\boldsymbol{\zeta}_{k-p}^{*} \tag{4.30}$$

The estimation law for $\boldsymbol{\psi}^*$ can be given by

$$\hat{\boldsymbol{\psi}}^*_{k+1} = \begin{cases} \hat{\boldsymbol{\psi}}^*_{k-p} + \varepsilon_k P_{k+1} \boldsymbol{\zeta}^*_{k-p} z_{k+1} & \forall k \in [p, \infty) \\ \hat{\boldsymbol{\psi}}^*_0 & \forall k \in [0, p) \end{cases} \tag{4.31}$$

$$P_{k+1} = \begin{cases} P_{k-p} - \varepsilon_k \frac{P_{k-p}\boldsymbol{\zeta}^*_{k-p}(\boldsymbol{\zeta}^*_{k-p})^\top P_{k-p}}{1+\varepsilon_k(\boldsymbol{\zeta}^*_{k-p})^\top P_{k-p}\boldsymbol{\zeta}^*_{k-p}} & \forall k \in [p, \infty) \\ P_0 > 0 & \forall k \in [0, p) \end{cases} \tag{4.32}$$

where $\varepsilon_k \in \Re$ is a positive coefficient that is used to ensure a nonsingular $\hat{\phi}_{\gamma,k}$ and $P_k \in \Re^{(p+2)\times(p+2)}$ is the symmetric positive-definite covariance matrix which similar to (3.6) has the properties, [96].

$$P^{-1}_{k+1} = P^{-1}_{k-p} + \varepsilon_k \boldsymbol{\zeta}_{k-p} \boldsymbol{\zeta}^\top_{k-p} \tag{4.33}$$

$$\boldsymbol{\zeta}^\top_{k-p} P_{k+1} \boldsymbol{\zeta}_{k-p} = \frac{\boldsymbol{\zeta}^\top_{k-p} P_{k-p} \boldsymbol{\zeta}_{k-p}}{1 + \varepsilon_k \boldsymbol{\zeta}^\top_{k-p} P_{k-p} \boldsymbol{\zeta}_{k-p}} \tag{4.34}$$

These properties will be used extensively in the stability analysis detailed in later sections. The choice of the values of ε_k is based on the requirement that $\hat{\phi}_{\gamma,k}$ is nonsingular as well as the asymptotic stability of (4.30). For the former case, to find out the allowable values of ε_k, consider (4.31) and let $\mathbf{s} = [0 \ \cdots \ 0 \ 1] \in \Re^{1\times(p+2)}$ such that

$$\mathbf{s}\hat{\boldsymbol{\psi}}^*_{k+1} = \hat{\phi}_{\gamma,k+1} = \hat{\phi}_{\gamma,k-p} + \varepsilon_k \mathbf{s} P_{k+1} \boldsymbol{\zeta}^*_{k-p} z_{k+1}. \tag{4.35}$$

Using the approach in [68], (4.35) can be rewritten as

$$\hat{\phi}_{\gamma,k+1} = \hat{\phi}_{\gamma,k-p} \varepsilon_k \left(\frac{1}{\varepsilon_k} + \hat{\phi}^{-1}_{\gamma,k-p} \mathbf{s} P_{k+1} \boldsymbol{\zeta}^*_{k-p} z_{k+1} \right). \tag{4.36}$$

If the initial choice of $\hat{\phi}_{\gamma,k}$ is nonsingular and $\varepsilon_k^{-1} \neq -\hat{\phi}^{-1}_{\gamma,k-p}\mathbf{s}P_{k+1}\boldsymbol{\zeta}^*_{k-p}z_{k+1}$ then, from (4.36), all computed values of $\hat{\phi}_{\gamma,k}$ will be nonsingular. Therefore, ε_k is selected as a constant value, for example $\varepsilon_k = 1$, and will remain at that value as long as the condition $\varepsilon_k^{-1} \neq -\hat{\phi}^{-1}_{\gamma,k-p}\mathbf{s}P_{k+1}\boldsymbol{\zeta}^*_{k-p}z_{k+1}$ is satisfied. If the condition is not satisfied then ε_k will be updated to another value.

Remark 4.2 Note that a semi-saturator approach similar to that in the Chap. 3 can also be used for this specific scalar system, however, for multivariable systems the semi-saturator approach is not applicable. Thus, the reader is introduced to this new approach for this simple case before adapting the approach to multivariable systems.

For the values of ε_k that ensure the asymptotic stability of (4.30) consider the following lemma:

Lemma 4.1 *For the system (4.17) with the control law (4.24) and adaptive laws (4.31), (4.32) the tracking error converges to zero asymptotically, i.e.,* $\lim_{k\to\infty} |e_k| = 0$ *if* $\varepsilon_k > 0$.

Proof Consider the following positive function

$$V_k = \sum_{i=0}^{p} \left[\left(\tilde{\boldsymbol{\psi}}^*_{k-i} \right)^\top P^{-1}_{k-i} \left(\tilde{\boldsymbol{\psi}}^*_{k-i} \right) \right]. \tag{4.37}$$

The forward difference $V_{k+1} - V_k$ can be found as,

$$V_{k+1} - V_k = \Delta V_k = -\frac{\varepsilon_k z^2_{k+1}}{1 + \varepsilon_k \left(\boldsymbol{\zeta}^*_{k-p} \right)^\top P_k \left(\boldsymbol{\zeta}^*_{k-p} \right)} \tag{4.38}$$

which implies that $\tilde{\boldsymbol{\psi}}^*_k$ is bounded and, therefore, $\hat{\boldsymbol{\psi}}^*_k$ is also bounded if $\varepsilon_k \in (0, 1]$. Note that for any $k \in [k_0, \infty)$ the following is true

$$V_{k+1} = V_{k_0} + \sum_{i=0}^{k-k_0} \Delta V_{k_0+i} \tag{4.39}$$

Substituting (4.38) in (4.39)

$$\lim_{k\to\infty} V_{k+1} \le \max_{k_0 \in [0,p)} V_{k_0} - \lim_{k\to\infty} \sum_{i=0}^{k-k_0} \frac{\varepsilon_k z^2_{k_0+i+1}}{1+\varepsilon_k \left(\boldsymbol{\zeta}^*_{k_0+i-p} \right)^\top P_{k_0+i-p} \left(\boldsymbol{\zeta}^*_{k_0+i-p} \right)}.$$

Consider that V_{k+1} is non-negative and V_{k_0} is finite in the interval $[0, p)$, thus, according to the convergence theorem of the sum of series

$$\lim_{k\to\infty} \frac{\varepsilon_k z^2_{k+1}}{1 + \varepsilon_k \left(\boldsymbol{\zeta}^*_{k-p} \right)^\top P_{k-p} \left(\boldsymbol{\zeta}^*_{k-p} \right)} = 0. \tag{4.40}$$

To guarantee that z_k is asymptotically stable, i.e. $\lim_{k\to\infty} |z_k| = 0$, it must be guaranteed that $\|\boldsymbol{\zeta}^*_k\| \le \mu_0 + \mu_1 \max_{i\in[0,k]} |z_{i+1}|$. Consider the definition of z_k, we have

$$e_{k+1} = a_{\mathrm{m}} e_k + b_{\mathrm{n}} z_{k+1} \tag{4.41}$$

using the fact that a_{m} is in the unit circle, $e_k = x_k - x_{\mathrm{m},k}$ and that $x_{\mathrm{m},k}$ is bounded then there exists constants γ_0 and γ_1 such that, [96],

$$|x_{k+1}| \leq \gamma_0 + \gamma_1 |z_{k+1}|. \tag{4.42}$$

Since the system is assumed to be minimum-phase, there exists

$$|u_k| \leq \kappa_0 + \kappa_1 |x_k|. \tag{4.43}$$

Looking at the signal growth rates, $\boldsymbol{\zeta}_k^*$ is a vector containing x_k and u_k, then there exists

$$\|\boldsymbol{\zeta}_k^*\| \leq \mu_0 + \mu_1 |z_k| \leq \mu_0 + \mu_1 \max_{i \in [0, k+1]} |z_i| \tag{4.44}$$

and, therefore, using the *Key Technical Lemma*, [68, 96], $\lim_{k\to\infty} |z_k| = 0$. Further, from (4.41) as $k \to \infty$, the following stable error dynamics is achieved

$$e_{k+1} = a_{\mathrm{m}} e_k \tag{4.45}$$

and, therefore, $\lim_{k\to\infty} |e_k| = 0$ is guaranteed.

4.3.3 Extension to Higher Order Systems

In this section the focus will be on higher order time-delay systems. Consider the sampled-data form of system (4.1) given by

$$\begin{aligned} \mathbf{x}_{k+1} &= \Phi \mathbf{x}_k + \Gamma \mathbf{u}_{k-p} \\ \mathbf{y}_k &= C \mathbf{x}_k \end{aligned} \tag{4.46}$$

where $\mathbf{x}_k \in \Re^n$, $\mathbf{u}_k \in \Re^m$, $\mathbf{y}_k \in \Re^m$ and $C \in \Re^{m\times n}$. The matrices $\Phi \in \Re^{n\times n}$, $\Gamma \in \Re^{n\times m}$ are considered uncertain.

Assumption 4.3 Similar to the continuous-time system, the time-delay steps p is known.

Assumption 4.4 Minimum-phaseness of the system (4.1) is not lost upon sampling.

Assumption 4.5 Since (4.46) is a sampled-data system, $C\Gamma_n$ is non-singular where Γ_n is the known nominal value of Γ.

Consider the sampled-data form of the reference model (4.2)

$$\begin{aligned} \mathbf{x}_{\mathrm{m},k+1} &= \Phi_{\mathrm{m}} \mathbf{x}_{\mathrm{m},k} + \Gamma_{\mathrm{m}} \mathbf{r}_{k-p} \\ \mathbf{y}_{\mathrm{m},k} &= C \mathbf{x}_{\mathrm{m},k}. \end{aligned} \tag{4.47}$$

As in the continuous-time problem, the objective is to force the system (4.46) to track the reference model (4.47) and thereby achieve $\lim_{k\to\infty} \mathbf{x}_k = \mathbf{x}_{\mathrm{m},k}$. The reference model (4.47) is designed by using the nominal values of the system parameters. In other words, assuming that there is a known Φ_{n} and Γ_{n} that are equal to Φ and Γ without uncertainty.

Consider initially that Φ and Γ are known, in order to derive the controller, subtract (4.47) from (4.46) to obtain

$$\mathbf{x}_{k+1} - \mathbf{x}_{\mathrm{m},k+1} = \Phi\mathbf{x}_k - \Phi_{\mathrm{m}}\mathbf{x}_{\mathrm{m},k} + \Gamma\mathbf{u}_{k-p} - \Gamma_{\mathrm{m}}\mathbf{r}_{k-p}. \tag{4.48}$$

Further, the term $\Phi_{\mathrm{m}}\mathbf{x}_k$ is added and subtracted on the right hand side of (4.48) to obtain

$$\mathbf{e}_{k+1} = \Phi_{\mathrm{m}}\mathbf{e}_k + (\Phi - \Phi_{\mathrm{m}})\,\mathbf{x}_k + \Gamma\mathbf{u}_{k-p} - \Gamma_{\mathrm{m}}\mathbf{r}_{k-p}. \tag{4.49}$$

where $\mathbf{e}_k = \mathbf{x}_k - \mathbf{x}_{\mathrm{m},k}$. The goal is to have $\lim_{k\to\infty} \mathbf{x}_k = \mathbf{x}_{\mathrm{m},k}$ or in other words $\lim_{k\to\infty} \mathbf{e}_k = 0$, therefore, assuming that there exists a $\Theta \in \Re^{m\times n}$ and a positive-definite $\Theta_\gamma \in \Re^{m\times m}$ such that

$$\Phi - \Gamma_{\mathrm{n}}\Theta = \Phi_{\mathrm{m}} \;\&\; \Gamma = \Gamma_{\mathrm{n}}\Theta_\gamma \tag{4.50}$$

then it is possible to construct a control law

$$\mathbf{u}_k = -\Theta_\gamma^{-1}\left(\Theta\mathbf{x}_{k+p} - \Theta_{\mathrm{r}}\mathbf{r}_k\right) \tag{4.51}$$

where the known matrix $\Theta_{\mathrm{r}} \in \Re^{m\times m}$ is selected such that $\Gamma_{\mathrm{m}} = \Gamma_{\mathrm{n}}\Theta_{\mathrm{r}}$. Since the controller (4.51) is non-causal, an approach similar to that in §4.3.2 is employed where the future $\mathbf{x}_{k+p}$ is computed as

$$\mathbf{x}_{k+p} = \Phi^p\mathbf{x}_k + \left(\Phi^{p-1}\Gamma\mathbf{u}_{k-p} + \Phi^{p-2}\Gamma\mathbf{u}_{k-p+1} + \cdots + \Gamma\mathbf{u}_{k-1}\right). \tag{4.52}$$

Substituting (4.52) in (4.51) and simplifying leads to a causal controller of the form

$$\mathbf{u}_k = -\Theta_\gamma^{-1}\left(\Theta_{\mathrm{x}}\mathbf{x}_k + \Theta_{\mathrm{u}}\boldsymbol{\xi}_k - \Theta_{\mathrm{r}}\mathbf{r}_k\right) \tag{4.53}$$

where the parameters $\Theta_{\mathrm{x}} = \Theta\Phi^p \in \Re^{m\times n}$, $\Theta_{\mathrm{u}} = \Theta\left[\Gamma \,\vdots\, \Phi\Gamma \,\vdots\, \Phi^2\Gamma \,\vdots\, \cdots \,\vdots\, \Phi^{p-1}\Gamma\right]$ $\in \Re^{m\times pm}$ are uncertain and $\boldsymbol{\xi}_k^\top = \left[\mathbf{u}_{k-1}^\top \,\vdots\, \mathbf{u}_{k-2}^\top \,\vdots\, \cdots \,\vdots\, \mathbf{u}_{k-p}^\top\right] \in \Re^{pm}$.

Consider (4.49), using (4.50) and (4.52) it is obtained that

$$\mathbf{e}_{k+1} = \Phi_{\mathrm{m}}\mathbf{e}_k + \Gamma_{\mathrm{n}}\left(\Theta_{\mathrm{x}}\mathbf{x}_{k-p} + \Theta_{\mathrm{u}}\boldsymbol{\xi}_{k-p}\right) + \Gamma_{\mathrm{n}}\Theta_\gamma\mathbf{u}_{k-p} - \Gamma_{\mathrm{n}}\Theta_{\mathrm{r}}\mathbf{r}_{k-p}. \tag{4.54}$$

Substitution of the control law (4.53) in the tracking error (4.54) it is obtained that

$$\mathbf{e}_{k+1} = \Phi_{\mathrm{m}}\mathbf{e}_k \tag{4.55}$$

which is stable.

Proceeding now with uncertain Φ and Γ matrices, the parameters Θ_{x}, Θ_{u} and Θ_{γ} also become uncertain. The control law (4.53) is, therefore, modified to the form

$$\mathbf{u}_k = -\hat{\Theta}_{\gamma,k}^{-1}\left(\hat{\Theta}_{\mathrm{x},k}x_k + \hat{\Theta}_{\mathrm{u},k}\boldsymbol{\xi}_k - \Theta_r\mathbf{r}_k\right) \tag{4.56}$$

where $\hat{\Theta}_{\mathrm{x},k}$, $\hat{\Theta}_{\mathrm{u},k}$ and $\hat{\Theta}_{\gamma,k}$ are the estimates of Θ_{x}, Θ_{u} and Θ_{γ} respectively. In order to derive the estimation law for $\hat{\Theta}_{\mathrm{x},k}$, $\hat{\Theta}_{\mathrm{u},k}$ and $\hat{\Theta}_{\gamma,k}$ it is necessary to derive the closed-loop system.

Consider the system (4.54), adding and subtracting the term $\Gamma_{\mathrm{n}}\hat{\Theta}_{\gamma,k-p}\mathbf{u}_{k-p}$ it is obtained that

$$\begin{aligned}\mathbf{e}_{k+1} = {} & \Phi_{\mathrm{m}}\mathbf{e}_k + \Gamma_{\mathrm{n}}\left(\Theta_{\mathrm{x}}\mathbf{x}_{k-p} + \Theta_{\mathrm{u}}\boldsymbol{\xi}_{k-p}\right) + \Gamma_{\mathrm{n}}\Theta_{\gamma}\mathbf{u}_{k-p} - \Gamma_{\mathrm{n}}\hat{\Theta}_{\gamma,k-p}\mathbf{u}_{k-p} \\ & + \Gamma_{\mathrm{n}}\hat{\Theta}_{\gamma,k-p}\mathbf{u}_{k-p} - \Gamma_{\mathrm{n}}\Theta_{\mathrm{r}}\mathbf{r}_{k-p}.\end{aligned} \tag{4.57}$$

Define the estimation errors as $\tilde{\Theta}_{\mathrm{x},k} = \Theta_{\mathrm{x}} - \hat{\Theta}_{\mathrm{x},k}$, $\tilde{\Theta}_{\mathrm{u},k} = \Theta_{\mathrm{u}} - \hat{\Theta}_{\mathrm{u},k}$ and $\tilde{\Theta}_{\gamma,k} = \Theta_{\gamma} - \hat{\Theta}_{\gamma,k}$. Using these definitions the system (4.57) can be simplified to the form

$$\begin{aligned}\mathbf{e}_{k+1} = {} & \Phi_{\mathrm{m}}\mathbf{e}_k + \Gamma_{\mathrm{n}}\left(\Theta_{\mathrm{x}}\mathbf{x}_{k-p} + \Theta_{\mathrm{u}}\boldsymbol{\xi}_{k-p}\right) + \Gamma_{\mathrm{n}}\tilde{\Theta}_{\gamma,k-p}\mathbf{u}_{k-p} \\ & + \Gamma_{\mathrm{n}}\hat{\Theta}_{\gamma,k-p}\mathbf{u}_{k-p} - \Gamma_{\mathrm{n}}\Theta_{\mathrm{r}}\mathbf{r}_{k-p}.\end{aligned}$$

Further, substitution of (4.56) into (4.58) it is obtained that

$$\mathbf{e}_{k+1} = \Phi_{\mathrm{m}}\mathbf{e}_k + \Gamma_{\mathrm{n}}\left(\tilde{\Theta}_{\mathrm{x},k-p}\mathbf{x}_{k-p} + \tilde{\Theta}_{\mathrm{u},k-p}\boldsymbol{\xi}_{k-p} + \tilde{\Theta}_{\gamma,k-p}\mathbf{u}_{k-p}\right) \tag{4.58}$$

which is the closed-loop dynamics of the system in terms of the parameter estimation errors. It is convenient to rewrite the error dynamics (4.58) in the augmented form

$$\mathbf{e}_{k+1} = \Phi_{\mathrm{m}}\mathbf{e}_k + \Gamma_{\mathrm{n}}\tilde{\Psi}_{k-p}^{\top}\boldsymbol{\zeta}_{k-p} \tag{4.59}$$

where $\tilde{\Psi}_k^{\top} = \left[\,\tilde{\Theta}_{\mathrm{x},k} \mid \tilde{\Theta}_{\mathrm{u},k} \mid \tilde{\Theta}_{\gamma,k}\,\right] \in \Re^{m\times(n+m(p+1))}$ is the augmented parameter estimation error and $\boldsymbol{\zeta}_k^{\top} = \left[\,\mathbf{x}_k^{\top} \mid \boldsymbol{\xi}_k^{\top} \mid \mathbf{u}_k^{\top}\,\right] \in \Re^{n+m(p+1)}$. In order to proceed with the formulation of the adaptation law define $\mathbf{z}_{k+1} = C_{\gamma}^{\top}(\mathbf{e}_{k+1} - \Phi_{\mathrm{m}}\mathbf{e}_k) \in \Re^m$ where $C_{\gamma}^{\top} = (C\Gamma_{\mathrm{n}})^{-1}C$ and substitute (4.59) to obtain

$$\mathbf{z}_{k+1} = \tilde{\Psi}_{k-p}^{\top}\boldsymbol{\zeta}_{k-p}. \tag{4.60}$$

The adaptation laws must be formulated with the objective of minimizing $\mathbf{z}_{k+1}$ so that the tracking error would follow the dynamics $\mathbf{e}_{k+1} = \Phi_{\mathrm{m}}\mathbf{e}_k$. Therefore, the adaptation laws are formulated as follows

$$\hat{\Psi}_{k+1} = \begin{cases} \hat{\Psi}_{k-p} + \varepsilon_k P_{k+1}\boldsymbol{\zeta}_{k-p}\mathbf{z}_{k+1}^{\top} & \forall k \in [p, \infty) \\ \hat{\Psi}_0 & \forall k \in [0, p) \end{cases} \tag{4.61}$$

$$P_{k+1} = \begin{cases} P_{k-p} - \varepsilon_k \frac{P_{k-p}\boldsymbol{\zeta}_{k-p}\boldsymbol{\zeta}_{k-p}^{\top}P_{k-p}}{1+\varepsilon_k\boldsymbol{\zeta}_{k-p}^{\top}P_{k-p}\boldsymbol{\zeta}_{k-p}} & \forall k \in [p, \infty) \\ P_0 > 0 & \forall k \in [0, p) \end{cases} \tag{4.62}$$

where $\varepsilon_k \in \Re$ is a positive coefficient used to prevent a singular $\hat{\Theta}_{\gamma,k}$ and the matrix $P_k \in \Re^{(n+m(p+1))\times(n+m(p+1))}$ is the symmetric, positive-definite covariance matrix.

Remark 4.3 Note that similar to the scalar case, in order for $\hat{\Theta}_{\gamma,k}$ not to be singular then, using the approach in [68], ε_k^{-1} must be selected such that it is not an eigenvalue of $-\hat{\Theta}_{\gamma,k-p}^{-1}SP_{k+1}\boldsymbol{\zeta}_{k-p}\mathbf{z}_{k+1}^{\top}$ where $S = [\mathbf{0} \;\cdots\; \mathbf{0}\; I_m] \in \Re^{m\times(n+m(p+1))}$.

4.3.4 Stability Analysis

In this section the stability analysis of the system under control (4.56) and adaptive law (4.61) are presented.

Theorem 4.1 *The system (4.46) and the adaptive laws (4.61) and (4.62) results in a closed-loop system with a bounded* $\tilde{\Psi}_k$ *and* $\lim_{k\to\infty}\|\mathbf{e}_k\| = 0$ *if* $\varepsilon_k > 0$.

Proof To proceed with the proof, define the vector $\mathbf{z}_k^{\top} = \left[\, z_{1,k} \mid z_{2,k} \mid \cdots \mid z_{m,k} \,\right]$ and $\tilde{\Psi}_k^{\top} = \left[\, \tilde{\boldsymbol{\psi}}_{1,k} \mid \tilde{\boldsymbol{\psi}}_{2,k} \mid \cdots \mid \tilde{\boldsymbol{\psi}}_{m,k} \,\right]$, where $\tilde{\boldsymbol{\psi}}_{j,k} \in \Re^{(n+m(p+1))\times 1}$ and $j = 1, \ldots, m$. Now, consider the following positive function

$$V_k = \sum_{j=1}^{m}\left(\sum_{i=0}^{p}\tilde{\boldsymbol{\psi}}_{j,k-i}^{\top}P_{k-i}^{-1}\tilde{\boldsymbol{\psi}}_{j,k-i}\right). \tag{4.63}$$

The forward difference of (4.63) is given by

$$\Delta V_k = V_{k+1} - V_k = \sum_{j=1}^{m}\left[\tilde{\boldsymbol{\psi}}_{j,k+1}^{\top}P_{k+1}^{-1}\tilde{\boldsymbol{\psi}}_{j,k+1} - \tilde{\boldsymbol{\psi}}_{j,k-p}^{\top}P_{k-p}^{-1}\tilde{\boldsymbol{\psi}}_{j,k-p}\right]. \tag{4.64}$$

Consider the update law (4.61), subtracting both sides from $\boldsymbol{\psi}_j$ it is possible to obtain

$$\boldsymbol{\psi}_j - \hat{\boldsymbol{\psi}}_{j,k+1} = \boldsymbol{\psi}_j - \hat{\boldsymbol{\psi}}_{j,k-p} - \varepsilon_k P_{k+1}\boldsymbol{\zeta}_{k-p}z_{j,k+1} \tag{4.65}$$

and defining $\tilde{\boldsymbol{\psi}}_{j,k} = \boldsymbol{\psi}_j - \hat{\boldsymbol{\psi}}_{j,k}$ we obtain

$$\tilde{\boldsymbol{\psi}}_{j,k+1} = \tilde{\boldsymbol{\psi}}_{j,k-p} - \varepsilon_k P_{k+1} \boldsymbol{\zeta}_{k-p} z_{j,k+1} \tag{4.66}$$

substitute (4.66) in (4.64) to obtain

$$\begin{aligned} \Delta V_k = \sum_{j=1}^{m} \Big[& \left(\tilde{\boldsymbol{\psi}}_{j,k-p} - \varepsilon_k P_{k+1} \boldsymbol{\zeta}_{k-p} z_{j,k+1} \right)^\top P_{k+1}^{-1} \left(\tilde{\boldsymbol{\psi}}_{j,k-p} - \varepsilon_k P_{k+1} \boldsymbol{\zeta}_{k-p} z_{j,k+1} \right) \\ & - \tilde{\boldsymbol{\psi}}_{j,k-p}^\top P_{k-p}^{-1} \tilde{\boldsymbol{\psi}}_{j,k-p} \Big] \end{aligned} \tag{4.67}$$

Grouping similar terms with each other leads to

$$\begin{aligned} \Delta V_k = \sum_{j=1}^{m} \Big[& \tilde{\boldsymbol{\psi}}_{j,k-p}^\top \left(P_{k+1}^{-1} - P_{k-p}^{-1} \right) \tilde{\boldsymbol{\psi}}_{j,k-p} - 2\varepsilon_k \tilde{\boldsymbol{\psi}}_{k-p}^\top \boldsymbol{\zeta}_{k-p} z_{j,k+1} \\ & + \varepsilon_k^2 \boldsymbol{\zeta}_{k-p}^\top P_{k+1} \boldsymbol{\zeta}_{k-p} z_{j,k+1}^2 \Big]. \end{aligned} \tag{4.68}$$

Substituting (4.33) into (4.68) and, since, $\varepsilon_k > 0$ it is obtained

$$\begin{aligned} \Delta V_k \leq \sum_{j=1}^{m} \Big[& \varepsilon_k \tilde{\boldsymbol{\psi}}_{j,k-p}^\top \boldsymbol{\zeta}_{k-p} \boldsymbol{\zeta}_{k-p}^\top \tilde{\boldsymbol{\psi}}_{j,k-p} - 2\varepsilon_k \tilde{\boldsymbol{\psi}}_{j,k-p}^\top \boldsymbol{\zeta}_{k-p} z_{j,k+1} \\ & + \varepsilon_k^2 \boldsymbol{\zeta}_{k-p}^\top P_{k+1} \boldsymbol{\zeta}_{k-p} z_{j,k+1}^2 \Big]. \end{aligned} \tag{4.69}$$

Further, note that $z_{j,k+1} = \boldsymbol{\psi}_{j,k-p}^\top \boldsymbol{\zeta}_{k-p}$. Using this substitution in (4.69) results in

$$\Delta V_k \leq \sum_{j=1}^{m} \left[\varepsilon_k z_{j,k+1}^2 - 2\varepsilon_k z_{j,k+1}^2 + \varepsilon_k^2 \boldsymbol{\zeta}_{k-p}^\top P_{k+1} \boldsymbol{\zeta}_{k-p} z_{j,k+1}^2 \right]. \tag{4.70}$$

Simplifying (4.70) and using (4.34), ΔV_k becomes

$$\Delta V_k = - \sum_{j=1}^{m} \left[\frac{\varepsilon_k z_{j,k+1}^2}{1 + \varepsilon_k \boldsymbol{\zeta}_{k-p}^\top P_{k-p} \boldsymbol{\zeta}_{k-p}} \right] \tag{4.71}$$

which can be rewritten in the form

$$\Delta V_k = - \frac{\varepsilon_k \mathbf{z}_{k+1}^\top \mathbf{z}_{k+1}}{1 + \varepsilon_k \boldsymbol{\zeta}_{k-p}^\top P_{k-p} \boldsymbol{\zeta}_{k-p}} \tag{4.72}$$

which implies that V_k is non-increasing and, thus, $\tilde{\Psi}_k$ is bounded. Similar to *Lemma* 4.1 it can be concluded that

$$\lim_{k\to\infty} \frac{\varepsilon_k \mathbf{z}_{k+1}^{\top} \mathbf{z}_{k+1}}{1 + \varepsilon_k \boldsymbol{\zeta}_{k-p}^{\top} P_{k-p} \boldsymbol{\zeta}_{k-p}} = 0. \tag{4.73}$$

Following the steps in *Lemma* 4.1, $\lim_{k\to\infty} \|\mathbf{e}_k\| = 0$.

4.4 Extension to More General Cases

In this section the discrete-time adaptive posicast control is extended to linear systems with uncertain but upper-bounded time-delay and nonlinear systems.

4.4.1 Uncertain Upper-Bounded Time-Delay

Consider the system (4.46), but, with an unknown input time-delay steps d such that

$$\begin{aligned} \mathbf{x}_{k+1} &= \Phi \mathbf{x}_k + \Gamma \mathbf{u}_{k-d} \\ \mathbf{y}_k &= C \mathbf{x}_k \end{aligned} \tag{4.74}$$

and the unknown time-delay is assumed to have a known upper-bound such that $d \le p$ for a known p. Subtracting the reference model (4.47) from (4.74) and deriving the error dynamics as

$$\mathbf{e}_{k+1} = \Phi_{\mathrm{m}} \mathbf{e}_k + \Gamma_{\mathrm{n}} \Theta \mathbf{x}_k + \Gamma_{\mathrm{n}} \Theta_{\gamma} \mathbf{u}_{k-d} - \Gamma_{\mathrm{n}} \Theta_{\mathrm{r}} \mathbf{r}_{k-p} \tag{4.75}$$

where $\mathbf{e}_k = \mathbf{x}_k - \mathbf{x}_{\mathrm{m},k}$. Note that $\mathbf{x}_{k+p}$ can be written as

$$\mathbf{x}_{k+p} = \Phi^p \mathbf{x}_k + \left(\Phi^{p-1} \Gamma \mathbf{u}_{k-d} + \Phi^{p-2} \Gamma \mathbf{u}_{k-d+1} + \cdots + \Gamma \mathbf{u}_{k+p-d-1} \right) \tag{4.76}$$

Substituting a p time steps delayed form of (4.76) into (4.75)

$$\begin{aligned} \mathbf{e}_{k+1} = {} & \Phi_{\mathrm{m}} \mathbf{e}_k + \Gamma_{\mathrm{n}} \Theta \Phi^p \mathbf{x}_{k-p} + \Gamma_{\mathrm{n}} \Theta \left(\Phi^{p-1} \Gamma \mathbf{u}_{k-p-d} + \Phi^{p-2} \Gamma \mathbf{u}_{k-p-d+1} + \cdots \right. \\ & \left. + \Gamma \mathbf{u}_{k-d-1} \right) + \Gamma_{\mathrm{n}} \Theta_{\gamma} \mathbf{u}_{k-d} - \Gamma_{\mathrm{n}} \Theta_{\mathrm{r}} \mathbf{r}_{k-p}. \end{aligned} \tag{4.77}$$

Let $\boldsymbol{\xi}_k^{\top} = \left[\mathbf{u}_{k-1}^{\top} \mid \mathbf{u}_{k-2}^{\top} \mid \cdots \mid \mathbf{u}_{k-p}^{\top}\right] \in \Re^{pm}$ and rewrite (4.77) as

$$\begin{aligned}\mathbf{e}_{k+1} &= \Phi_{\mathrm{m}}\mathbf{e}_k + \Gamma_{\mathrm{n}}\Theta\Phi^{p}\mathbf{x}_{k-p} + \Gamma_{\mathrm{n}}(0\cdot\mathbf{u}_{k-2p} + \cdots + 0\cdot\mathbf{u}_{k-p-d-1})\\ &\quad + \Gamma_{\mathrm{n}}\Theta\left(\Phi^{p-1}\Gamma\mathbf{u}_{k-p-d} + \Phi^{p-2}\Gamma\mathbf{u}_{k-p-d+1} + \cdots + \Gamma\mathbf{u}_{k-d-1}\right)\\ &\quad + \Gamma_{\mathrm{n}}\Theta_{\gamma}\mathbf{u}_{k-d} + \Gamma_{\mathrm{n}}(0\cdot\mathbf{u}_{k-d+1} + \cdots + 0\cdot\mathbf{u}_{k-1}) - \Gamma_{\mathrm{n}}\Theta_{\mathrm{r}}\mathbf{r}_{k-p}. \end{aligned} \tag{4.78}$$

It is possible to simplify (4.78) further to the form

$$\mathbf{e}_{k+1} = \Phi_{\mathrm{m}}\mathbf{e}_k + \Gamma_{\mathrm{n}}\Theta_{\mathrm{x}}\mathbf{x}_{k-p} + \Gamma_{\mathrm{n}}\Theta_{\mathrm{u}}\boldsymbol{\xi}_{k-p} - \Gamma_{\mathrm{n}}\Theta_{\mathrm{r}}\mathbf{r}_{k-p} + \Gamma_{\mathrm{n}}\Theta_{\mathrm{p}}\mathbf{u}_{k-p} + \Gamma_{\mathrm{n}}\Omega_{\mathrm{u}}\boldsymbol{\xi}_k \tag{4.79}$$

where the uncertain parameters are now given as $\Theta_{\mathrm{x}} \in \Re^{m\times n}$, $\Theta_{\mathrm{p}} = \Theta\Phi^{p-d-1}\Gamma \in \Re^{m\times m}$, $\Theta_{\mathrm{u}} = \left[[0]_{m\times m(p-d)} \mid \Theta\Phi^{p-1}\Gamma \mid \Theta\Phi^{p-2}\Gamma \mid \cdots \mid \Theta\Phi^{p-d}\Gamma\right] \in \Re^{m\times m}$ and $\Omega_{\mathrm{u}} = \left[\Theta\Phi^{p-d-2}\Gamma \mid \Theta\Phi^{p-d-3}\Gamma \mid \cdots \mid \Theta\Gamma \mid [0]_{m\times md}\right] \in \Re^{m\times pm}$. It should be noted that some of the elements of Θ_{u} and Ω_{u} matrices are zero as seen from (4.78).

The reason (4.77) is rewritten in the form (4.78) is to eliminate the dependency on the unknown time-delay steps d. From (4.79) it seen that the system is written in terms of the known upperbound p rather than the unknown time-delay steps d. Proceeding further, assume a controller of the form

$$\mathbf{u}_k = -\hat{\Theta}_{\mathrm{p},k}^{-1}\left(\hat{\Theta}_{\mathrm{x},k}\mathbf{x}_k + \hat{\Theta}_{\mathrm{u},k}\boldsymbol{\xi}_k - \Theta_{\mathrm{r}}\mathbf{r}_k\right). \tag{4.80}$$

Substitution of (4.80) into (4.79) and after performing some simplifications it is obtained that

$$\begin{aligned}\mathbf{e}_{k+1} &= \Phi_{\mathrm{m}}\mathbf{e}_k + \Gamma_{\mathrm{n}}\tilde{\Theta}_{\mathrm{x},k-p}\mathbf{x}_{k-p} + \Gamma_{\mathrm{n}}\tilde{\Theta}_{\mathrm{u},k-p}\boldsymbol{\xi}_{k-p}\\ &\quad + \Gamma_{\mathrm{n}}\left(\Theta_{\mathrm{p}} - \hat{\Theta}_{\mathrm{p},k-p}\right)\mathbf{u}_{k-p} + \Gamma_{\mathrm{n}}\Omega_{\mathrm{u}}\boldsymbol{\xi}_k\\ &= \Phi_{\mathrm{m}}\mathbf{e}_k + \Gamma_{\mathrm{n}}\tilde{\Theta}_{\mathrm{x},k-p}\mathbf{x}_{k-p} + \Gamma_{\mathrm{n}}\tilde{\Theta}_{\mathrm{u},k-p}\boldsymbol{\xi}_{k-p}\\ &\quad + \Gamma_{\mathrm{n}}\tilde{\Theta}_{\mathrm{p},k-p}\mathbf{u}_{k-p} + \Gamma_{\mathrm{n}}\Omega_{\mathrm{u}}\boldsymbol{\xi}_k. \end{aligned} \tag{4.81}$$

Consider now the inclusion of the terms $\Gamma_{\mathrm{n}}\cdot 0\cdot\mathbf{x}_k + \Gamma_{\mathrm{n}}\cdot 0\cdot\mathbf{u}_k$ in (4.81) such that

$$\begin{aligned}\mathbf{e}_{k+1} &= \Phi_{\mathrm{m}}\mathbf{e}_k + \Gamma_{\mathrm{n}}\tilde{\Theta}_{\mathrm{x},k-p}\mathbf{x}_{k-p} + \Gamma_{\mathrm{n}}\tilde{\Theta}_{\mathrm{u},k-p}\boldsymbol{\xi}_{k-p} + \Gamma_{\mathrm{n}}\tilde{\Theta}_{\mathrm{p},k-p}\mathbf{u}_{k-p} + \Gamma_{\mathrm{n}}\cdot 0\cdot\mathbf{x}_k\\ &\quad + \Gamma_{\mathrm{n}}\cdot 0\cdot\mathbf{u}_k + \Gamma_{\mathrm{n}}\Omega_{\mathrm{u}}\boldsymbol{\xi}_k. \end{aligned} \tag{4.82}$$

and let $\boldsymbol{\zeta}_k^{\top} = \left[\mathbf{x}_k^{\top} \mid \boldsymbol{\xi}_k^{\top} \mid \mathbf{u}_k^{\top}\right] \in \Re^{n+m(p+1)}$, $\tilde{\Psi}_k^{\top} = \left[\tilde{\Theta}_{\mathrm{x},k} \mid \tilde{\Theta}_{\mathrm{u},k} \mid \tilde{\Theta}_{\mathrm{p},k}\right] \in \Re^{m\times(n+m(p+1))}$ and $\Omega^{\top} = \left[[0]_{m\times n} \mid \Omega_{\mathrm{u}} \mid [0]_{m\times m}\right] \in \Re^{m\times(n+m(p+1))}$ then it is possible to obtain the compact error dynamics of the form

$$\mathbf{e}_{k+1} = \Phi_{\mathrm{m}}\mathbf{e}_k + \Gamma_{\mathrm{n}}\tilde{\Psi}_{k-p}^{\top}\boldsymbol{\zeta}_{k-p} + \Gamma_{\mathrm{n}}\Omega^{\top}\boldsymbol{\zeta}_k. \tag{4.83}$$

Note that the error dynamics (4.83) is similar to (4.59) with the only difference being the extra term $\Gamma_{\rm n}\Omega^\top \boldsymbol{\zeta}_k$ which exists due to the uncertainty in the time-delay. If the time-delay steps d is known and $d = p$ then Ω would be a null matrix. Proceeding further, using $\mathbf{z}_{k+1} = (C\Gamma_{\rm n})^{-1}C\,(\mathbf{e}_{k+1} - \Phi_{\rm m}\mathbf{e}_k)$ to obtain

$$\mathbf{z}_{k+1} = \tilde{\Psi}_{k-p}^\top \boldsymbol{\zeta}_{k-p} + \Omega^\top \boldsymbol{\zeta}_k \tag{4.84}$$

where $\mathbf{z}_{k+1} \in \Re^m$. The adaptation law will be formulated in such a way as to be robust to the term $\Omega^\top \boldsymbol{\zeta}_k$. Based on (4.84), the adaptation law is proposed as

$$\hat{\Psi}_{k+1} = \begin{cases} \hat{\Psi}_{k-p} + \varepsilon_k \frac{\beta_k}{\varphi_k} Q \boldsymbol{\zeta}_{k-p}\mathbf{z}_{k+1}^\top & \forall k \in [p, \infty) \\ \hat{\Psi}_0 & \forall k \in [0, p) \end{cases} \tag{4.85}$$

where the scalar function $\varphi_k = 1 + \varepsilon_k \boldsymbol{\zeta}_{k-p}^\top Q \boldsymbol{\zeta}_{k-p} + \varepsilon_k \gamma \lambda^2 \|\boldsymbol{\zeta}_k\|^2$, the matrix Q is a constant positive definite matrix of dimension $n + m(p+1)$, γ, λ are positive tuning constants, β_k is a positive scalar weighing coefficient and $\varepsilon_k > 0$ is a coefficient used to ensure a nonsingular $\hat{\Theta}_{{\rm p},k}$.

Remark 4.4 Note that similar to the previous cases ε_k must be selected such that ε_k^{-1} is not an eigenvalue of $-\hat{\Theta}_{{\rm p},k-p}^{-1} S \frac{\beta_k}{\varphi_k} Q \boldsymbol{\zeta}_{k-p}\mathbf{z}_{k+1}^\top$ where $S = (C\Gamma_{\rm n})^{-1}C[\mathbf{0}\ \cdots\ \mathbf{0}\ I_m] \in \Re^{m\times(n+m(p+1))}$.

Consider the constant uncertainty Ω and assume that $\|\Omega\| = \lambda\rho$ where ρ is an uncertain positive constant, it is easy to see that $\|\Omega^\top \boldsymbol{\zeta}_k\| \le \lambda\rho\|\boldsymbol{\zeta}_k\|$. Further, the weighing coefficient β_k can be defined as,

$$\beta_k = \begin{cases} 1 - \frac{\lambda\hat{\rho}_k\|\boldsymbol{\zeta}_k\|}{\|\mathbf{z}_{k+1}\|} & \text{if } \|\mathbf{z}_{k+1}\| \ge \lambda\hat{\rho}_k\|\boldsymbol{\zeta}_k\| \\ 0 & \text{if } \|\mathbf{z}_{k+1}\| < \lambda\hat{\rho}_k\|\boldsymbol{\zeta}_k\| \end{cases} \tag{4.86}$$

where $\hat{\rho}_k$ is the estimate of ρ and λ can be chosen as any constant as long as it satisfies $0 < \lambda < \lambda_{\max}$, with $\lambda_{\max}$ being defined later. The estimation law for ρ can be given as

$$\hat{\rho}_{k+1} = \hat{\rho}_k + \varepsilon_k \frac{\beta_k \lambda \gamma \|\boldsymbol{\zeta}_k\| \cdot \|\mathbf{z}_{k+1}\|}{\varphi_k}. \tag{4.87}$$

Note that from (4.86) if $\|\mathbf{z}_{k+1}\| \ge \lambda\hat{\rho}_k\|\boldsymbol{\zeta}_k\|$ it is obtained that

$$\beta_k^2 \mathbf{z}_{k+1}^\top \mathbf{z}_{k+1} = \beta_k \mathbf{z}_{k+1}^\top \mathbf{z}_{k+1} - \lambda\hat{\rho}_k\beta_k\|\boldsymbol{\zeta}_k\| \cdot \|\mathbf{z}_{k+1}\|. \tag{4.88}$$

The validity of the above adaption law is verified by the following theorem:

Theorem 4.2 *Under the adaptation law (4.85) and the closed-loop dynamics (4.84) the tracking error* $\mathbf{e}_k$ *is bounded.*

Proof The convergence analysis is analogous to the preceding cases. Selecting a nonnegative function

$$V_k = \sum_{j=1}^{m}\left(\sum_{i=k-p}^{k} \tilde{\boldsymbol{\psi}}_{j,i}^{\top} Q^{-1}\tilde{\boldsymbol{\psi}}_{j,i}\right) + \frac{1}{\gamma}\tilde{\rho}_k^2 \tag{4.89}$$

where $\tilde{\Psi}_k^{\top} = \left[\, \tilde{\boldsymbol{\psi}}_{1,k} \,\vdots\, \tilde{\boldsymbol{\psi}}_{2,k} \,\vdots\, \cdots \,\vdots\, \tilde{\boldsymbol{\psi}}_{m,k} \,\right]$. The difference of V_k with respect to a single time step is

$$\begin{aligned}
\Delta V_k &= V_{k+1} - V_k \\
&= \sum_{j=1}^{m}\left(\tilde{\boldsymbol{\psi}}_{j,k+1}^{\top} Q^{-1}\tilde{\boldsymbol{\psi}}_{j,k+1} - \tilde{\boldsymbol{\psi}}_{j,k-p}^{\top} Q^{-1}\tilde{\boldsymbol{\psi}}_{j,k-p}\right) + \frac{1}{\gamma}\left(\tilde{\rho}_{k+1}^2 - \tilde{\rho}_k^2\right).
\end{aligned} \tag{4.90}$$

Defining $\tilde{\rho}_k = \rho - \hat{\rho}_k$, then it is possible to obtain

$$\tilde{\rho}_{k+1} = \tilde{\rho}_k - \varepsilon_k \frac{\beta_k \lambda \gamma \|\boldsymbol{\zeta}_k\| \cdot \|\mathbf{z}_{k+1}\|}{\varphi_k}. \tag{4.91}$$

Substituting (4.85) and (4.91) in (4.90) and simplifying further it is obtained that

$$\begin{aligned}
\Delta V_k &= \sum_{j=1}^{m}\left(\tilde{\boldsymbol{\psi}}_{j,k-p} - \varepsilon_k \frac{\beta_k}{\varphi_k} Q \boldsymbol{\zeta}_{k-p} z_{j,k+1}\right)^{\top} Q^{-1}\left(\tilde{\boldsymbol{\psi}}_{j,k-p} - \varepsilon_k \frac{\beta_k}{\varphi_k} Q \boldsymbol{\zeta}_{k-p} z_{j,k+1}\right) \\
&\quad - \sum_{j=1}^{m} \tilde{\boldsymbol{\psi}}_{j,k-p}^{\top} Q^{-1}\tilde{\boldsymbol{\psi}}_{j,k-p} + \frac{1}{\gamma}\left(\tilde{\rho}_k - \frac{\varepsilon_k \beta_k \lambda \gamma \|\boldsymbol{\zeta}_{k-p}\| \cdot \|\mathbf{z}_{k+1}\|}{\varphi_k}\right)^2 - \frac{1}{\gamma}\tilde{\rho}_k^2 \\
&\le \sum_{j=1}^{m}\left(\frac{\varepsilon_k^2 \beta_k^2}{\varphi_k^2} \boldsymbol{\zeta}_{k-p}^{\top} Q \boldsymbol{\zeta}_{k-p} z_{j,k+1}^2 - 2\frac{\varepsilon_k \beta_k}{\varphi_k} \tilde{\boldsymbol{\psi}}_{j,k-p}^{\top} \boldsymbol{\zeta}_{k-p} z_{j,k+1}\right) \\
&\quad + \frac{\varepsilon_k^2 \beta_k^2}{\varphi_k^2} \gamma \lambda^2 \|\boldsymbol{\zeta}_k\|^2 \mathbf{z}_{k+1}^{\top} \mathbf{z}_{k+1} - 2\frac{\varepsilon_k \beta_k}{\varphi_k} \lambda \tilde{\rho}_k \|\boldsymbol{\zeta}_k\| \cdot \|\mathbf{z}_{k+1}\| \\
&\le -2\frac{\varepsilon_k \beta_k}{\varphi_k} \mathbf{z}_{k+1}^{\top} \mathbf{z}_{k+1} + 2\frac{\varepsilon_k \beta_k}{\varphi_k} \lambda \tilde{\rho}_k \|\boldsymbol{\zeta}_k\| \cdot \|\mathbf{z}_{k+1}\| - 2\frac{\varepsilon_k \beta_k}{\varphi_k} \lambda \tilde{\rho}_k \|\boldsymbol{\zeta}_k\| \cdot \|\mathbf{z}_{k+1}\| \\
&\quad + \frac{\varepsilon_k^2 \beta_k^2}{\varphi_k^2}\left(\boldsymbol{\zeta}_{k-p}^{\top} Q \boldsymbol{\zeta}_{k-p} + \gamma \lambda^2 \|\boldsymbol{\zeta}_k\|^2\right) \mathbf{z}_{k+1}^{\top} \mathbf{z}_{k+1} \\
&\le -2\frac{\varepsilon_k \beta_k}{\varphi_k} \mathbf{z}_{k+1}^{\top} \mathbf{z}_{k+1} + 2\frac{\varepsilon_k \beta_k \lambda}{\varphi_k} \hat{\rho}_k \|\boldsymbol{\zeta}_k\| \cdot \|\mathbf{z}_{k+1}\| \\
&\quad + \frac{\varepsilon_k^2 \beta_k^2}{\varphi_k^2}\left(\boldsymbol{\zeta}_{k-p}^{\top} Q \boldsymbol{\zeta}_{k-p} + \gamma \lambda^2 \|\boldsymbol{\zeta}_k\|^2\right) \mathbf{z}_{k+1}^{\top} \mathbf{z}_{k+1}.
\end{aligned} \tag{4.92}$$

Using the fact that $\frac{\varepsilon_k \boldsymbol{\zeta}_{k-p}^{\top} Q \boldsymbol{\zeta}_{k-p} + \varepsilon_k \gamma \lambda^2 \|\boldsymbol{\zeta}_k\|^2}{\varphi_k} < 1$ and (4.88) it is possible to obtain

$$\Delta V_k < -\varepsilon_k \frac{\beta_k^2}{\varphi_k} \mathbf{z}_{k+1}^\top \mathbf{z}_{k+1}. \tag{4.93}$$

Following the same steps that lead to (4.73) in *Theorem* 4.1, it is concluded that

$$\lim_{k\to\infty} \varepsilon_k \frac{\beta_k^2}{\varphi_k} \mathbf{z}_{k+1}^\top \mathbf{z}_{k+1} = 0. \tag{4.94}$$

The result (4.93) shows that $\tilde{\Theta}_{\mathrm{x},k}$, $\tilde{\Theta}_{p,k}$, $\tilde{\Theta}_{\mathrm{u},k}$ and $\tilde{\rho}_k$ are bounded because V_k is non-increasing. Using arguments similar to that in *Theorem* 4.1 and *Lemma* 4.1, then it is possible to establish that $\|\boldsymbol{\zeta}_k\| \le d_1 + d_2\|\mathbf{z}_{k+1}\|$ for some positive constants d_1 and d_2. Thus, establishing the condition required by the *Key Technical Lemma* that guarantees that $\lim_{k\to\infty} \beta_k \|\mathbf{z}_{k+1}\| = 0$.

Further, there must exist a positive constant μ such that $\max_{i\in[0,k]}\{\beta_i \|\mathbf{z}_{i+1}\|\} \le \mu$. Then according to the definition of β_k in (4.86)

$$\max_{i\in[0,k]} \|\mathbf{z}_{i+1}\| \le \mu + \lambda\bar{\rho} \max_{i\in[0,k]} \|\boldsymbol{\zeta}_i\| \tag{4.95}$$

where $\max_{i\in[0,k]} \hat{\rho}_i \le \bar{\rho}$. Following the analysis in *Theorem* 4.1, *Lemma* 4.1 and the bound on $\boldsymbol{\zeta}_k$ the maximum tracking error bound is found to be

$$\max_{i\in[0,k]} \|\mathbf{z}_{i+1}\| \le \mu + \lambda\bar{\rho} \max_{i\in[0,k]} \{d_1 + d_2\|\mathbf{z}_i\|\} \tag{4.96}$$

which results in

$$\|\mathbf{z}_{k+k_0}\| \le (\lambda\bar{\rho}d_2)^k \|\mathbf{z}_{k_0}\| + (\mu + \lambda\bar{\rho}d_1) \sum_{i=0}^{k-1} (\lambda\bar{\rho}d_2)^i \tag{4.97}$$

implying that to guarantee boundedness of $\|\mathbf{z}_k\|$ then $|\lambda| < \lambda_{\max} < \frac{1}{\bar{\rho}d_2}$. Using the boundedness condition on $\|\mathbf{z}_k\|$ and the analysis in *Lemma* 4.1, the boundedness of $\|\mathbf{e}_k\|$ is established.

4.4.2 Extension to Nonlinear Systems

Consider the nth order feedback linearizable nonlinear system given as

$$\begin{aligned} \mathbf{x}_{k+1} &= \Theta^\top \boldsymbol{\chi}(\mathbf{x}_k) + \Gamma \mathbf{u}_{k-p} \\ \mathbf{y}_k &= C\mathbf{x}_k \end{aligned} \tag{4.98}$$

where $\Theta \in \Re^{q\times n}$ is a matrix of uncertain parameters, $\Gamma \in \Re^{n\times m}$ is the uncertain input gain and $C \in \Re^{n\times m}$ is the output matrix. The function $\boldsymbol{\chi}(\mathbf{x}_k) \in \Re^q$ is assumed to be a vector of known nonlinear functions that satisfies $\|\boldsymbol{\chi}(\mathbf{x}_k)\| \leq c_1 + c_2\|\mathbf{x}_k\|$ for some constants c_1 and c_2.

Assumption 4.6 The control input $\mathbf{u}_k$ is bounded for a bounded $\mathbf{x}_k$.

The objective is to force the system (4.98) to follow the reference model (4.47).

Similar to the linear case, assume that there exists a positive definite $\Theta_\gamma \in \Re^{m\times m}$ such that $\Gamma = \Gamma_\mathrm{n}\Theta_\gamma$ where Γ_n is the known nominal value of Γ and assume that the nonlinearity is matched by the control input such that there exists a $\Theta_\mathrm{x} \in \Re^{q\times m}$ and that $\Theta = \Theta_\mathrm{x}\Gamma_\mathrm{n}^\top$. Subtracting (4.47) from (4.98) and rearranging the terms it is obtained that

$$\mathbf{e}_{k+1} = \Phi_\mathrm{m}\mathbf{e}_k + \Theta^\top\boldsymbol{\chi}(\mathbf{x}_k) - \Phi_\mathrm{m}\mathbf{x}_k - \Gamma_\mathrm{n}\Theta_\mathrm{r}\mathbf{r}_{k-p} + \Gamma_\mathrm{n}\Theta_\gamma\mathbf{u}_{k-p} \tag{4.99}$$

where $\mathbf{e}_k = \mathbf{x}_k - \mathbf{x}_{\mathrm{m},k}$. Note that the linear term $\Phi\mathbf{x}_k$ in (4.46) is not present in the nonlinear system (4.98) and, therefore, $\Phi_\mathrm{m} = \Gamma_\mathrm{n}\Theta_\mathrm{m}$ where $\Theta_\mathrm{m} \in \Re^{m\times n}$ is a known constant matrix and Φ_m will have $n - m$ deadbeat poles. Using $\Theta = \Theta_\mathrm{x}\Gamma_\mathrm{n}^\top$ and $\Phi_\mathrm{m} = \Gamma_\mathrm{n}\Theta_\mathrm{m}$ in (4.99) it is obtained that

$$\mathbf{e}_{k+1} = \Phi_\mathrm{m}\mathbf{e}_k + \Gamma_\mathrm{n}\Theta_\mathrm{x}^\top\boldsymbol{\chi}(\mathbf{x}_k) - \Gamma_\mathrm{n}\Theta_\mathrm{m}\mathbf{x}_k - \Gamma_\mathrm{n}\Theta_\mathrm{r}\mathbf{r}_{k-p} + \Gamma_\mathrm{n}\Theta_\gamma\mathbf{u}_{k-p} \tag{4.100}$$

which can be simplified further to the form

$$\mathbf{e}_{k+1} = \Phi_\mathrm{m}\mathbf{e}_k + \Gamma_\mathrm{n}\left(\Theta_\mathrm{x}^\top\boldsymbol{\chi}(\mathbf{x}_k) - \Theta_\mathrm{m}\mathbf{x}_k - \Theta_\mathrm{r}\mathbf{r}_{k-p} + \Theta_\gamma\mathbf{u}_{k-p}\right). \tag{4.101}$$

From (4.101) the control law can be selected as

$$\mathbf{u}_k = -\Theta_\gamma{}^{-1}\left(\Theta_\mathrm{x}^\top\boldsymbol{\chi}(\mathbf{x}_{k+p}) - \Theta_\mathrm{m}\mathbf{x}_{k+p} - \Theta_\mathrm{r}\mathbf{r}_k\right). \tag{4.102}$$

Proceeding further, the parameters Θ_x and Θ_γ are assumed to be uncertain, therefore, the controller (4.102) is modified to the form

$$\mathbf{u}_k = -\hat{\Theta}_{\gamma,k}^{-1}\left(\hat{\Theta}_{\mathrm{x},k}^\top\boldsymbol{\chi}(\mathbf{x}_{k+p}) - \Theta_\mathrm{m}\mathbf{x}_{k+p} - \Theta_\mathrm{r}\mathbf{r}_k\right) \tag{4.103}$$

where $\hat{\Theta}_{\gamma,k}$ and $\hat{\Theta}_{\mathrm{x},k}$ are the estimates of Θ_γ and Θ_x respectively. Note that from Θ we have $\Theta_\mathrm{x}^\top = (C\Gamma_\mathrm{n})^{-1}C\Theta^\top$ and, similarly, if $\hat{\Theta}_k$ is available then $\hat{\Theta}_{\mathrm{x},k}^\top = (C\Gamma_\mathrm{n})^{-1}C\hat{\Theta}_k^\top$.

Proceeding further, controller (4.103) is not causal since it requires $\mathbf{x}_{k+p}$ and $\boldsymbol{\chi}(\mathbf{x}_{k+p})$. In order to solve this problem define the states

$$\begin{aligned}
\eta_{1,k} &= \mathbf{x}_k \\
\eta_{2,k} &= \hat{\Theta}_{k-p}^\top\boldsymbol{\chi}(\eta_{1,k}) + \hat{\Gamma}_{k-p}\mathbf{u}_{k-p} \\
\eta_{i,k} &= \hat{\Theta}_{k-p+i-2}^\top\boldsymbol{\chi}(\eta_{i-1,k}) + \hat{\Gamma}_{k-p+i-2}\mathbf{u}_{k-p+i-2}
\end{aligned} \tag{4.104}$$

where $i = 3, \ldots, p+1$. The state $\boldsymbol{\eta}_{2,k}$ represents the estimate of $\mathbf{x}_{k+1}$ and similarly $\boldsymbol{\eta}_{i,k}$ represents the estimate of $\mathbf{x}_{k+i-1}$. In order to design the adaptive law consider (4.104), it is possible to obtain

$$\begin{aligned}\boldsymbol{\eta}_{1,k+1} &= \boldsymbol{\eta}_{2,k} + \tilde{\Theta}_{k-p}^{\top}\boldsymbol{\chi}(\boldsymbol{\eta}_{1,k}) + \tilde{\Gamma}_{k-p}\mathbf{u}_{k-p} \\ \boldsymbol{\eta}_{2,k+1} &= \boldsymbol{\eta}_{3,k} + \hat{\Theta}_{k-p+1}^{\top}\left(\boldsymbol{\chi}(\boldsymbol{\eta}_{1,k+1}) - \boldsymbol{\chi}(\boldsymbol{\eta}_{2,k})\right) \\ \boldsymbol{\eta}_{i-1,k+1} &= \boldsymbol{\eta}_{i,k} + \hat{\Theta}_{k-p+i-2}^{\top}\left(\boldsymbol{\chi}(\boldsymbol{\eta}_{i-2,k+1}) - \boldsymbol{\chi}(\boldsymbol{\eta}_{i-1,k})\right)\end{aligned} \tag{4.105}$$

where $\tilde{\Theta}_k = \Theta - \hat{\Theta}_k$ and $\tilde{\Gamma}_k = \Gamma - \hat{\Gamma}_k$. Rewriting the dynamics

$$\boldsymbol{\eta}_{1,k+1} = \boldsymbol{\eta}_{2,k} + \tilde{\Theta}_{k-p}\boldsymbol{\chi}(\boldsymbol{\eta}_{1,k}) + \tilde{\Gamma}_{k-p}\mathbf{u}_{k-p} \tag{4.106}$$

as

$$\tilde{\boldsymbol{\eta}}_{1,k+1} = \boldsymbol{\eta}_{1,k+1} - \boldsymbol{\eta}_{2,k} = \tilde{\boldsymbol{\Psi}}_{k-p}^{\top}\boldsymbol{\zeta}_k \tag{4.107}$$

where $\tilde{\Psi}_k^{\top} = \left[\tilde{\Theta}_k^{\top} \,\vdots\, \tilde{\Gamma}_k\right] \in \Re^{n\times(q+m)}$ and $\boldsymbol{\zeta}_k^{\top} = \left[\boldsymbol{\chi}^{\top}(\boldsymbol{\eta}_{1,k}) \,\vdots\, \mathbf{u}_{k-p}^{\top}\right] \in \Re^{q+m}$. Using (4.107), it is possible to formulate the adaptation law as follows

$$\hat{\Psi}_{k+1} = \begin{cases} \hat{\Psi}_{k-p} + \varepsilon_k P_{k+1}\boldsymbol{\zeta}_k^{\top}\tilde{\boldsymbol{\eta}}_{1,k+1}^{\top} & \forall k \in [p,\infty) \\ \hat{\Psi}_0 & \forall k \in [0,p) \end{cases} \tag{4.108}$$

$$P_{k+1} = \begin{cases} P_{k-p} - \varepsilon_k \frac{P_{k-p}\boldsymbol{\zeta}_k\boldsymbol{\zeta}_k^{\top}P_{k-p}}{1+\varepsilon_k\boldsymbol{\zeta}_k^{\top}P_{k-p}\boldsymbol{\zeta}_k} & \forall k \in [p,\infty) \\ P_0 > 0 & \forall k \in [0,p) \end{cases} \tag{4.109}$$

where ε_k is a positive coefficient used to prevent a singular $\hat{\Theta}_{\gamma,k}$ and $P_k \in \Re^{(q+m)\times(q+m)}$ is the symmetric positive-definite covariance matrix.

Remark 4.5 As in the coefficient used in (4.61) and (4.62), in order for $\hat{\Theta}_{\gamma,k}$ not to be singular then ε_k^{-1} must selected such that it is not an eigenvalue of $-\hat{\Theta}_{\gamma,k-p}^{-1} S P_{k+1}\boldsymbol{\zeta}_k\tilde{\boldsymbol{\eta}}_{1,k+1}$ where $S = (C\Gamma_{\text{n}})^{-1}C[\mathbf{0} \ \cdots \ \mathbf{0} \ I_m] \in \Re^{m\times(q+m)}$.

Using (4.104), the control law (4.103) can be modified to the form

$$\mathbf{u}_k = -\hat{\Theta}_{\gamma,k}^{-1}\left(\hat{\Theta}_{\text{x},k}^{\top}\boldsymbol{\chi}(\boldsymbol{\eta}_{p+1,k}) - \Theta_{\text{m}}\boldsymbol{\eta}_{p+1,k} - \Theta_{\text{r}}\mathbf{r}_k\right) \tag{4.110}$$

where $\hat{\Theta}_{\gamma,k} = (C\Gamma_{\text{n}})^{-1}C\hat{\Gamma}_k$ can be derived easily from (4.50). Proceeding further, consider (4.99) and adding and subtracting the term $\hat{\Theta}_{k-p}^{\top}\boldsymbol{\chi}(\mathbf{x}_k) + \hat{\Gamma}_{k-p}\mathbf{u}_{k-p}$, where $\hat{\Gamma}_k$ is the estimate of Γ, to the right hand side of (4.99) it is obtained that

$$\begin{aligned}\mathbf{e}_{k+1} &= \Phi_{\mathrm{m}}\mathbf{e}_k + \left(\Theta^\top - \hat{\Theta}^\top_{k-p}\right)\boldsymbol{\chi}(\mathbf{x}_k) - \Gamma_{\mathrm{n}}\Theta_{\mathrm{m}}\mathbf{x}_k - \Gamma_{\mathrm{n}}\Theta_{\mathrm{r}}\mathbf{r}_{k-p} + \hat{\Gamma}_{k-p}\mathbf{u}_{k-p}\\ &\quad + \left(\Gamma - \hat{\Gamma}_{k-p}\right)\mathbf{u}_{k-p}\\ &= \Phi_{\mathrm{m}}\mathbf{e}_k + \tilde{\Theta}^\top_{k-p}\boldsymbol{\chi}(\mathbf{x}_k) + \hat{\Theta}^\top_{k-p}\boldsymbol{\chi}(\mathbf{x}_k) - \Gamma_{\mathrm{n}}\Theta_{\mathrm{m}}\mathbf{x}_k - \Gamma_{\mathrm{n}}\Theta_{\mathrm{r}}\mathbf{r}_{k-p} + \hat{\Gamma}_{k-p}\mathbf{u}_{k-p}\\ &\quad + \tilde{\Gamma}_{k-p}\mathbf{u}_{k-p}\end{aligned} \tag{4.111}$$

Further, (4.111) can be written as

$$\begin{aligned}\mathbf{e}_{k+1} &= \Phi_{\mathrm{m}}\mathbf{e}_k + \tilde{\Theta}^\top_{k-p}\boldsymbol{\chi}(\mathbf{x}_k) + \tilde{\Gamma}_k\mathbf{u}_{k-p}\\ &\quad + \Gamma_{\mathrm{n}}\left(\hat{\Theta}^\top_{\mathrm{x},k-p}\boldsymbol{\chi}(\mathbf{x}_k) - \Theta_{\mathrm{m}}\mathbf{x}_k - \Theta_{\mathrm{r}}\mathbf{r}_{k-p} + \Gamma_n\hat{\Theta}_{\gamma,k-p}\mathbf{u}_{k-p}\right)\end{aligned} \tag{4.112}$$

substitution of (4.110) into (4.112) and simplifying it is obtained that

$$\begin{aligned}\mathbf{e}_{k+1} &= \Phi_{\mathrm{m}}\mathbf{e}_k + \tilde{\Theta}^\top_{k-p}\boldsymbol{\chi}(\mathbf{x}_k) + \tilde{\Gamma}_{k-p}\mathbf{u}_{k-p}\\ &\quad - \Gamma_{\mathrm{n}}\left[\Theta_{\mathrm{m}}(\mathbf{x}_k - \boldsymbol{\eta}_{p+1,k-p}) - \hat{\Theta}^\top_{\mathrm{x},k-p}\left(\boldsymbol{\chi}(\mathbf{x}_k) - \boldsymbol{\chi}(\boldsymbol{\eta}_{p+1,k-p})\right)\right]\\ &= \Phi_{\mathrm{m}}\mathbf{e}_k + \tilde{\boldsymbol{\eta}}_{1,k+1}\\ &\quad - \Gamma_{\mathrm{n}}\left[\Theta_{\mathrm{m}}(\mathbf{x}_k - \boldsymbol{\eta}_{p+1,k-p}) - \hat{\Theta}^\top_{\mathrm{x},k-p}\left(\boldsymbol{\chi}(\mathbf{x}_k) - \boldsymbol{\chi}(\boldsymbol{\eta}_{p+1,k-p})\right)\right].\end{aligned} \tag{4.113}$$

From (4.113), the asymptotic stability of $\|\mathbf{e}_k\|$ is guaranteed if $\lim_{k\to\infty}\|\tilde{\boldsymbol{\eta}}_{1,k+1}\| = 0$, $\lim_{k\to\infty}\|\mathbf{x}_k - \boldsymbol{\eta}_{p+1,k-p}\| = 0$, $\lim_{k\to\infty}\|\boldsymbol{\chi}(\mathbf{x}_k) - \boldsymbol{\chi}(\boldsymbol{\eta}_{p+1,k-p})\| = 0$ and $\hat{\Theta}_{x,k}$ or $\hat{\Theta}_k$ is bounded. These are summarized in the following theorem:

Theorem 4.3 *For the system (4.105) and the adaptative laws (4.108) and (4.109), the difference $\|\tilde{\boldsymbol{\eta}}_{1,k+1}\|$ is asymptotically stable and $\|\hat{\Theta}_k\|$ is bounded, furthermore, the following conditions are true*

$$\begin{aligned}\lim_{k\to\infty}\|\mathbf{x}_k - \boldsymbol{\eta}_{p+1,k-p}\| &= 0\\ \lim_{k\to\infty}\|\boldsymbol{\chi}(\mathbf{x}_k) - \boldsymbol{\chi}(\boldsymbol{\eta}_{p+1,k-p})\| &= 0\end{aligned} \tag{4.114}$$

Proof Consider the system (4.105) and the adaptative laws (4.108) and (4.109). Propose a positive function of the form

$$V_k = \sum_{j=1}^{n}\left(\sum_{i=0}^{p}\tilde{\boldsymbol{\psi}}^\top_{j,k-i}P^{-1}_{k-i}\tilde{\boldsymbol{\psi}}_{j,k-i}\right) \tag{4.115}$$

where $\tilde{\Psi}^\top_k = \left[\tilde{\boldsymbol{\psi}}_{1,k} \,\vdots\, \tilde{\boldsymbol{\psi}}_{2,k} \,\vdots\, \cdots \,\vdots\, \tilde{\boldsymbol{\psi}}_{n,k}\right]^\top$. The difference $\Delta V_k = V_{k+1} - V_{k-p}$ is given as

$$\Delta V_k = \sum_{j=1}^{n}\left(\tilde{\boldsymbol{\psi}}^\top_{j,k+1}P^{-1}_{k+1}\tilde{\boldsymbol{\psi}}_{j,k+1} - \tilde{\boldsymbol{\psi}}^\top_{j,k-p}P^{-1}_{k-p}\tilde{\boldsymbol{\psi}}_{j,k-p}\right). \tag{4.116}$$

Substitution of (4.108) and (4.109) into (4.116) and proceeding as in *Theorem* 4.1 it is obtained that

$$\Delta V_k = -\frac{\varepsilon_k}{1+\varepsilon_k \boldsymbol{\zeta}_k^\top P_{k-p}\boldsymbol{\zeta}_k}\tilde{\boldsymbol{\eta}}_{1,k+1}^\top \tilde{\boldsymbol{\eta}}_{1,k+1} \leq 0. \tag{4.117}$$

which implies that $\hat{\Psi}_k$ is bounded and, consequently, $\hat{\Theta}_k$ and $\hat{\Gamma}_k$ are also bounded. Using the analysis in *Lemma* 4.1 it can also be shown that $\lim_{k\to\infty} \Delta V_k = 0$. To guarantee that $\lim_{k\to\infty} \|\tilde{\boldsymbol{\eta}}_{1,k+1}\|$, certain bounds on $\boldsymbol{\zeta}_k$ must be established.

Consider the vector $\boldsymbol{\zeta}_k$, since $\boldsymbol{\zeta}_k^\top = \left[\boldsymbol{\chi}^\top(\boldsymbol{\eta}_{1,k}) \,\vdots\, \mathbf{u}_{k-p}^\top\right]$, using the bound on $\boldsymbol{\chi}(\mathbf{x}_k) = \boldsymbol{\chi}(\boldsymbol{\eta}_{1,k})$ and *Assumption* 4.6 then

$$\|\boldsymbol{\zeta}_k\| \leq g_1 + g_2\|\mathbf{x}_k\| = g_1 + g_2\|\boldsymbol{\eta}_{1,k}\| \leq g_1 + g_2 \max_{i\in[0,k]} \|\boldsymbol{\eta}_{1,i+1}\|. \tag{4.118}$$

Further, using the definition of $\boldsymbol{\eta}_{2,k}$ given as

$$\boldsymbol{\eta}_{2,k} = \hat{\Theta}_{k-p}^\top \boldsymbol{\chi}(\boldsymbol{\eta}_{1,k}) + \hat{\Gamma}_{k-p}\mathbf{u}_{k-p} \tag{4.119}$$

and using the fact that $\hat{\Theta}_k$, $\hat{\Gamma}_k$ are bounded and using the bounds on $\boldsymbol{\chi}(\boldsymbol{\eta}_{1,k})$ and *Assumption* 4.6 it is obtained that

$$\|\tilde{\boldsymbol{\eta}}_{1,k+1}\| = \|\boldsymbol{\eta}_{1,k+1} - \boldsymbol{\eta}_{2,k}\| \leq h_1 + h_2 \max_{i\in[0,k]} \|\boldsymbol{\eta}_{1,i+1}\|. \tag{4.120}$$

Thus, it is shown that both the numerator and denominator terms of (4.117) are similarly upperbounded and, therefore, the term $\varepsilon_k \boldsymbol{\zeta}_k^\top P_k \boldsymbol{\zeta}_k$ cannot grow faster than $\tilde{\boldsymbol{\eta}}_{1,k+1}^\top \tilde{\boldsymbol{\eta}}_{1,k+1}$. This implies that if $\lim_{k\to\infty} \Delta V_k = 0$ then $\lim_{k\to\infty} \|\tilde{\boldsymbol{\eta}}_{1,k+1}\| = 0$.

Proceeding with the two conditions in *Theorem* 4.3, consider (4.105) and based on the first result of this theorem it is concluded that if $\lim_{k\to\infty} \|\tilde{\boldsymbol{\eta}}_{1,k+1}\| = 0$ then $\lim_{k\to\infty} \|\boldsymbol{\chi}(\boldsymbol{\eta}_{1,k+1}) - \boldsymbol{\chi}(\boldsymbol{\eta}_{2,k})\| = 0$. Proceeding further, consider from (4.105) we have

$$\boldsymbol{\eta}_{2,k+1} - \boldsymbol{\eta}_{3,k} = \hat{\Theta}_{k-p+1}^\top \left(\boldsymbol{\chi}(\boldsymbol{\eta}_{1,k+1}) - \boldsymbol{\chi}(\boldsymbol{\eta}_{2,k})\right) \tag{4.121}$$

and since $\hat{\Theta}_k$ is bounded and $\lim_{k\to\infty} \|\boldsymbol{\eta}_{1,k+1} - \boldsymbol{\eta}_{2,k}\| = 0$ then it is obtained that

$$\lim_{\to\infty} \|\boldsymbol{\eta}_{2,k+1} - \boldsymbol{\eta}_{3,k}\| = 0. \tag{4.122}$$

Using this logic it can be concluded that $\lim_{k\to\infty} \|\boldsymbol{\eta}_{i-1,k+1} - \boldsymbol{\eta}_{i,k}\| = 0$ which implies that $\lim_{k\to\infty} \|\boldsymbol{\chi}(\boldsymbol{\eta}_{i-1,k+1}) - \boldsymbol{\chi}(\boldsymbol{\eta}_{i,k})\| = 0$. Using successive substitutions it is found that $\lim_{k\to\infty} \|\boldsymbol{\eta}_{1,k} - \boldsymbol{\eta}_{p+1,k-p}\| = \lim_{k\to\infty} \|\mathbf{x}_k - \boldsymbol{\eta}_{p+1,k-p}\| = 0$ and, consequently, $\lim_{k\to\infty} \|\boldsymbol{\chi}(\mathbf{x}_k) - \boldsymbol{\chi}(\boldsymbol{\eta}_{p+1,k-p})\| = 0$.

4.5 Illustrative Examples

4.5.1 *Linear Systems*

To illustrate the advantages of the discrete-time APC, a flight control example with the longitudinal dynamics of a four-engine jet transport aircraft, [32], was used. The aircraft flies straight and level flight at 40,000 ft with a velocity of 600 ft/s. Under these conditions, the nominal short period dynamics is given by

$$\begin{bmatrix} \dot{\alpha}(t) \\ \dot{q}(t) \end{bmatrix} = \begin{bmatrix} -0.3233 & 1 \\ -1.1692 & -0.4802 \end{bmatrix} \begin{bmatrix} \alpha(t) \\ q(t) \end{bmatrix} + \begin{bmatrix} -0.0178 \\ -1.3789 \end{bmatrix} \sigma_e(t-\tau) \tag{4.123}$$

where $\alpha(t)$ is the angle of attack in radians, $q(t)$ is the pitch rate in radians per second and σ_e is the elevator deflection also in radians. The time-delay value used in the simulation is given as $\tau = 0.3\,\text{s}$. Eigenvalues are $-0.4017 \pm 1.0785i$, giving a nominal short period natural frequency of $\omega_\text{n} = 1.1423\,\text{rad/s}$ and a nominal damping ratio of $\zeta = 0.3517$.

To obtain a challenging scenerio, control effectiveness uncertainty was introduced resulting in a 30 % decrease in elevator effectiveness. In addition, by adding further uncertainty to the state matrix, proximity of the open loop poles to the imaginary axis was halved and the damping ratio was reduced by 48 %. The resulting uncertain system becomes

$$\begin{bmatrix} \dot{\alpha}(t) \\ \dot{q}(t) \end{bmatrix} = \begin{bmatrix} -0.3233 & 1.0052 \\ -1.1755 & -0.0766 \end{bmatrix} \begin{bmatrix} \alpha(t) \\ q(t) \end{bmatrix} + \begin{bmatrix} -0.0089 \\ -0.6894 \end{bmatrix} \sigma_e(t-\tau). \tag{4.124}$$

In order to implement the controller (4.53) the reference model needs to be computed in discrete-time. To do this the nominal system (4.123) will be sampled at $T_\text{s} = 0.02\,\text{s}$ resulting in the sampled-data system

$$\begin{bmatrix} \alpha_{k+1} \\ q_{k+1} \end{bmatrix} = \begin{bmatrix} 0.9933 & 0.0198 \\ -0.0232 & 0.9902 \end{bmatrix} \begin{bmatrix} \alpha_k \\ q_k \end{bmatrix} + \begin{bmatrix} -0.0006 \\ -0.0274 \end{bmatrix} \sigma_{e,k-p} \tag{4.125}$$

where $p = \frac{\tau}{T_\text{s}} = 15$. The reference model is designed using the LQR method ignoring the time-delay. The feedback matrix is calculated by selecting $Q_\text{x} = diag(10, 10)$ and $R = 1$ resulting in a reference model of the form

$$\begin{bmatrix} \alpha_{\text{m},k+1} \\ q_{\text{m},k+1} \end{bmatrix} = \begin{bmatrix} 0.9924 & 0.0179 \\ -0.0622 & 0.9078 \end{bmatrix} \begin{bmatrix} \alpha_{\text{m},k} \\ q_{\text{m},k} \end{bmatrix} + \begin{bmatrix} 0.0021 \\ 0.0905 \end{bmatrix} r_{k-p}. \tag{4.126}$$

The response of the reference model with respect to a particular reference signal r_k is shown in Fig. 4.1.

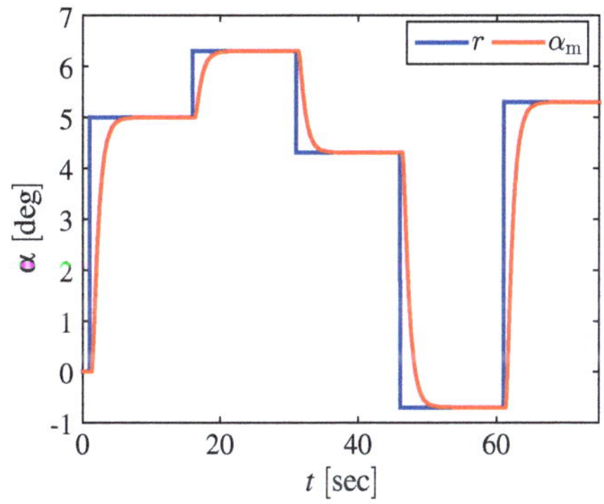

Fig. 4.1 Tracking performance of the reference model

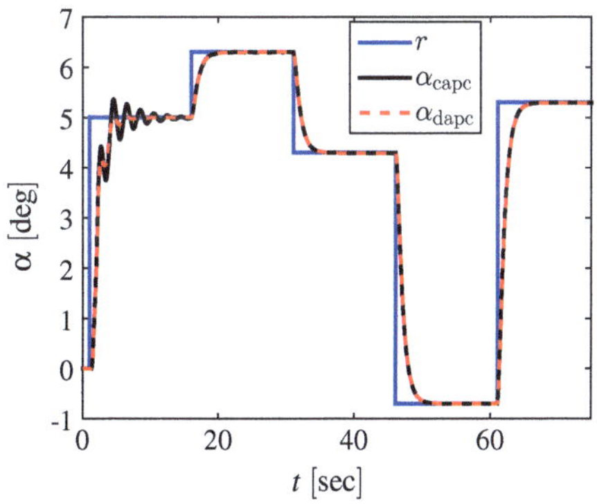

Fig. 4.2 Tracking performance of the discrete approximation of continuous-time APC versus discrete-time APC with $\tau = 0.3\,\text{s}$

4.5.1.1 Discrete APC Versus Discrete Approximation of Continuous APC

The adaptive gains of the continuous-time APC are calculated as $\Psi_{\text{x}} = \text{diag}(3.7, 8.3) \times 10^3$ and $\Psi_{\text{r}} = 8.4 \times 10^3$. The gains used in the integral approximation are tuned to get the best performance. As for the discrete-time APC, the parameter values for $P_0 = \text{diag}(P_{\text{x},0}, P_{\text{u},0}, P_{\gamma,0})$ where $P_{\text{x},0} = \text{diag}(44, 105)$, $P_{\text{u},0} = 5.1 I_p$, $P_{\gamma,0} = 0.10$. The performance of the two controllers is shown in Fig. 4.2. The performance of the continuous-time APC is oscillatory at certain intervals while the discrete-time APC has a short oscillatory period and then dampens quickly by the time the second pulse reference comes into effect. To push the comparison further the time-delay is increased to $\tau = 0.35\,\text{s}$. In Fig. 4.3 it is seen that the continuous-time APC is oscillatory while the discrete-time APC has a short oscillatory period after which it is smooth throughout.

4.5.1.2 Discrete APC Versus MRAC

The structure of the MRAC in discrete-time is similar to that of the discrete-time APC with the main difference being that the term $\hat{\Theta}_{\text{u},k}\boldsymbol{\xi}_k$ is absent from the controller

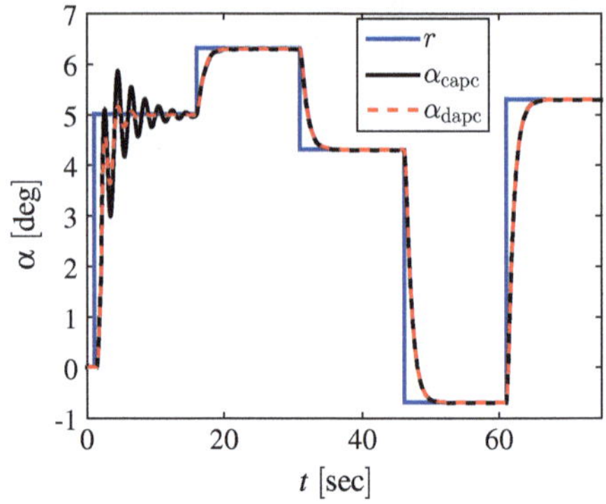

Fig. 4.3 Tracking performance of the discrete approximation of continuous-time APC versus discrete-time APC with $\tau = 0.4$ s

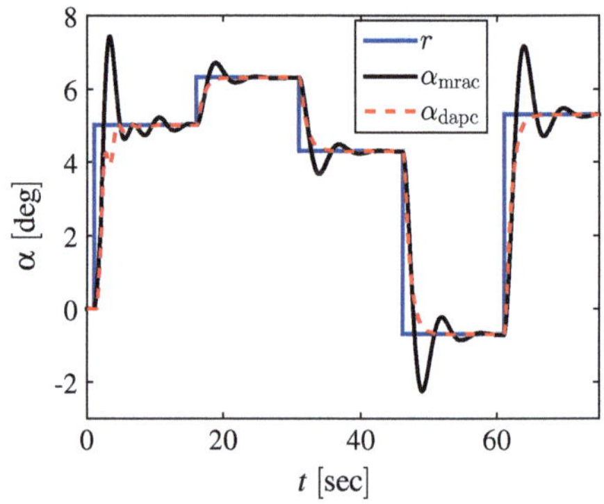

Fig. 4.4 Tracking performance of the discrete-time APC versus MRAC with $\tau = 0.35$ s

(4.56). Figure 4.4 shows that the MRAC is very oscillatory when an input time-delay of 0.35 s is introduced to the system. Even though it is well known that MRAC works well when there is no time-delay in the system its performance degrades considerably in the presence of time-delay. On the other hand the discrete-time APC is stable similar to the previous example. This example clearly presents the advantages discrete-time APC has over the conventional MRAC design.

4.5.1.3 Unknown but Upper-Bounded Time-Delay

Consider the system (4.123) where the time-delay is assumed to be $\tau_{\rm p} = 0.4$ s while the actual time-delay is $\tau_{\rm d} = 0.3$ s. Selecting $\lambda = 0.015$, $\gamma = 100$ and $Q = \text{diag}(300, 150, 60, I_p)$. The system is simulated under these conditions and the results can be seen in Fig. 4.5. The results show that the system converges within a reasonable error bound around the desired trajectory. Furthermore, the actual time is changed to $\tau_{\rm d} = 0.2$ s while the remaining parameters remain unchanged and the system is simulated once more. The results from Fig. 4.6 show that inspite of a 50 % uncertainty in the time-delay, very good performance is still possible using this approach.

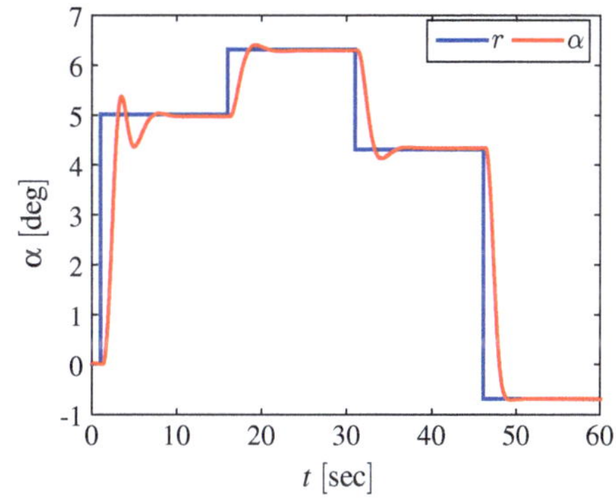

Fig. 4.5 Tracking performance of the discrete-time APC with $\tau_{\mathrm{d}} = 0.3\,\mathrm{s}$ and $\tau_{\mathrm{p}} = 0.4\,\mathrm{s}$

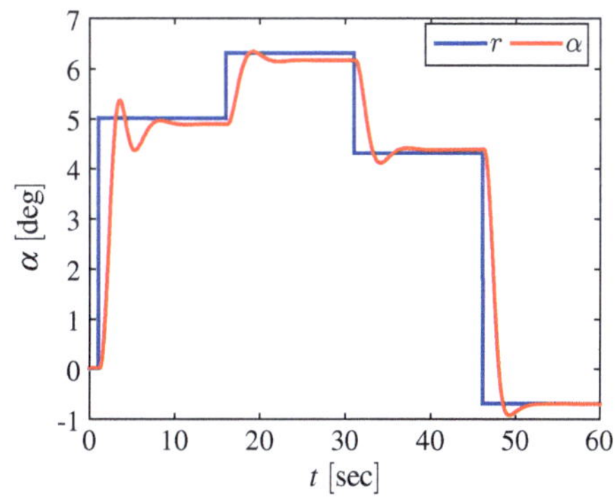

Fig. 4.6 Tracking performance of the discrete-time APC with $\tau_{\mathrm{d}} = 0.2\,\mathrm{s}$ and $\tau_{\mathrm{p}} = 0.4\,\mathrm{s}$

4.5.2 Nonlinear Systems

Consider a nonlinear discrete-time system of the form

$$\begin{bmatrix} \alpha_{k+1} \\ q_{k+1} \end{bmatrix} = \begin{bmatrix} 0.9 & 0 \\ 0.02 & 0 \\ 0 & -0.04 \\ 0 & 0.9 \end{bmatrix}^{\top} \begin{bmatrix} \alpha_k\,(sin(q_k+1)) \\ \alpha_k \left(\frac{q_k}{1+\|\alpha_k\|}\right) \\ \alpha_k \\ q_k\,(cos(\alpha_k+1)) \end{bmatrix} + \begin{bmatrix} -0.0006 \\ -0.03 \end{bmatrix} \sigma_{\mathrm{e},k-p} \tag{4.127}$$

Using the same reference model (4.126) the controller parameter is selected as $P_0 = 300 I_5$. Under a time-delay of $\tau = 0.4\,\mathrm{s}$ the system is simulated and the tracking performance is shown in Fig. 4.7. Furthermore, the estimated states are also shown in Figs. 4.8 and 4.9. As it can be seen. The proposed adaptive controller performs well for nonlinear systems with time-delay.

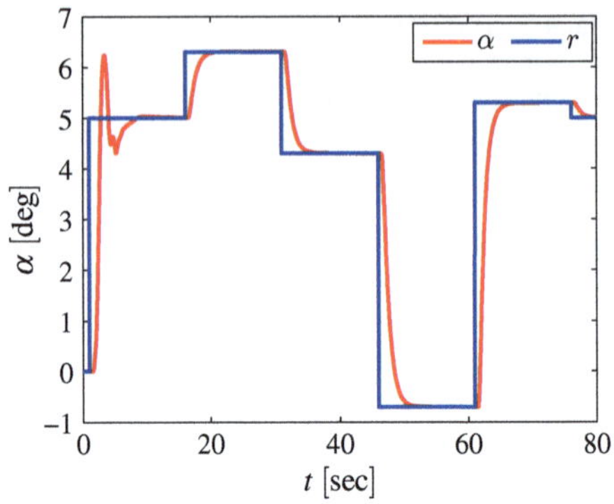

Fig. 4.7 Tracking performance of the discrete-time APC with $\tau = 0.4\,\mathrm{s}$

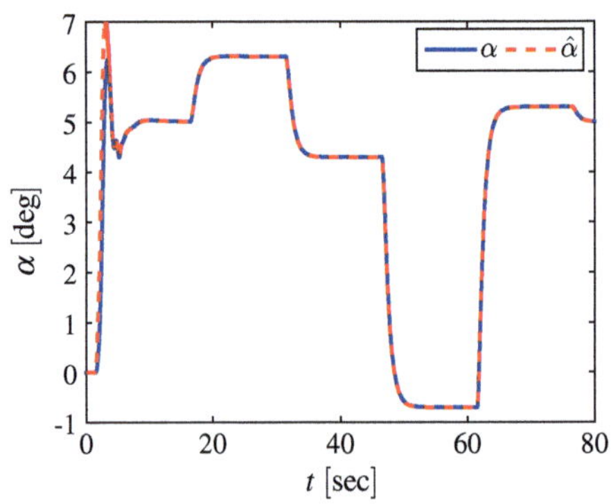

Fig. 4.8 α and its estimate $\hat{\alpha}$ with $\tau = 0.4\,\mathrm{s}$

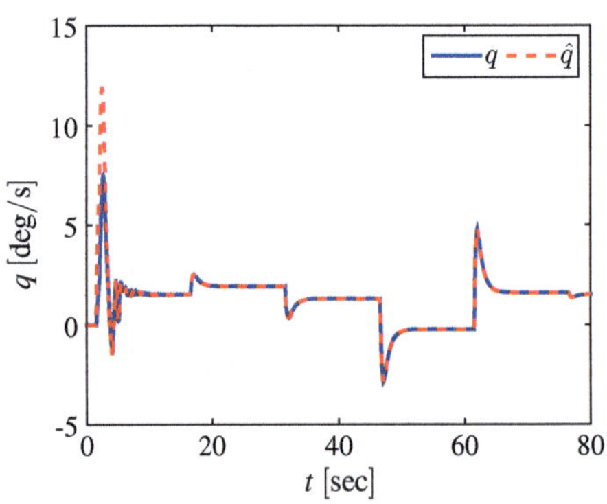

Fig. 4.9 q and its estimate $\hat{q}$ with $\tau = 0.4\,\mathrm{s}$

4.6 Conclusion

In this chapter, a discrete-time Adaptive Posicast Control (APC) method for time-delay systems has been derived. The method is extended to nonlinear systems and linear systems with unknown but upper bounded time-delay. The method is simulated and compared to a discrete-time approximation of the continuous-time APC and a

MRAC by applying each method to a flight control problem, where the short period dynamics of a jet transport aircraft were used as the system model. Further simulation results are shown for nonlinear and unknown upper bounded time-delay cases. A potential for the discrete-time APC to outperform both the continuous-time APC and the conventional MRAC has been highlighted. The stability of the closed loop system under different scenarios has been discussed.

Chapter 5
Discrete-Time Iterative Learning Control

Abstract In this study the convergence properties of iterative learning control (ILC) algorithms are discussed. The analysis is carried out in a framework using linear iterative systems, which enables several results from the theory of linear systems to be applied. This makes it possible to analyse both first-order and high-order ILC algorithms in both the time and frequency domains. The time and frequency domain results can also be tied together in a clear way. Illustrative examples are presented to support the analytical results.

5.1 Introduction

The idea of using an iterative method to compensate for a repetitive error is not new. For example, physical feats such as dance routines need to be repeated iteratively to be perfected. During each repetition, a dancers observes how he/she correctly executes the required motions correcting any errors each time. As the dancer continues to practice, the correct motion is learned and becomes ingrained into the muscle memory so that the execution of the routine is iteratively improved. The converged muscle motion profile is an open-loop control generated through repetition and learning. This type of learned open-loop control strategy is the essence of ILC.

We consider learning controllers for systems that perform the same operation repeatedly and under the same operating conditions. For such systems, a nonlearning controller yields the same tracking error on each pass. Although error signals from previous iterations are information rich, they are unused by a nonlearning controller. The objective of ILC is to improve performance by incorporating error information into the control for subsequent iterations. In doing so, high performance can be achieved with low transient tracking error despite large model uncertainties and repeating disturbances.

ILC differs from other learning-type control strategies, such as adaptive control, neural networks, and repetitive control (RC). Adaptive control strategies modify the controller, which is a system, whereas ILC modifies the control input, which is a signal, [118]. Additionally, adaptive controllers typically do not take advantage of

K. Abidi and J.-X. Xu, *Advanced Discrete-Time Control*, Studies in Systems, Decision and Control 23, DOI 10.1007/978-981-287-478-8_5

the information contained in repetitive command signals. Similarly, neural network learning involves the modification of controller parameters rather than a control signal; in this case, large networks of nonlinear neurons are modified. These large networks require extensive training data, and fast convergence may be difficult to guarantee, [86], whereas ILC usually converges adequately in just a few iterations.

ILC is perhaps most similar to RC, [79], except that RC is intended for continuous operation, whereas ILC is intended for discontinuous operation. For example, an ILC application might be to control a robot that performs a task, returns to its home position, and comes to a rest before repeating the task. On the other hand, an RC application might be the control of a conveyer system in a mass-production line moving items at periodic intervals where the next iteration immediately follows the current iteration. The difference between RC and ILC is the setting of the initial conditions for each trial, [104]. In ILC, the initial conditions are set to the same value on each trial. In RC, the initial conditions are set to the final conditions of the previous trial. The difference in initial-condition resetting leads to different analysis techniques and results, [104].

Traditionally, the focus of ILC has been on improving the performance of systems that execute a single, repeated operation. This focus includes many practical industrial systems in manufacturing, robotics, and chemical processing, where mass production on an assembly line entails repetition. ILC has been successfully applied to industrial robots [11, 36, 56, 114, 126], computer numerical control (CNC) machine tools, [94], wafer stage motion systems, [139], injection-molding machines, [63, 77], and many more.

The basic ideas of ILC can be found in a U.S. patent, [64], filed in 1967 as well as the 1978 journal publication, [153], written in Japanese. However, these ideas stayed dormant until a series of articles in 1984, [11, 42, 48, 92], sparked widespread interests in ILC. Since then, the number of publications on ILC has been growing rapidly, including two special issues, [4, 120], several books, [31, 45, 118, 166], and three surveys, [37, 83, 119].

The aim of this Chapter is to present a summary of theoretical analysis and design of ILC algorithms in the time and frequency domain. The analysis is based on the use of linear iterative systems. By incorporating the ILC algorithm and the system to be controlled into this class of systems many useful results from linear systems theory can be applied. As a consequence both first-order and high-order ILC algorithms can be analysed.

5.2 Preliminaries

In this section we shall present the problem description as well as highlight the basic differences between continuous-time ILC and sampled-data ILC.

5.2.1 Problem Formulation

Considering a tracking task that ends in a finite interval $[0, N]$ and repeats. Let the desired trajectory be $y_{m,k}$ $\forall k \in [0, N]$. Consider the ILC law

$$u_{i+1,k} = Q(\mathrm{q})[u_{i,k} + \beta L(\mathrm{q})e_{i,k+1}] \tag{5.1}$$

where $u_{i,k} \in \Re$ is the control input, $e_{i,k} = y_{m,k} - y_{i,k} \in \Re$ is the output tracking error, $Q(\mathrm{q})$ is a filter function, $L(\mathrm{q})$ is a learning function, q is the time-shift operator, $\beta > 0$ is a scalar learning gain, $i \in Z^+$ (Z^+ being the set of positive integers) denotes the iteration number, and $y_{i,k}$ is the output of the sampled-data system

$$\begin{aligned} \mathbf{x}_{i,k+1} &= \Phi \mathbf{x}_{i,k} + \Gamma u_{i,k} \\ y_{i,k} &= C\mathbf{x}_{i,k} \end{aligned} \tag{5.2}$$

where $\mathbf{x}_{i,k} \in \Re^n$ is the state vector, $\Phi \in \Re^{n\times n}$ is the state transition matrix, $\Gamma \in \Re^n$ is the gain matrix, and the initial state vector is $\mathbf{x}_{i,0}$. Define

$$\begin{aligned} \mathbf{x}^{\mathrm{d}}_{k+1} &= \Phi \mathbf{x}^{\mathrm{d}}_k + \Gamma u^{\mathrm{d}}_k \\ y^{\mathrm{d}}_k &= C\mathbf{x}^{\mathrm{d}}_k \end{aligned} \tag{5.3}$$

where $\mathbf{x}^{\mathrm{d}}_k \in \Re^n$, $y^{\mathrm{d}}_k \in \Re$ are the desired state and output trajectories and $u^{\mathrm{d}}_k \in \Re$ is the required control input to achieve those trajectories. Subtracting (5.2) from (5.3) leads to

$$\begin{aligned} \Delta\mathbf{x}_{i,k+1} &= \Phi \Delta\mathbf{x}_{i,k} + \Gamma \Delta u_{i,k} \\ e_{i,k} &= C\Delta\mathbf{x}_{i,k}. \end{aligned} \tag{5.4}$$

In all the analysis it is assumed that the identical initial condition *i.i.c*, $\mathbf{x}_{i,0} = \mathbf{x}^{\mathrm{d}}_0$ & $e_{i,0} = 0$, holds.

The control problem is to design the learning function $L(\mathrm{q})$ and filter $Q(\mathrm{q})$ such that the maximum bound of the tracking error in an iteration converges to zero asymptotically with respect to iteration, i.e.,

$$\lim_{i\to\infty}\left(\sup_{k\in(0,N)} |e_{i,k}|\right) = 0. \tag{5.5}$$

5.2.2 Difference with Continuous-Time Iterative Learning Control

In the derivation of the ILC, the question of the relative degree of the system is very important. For example, consider the nth order SISO system with a relative degree of n

$$\begin{aligned}\dot{\mathbf{x}}(t) &= A\mathbf{x}(t) + Bu(t)\\ y(t) &= C\mathbf{x}(t).\end{aligned} \tag{5.6}$$

Since the system is of relative degree n, the term $CA^{n-1}B$ is non-zero which is inferred from

$$\frac{d^n}{dt^n}y(t) = CA^n\mathbf{x}(t) + CA^{n-1}Bu(t) \tag{5.7}$$

while CB upto $CA^{n-2}B$ are 0.

Now, consider that (5.6) is sampled with time T resulting in the system

$$\begin{aligned}\mathbf{x}_{k+1} &= \Phi\mathbf{x}_k + \Gamma u_k\\ y_k &= C\mathbf{x}_k.\end{aligned} \tag{5.8}$$

Recall from Chap. 2 that the input gain for the sampled data system (5.8) can be written as

$$\begin{aligned}\Gamma &= \int_0^T e^{A\tau}Bd\tau\\ &= \int_0^T \left(B + AB\tau\frac{1}{2!}A^2B\tau^2 + \cdots + \frac{1}{(n-1)!}A^{n-1}B\tau^{n-1} + O(T^n)\right)d\tau\\ &= BT + \frac{1}{2!}ABT^2 + \cdots + \frac{1}{n!}A^{n-1}BT^n + O(T^{n+1})\end{aligned} \tag{5.9}$$

premultiply (5.9) with C results in

$$C\Gamma = \frac{1}{n!}CA^{n-1}BT^n + O(T^{n+1}). \tag{5.10}$$

This has an important implication, which is that the relative degree of the system has changed from n to 1 upon sampling. This can be seen from

$$y_{k+1} = C\Phi\mathbf{x}_k + C\Gamma u_k. \tag{5.11}$$

This result shows that in the continuous-time, the ILC would require the nth derivative of the output signal while it is not so in the sampled-data ILC. This greatly simplifies the ILC derivation for sampled-data systems irrespective of order and relative degree in continuous-time. However, note that the size of the term $C\Gamma$ depends on the sampling period and order of the system. Based on this result we can proceed to the analysis of the discrete-time ILC.

Remark 5.1 The above discussion may also be extended to LTV systems provided that the condition $C\Gamma \neq 0$ is satisfied.

5.3 General Iterative Learning Control: Time Domain

Considering (5.4) and define $\Delta\bar{\mathbf{x}}_i^\top = \left[\Delta\mathbf{x}_{i,1} \,\vdots\, \Delta\mathbf{x}_{i,2} \,\vdots\, \cdots \,\vdots\, \Delta\mathbf{x}_{i,N}\right]$ as the lifted state tracking error, $\mathbf{e}_i^\top = \left[e_{i,1} \,\vdots\, e_{i,2} \,\vdots\, \cdots \,\vdots\, e_{i,N}\right]$ as the lifted output tracking error, and $\Delta\mathbf{u}_i^\top = \left[\Delta u_{i,0} \,\vdots\, \Delta u_{i,1} \,\vdots\, \cdots \,\vdots\, \Delta u_{i,N-1}\right]$ as the lifted control input tracking error. Assuming that the *i.i.c is satisfied*, (5.4) can be written as

$$\Delta\bar{\mathbf{x}}_i = \begin{bmatrix} \Gamma & \mathbf{0} & \cdots & \mathbf{0} \\ \Phi\Gamma & \Gamma & \cdots & \mathbf{0} \\ \vdots & \vdots & \ddots & \vdots \\ \Phi^{N-1}\Gamma & \Phi^{N-2}\Gamma & \cdots & \Gamma \end{bmatrix} \Delta\mathbf{u}_i$$

$$\mathbf{e}_i = \underbrace{\begin{bmatrix} C\Gamma & 0 & \cdots & 0 \\ C\Phi\Gamma & C\Gamma & \cdots & 0 \\ \vdots & \vdots & \ddots & \vdots \\ C\Phi^{N-1}\Gamma & C\Phi^{N-2}\Gamma & \cdots & C\Gamma \end{bmatrix}}_{\mathbf{P}} \Delta\mathbf{u}_i. \tag{5.12}$$

If the rational functions $Q(\mathrm{q})$ and $L(\mathrm{q})$ are assumed causal and expanded as infinite series by dividing the denominator into its numerator, yielding

$$Q(\mathrm{q}) = q_0 + q_1\mathrm{q}^{-1} + q_2\mathrm{q}^{-2} + \cdots$$

and

$$L(\mathrm{q}) = l_0 + l_1\mathrm{q}^{-1} + l_2\mathrm{q}^{-2} + \cdots$$

respectively. In the lifted form the matrices $\mathbf{Q}$ and $\mathbf{L}$ are lower-triangular Toeplitz matrices as shown below

$$\mathbf{Q} = \begin{bmatrix} q_0 & 0 & \cdots & 0 \\ q_1 & \ddots & \ddots & \vdots \\ \vdots & \ddots & \ddots & 0 \\ q_{N-1} & \cdots & q_1 & q_0 \end{bmatrix} \quad \& \quad \mathbf{L} = \begin{bmatrix} l_0 & 0 & \cdots & 0 \\ l_1 & \ddots & \ddots & \vdots \\ \vdots & \ddots & \ddots & 0 \\ l_{N-1} & \cdots & l_1 & l_0 \end{bmatrix}. \tag{5.13}$$

Subtracting both sides of the control law (5.1) from u_k^{d} and writing the result in lifted form

$$\Delta \mathbf{u}_{i+1} = \mathbf{Q}\Delta \mathbf{u}_i - \beta \mathbf{Q}\mathbf{L}\mathbf{e}_i + (I - \mathbf{Q})\mathbf{u}^{\mathrm{d}} \tag{5.14}$$

where $\mathbf{u}^{\mathrm{d}} = \left[u_0^{\mathrm{d}} \,\vdots\, u_1^{\mathrm{d}} \,\vdots\, \cdots \,\vdots\, u_{N-1}^{\mathrm{d}} \right]^{\mathsf{T}}$. Substituting (5.12) into (5.14) gives

$$\Delta \mathbf{u}_{i+1} = \underbrace{\mathbf{Q}\,(I - \beta \mathbf{L}\mathbf{P})}_{\mathbf{F}}\,\Delta \mathbf{u}_i + (I - \mathbf{Q})\,\mathbf{u}^{\mathrm{d}}. \tag{5.15}$$

It can easily be shown that $\mathbf{F}$ has $q_0(1 - \beta l_0 C\Gamma)$ as a repeated eigenvalue. Stability of (5.15) is guaranteed if

$$|q_0(1 - \beta l_0 C\Gamma)| < 1. \tag{5.16}$$

Remark 5.2 Condition (5.16) is possible since $C\Gamma \neq 0$. This is true for sampled-data systems as was shown previously. However, depending on the order and degree of the continuous system, $C\Gamma$ can be quite small requiring a large learning gain.

Remark 5.3 Condition (5.16) is sufficient for BIBO stability, but, does not necessarily guarantee a monotonically decreasing $\|\mathbf{e}_i\|$, [127].

5.3.1 Convergence Properties

We can further investigate the convergence properties of the ILC system by looking at the matrix $\mathbf{F}$ in (5.15). Now consider (5.1) with $Q(\mathrm{q}) = L(\mathrm{q}) = 1$, (5.15) becomes

$$\Delta \mathbf{u}_{i+1} = (I - \beta \mathbf{P})\Delta \mathbf{u}_i \tag{5.17}$$

which can be expanded to the form

$$\Delta \mathbf{u}_{i+1} = \underbrace{\begin{bmatrix} (1-\beta C\Gamma) & 0 & \cdots & 0 \\ -\beta C\Phi\Gamma & (1-\beta C\Gamma) & \ddots & \vdots \\ \vdots & \vdots & \ddots & 0 \\ -\beta C\Phi^{N-1}\Gamma & -\beta C\Phi^{N-2}\Gamma & \cdots & (1-\beta C\Gamma) \end{bmatrix}}_{\mathbf{F}} \Delta \mathbf{u}_i. \tag{5.18}$$

The stability of (5.18) is now reduced to

$$|(1 - \beta C\Gamma)| < 1. \tag{5.19}$$

Assume that

$$\alpha = \sup_{1 \le k \le N-1} |\beta C \Phi^k \Gamma| \tag{5.20}$$

and let $\gamma = 1 - \beta C \Gamma$. Thus, we approximate the worst case matrix $\mathbf{F}$ as follows

$$\mathbf{F}_{\text{wc}} = \begin{bmatrix} \gamma & 0 & \cdots & 0 \\ -\alpha & \ddots & \ddots & \vdots \\ \vdots & \ddots & \ddots & 0 \\ -\alpha & \cdots & -\alpha & \gamma \end{bmatrix}. \tag{5.21}$$

Note that a general solution of (5.18) is given by

$$\Delta \mathbf{u}_{i+1} = \mathbf{F}^i \Delta \mathbf{u}_0 \tag{5.22}$$

where $\Delta \mathbf{u}_0$ is the error in control at the 0th iteration. If we use the approximation of $\mathbf{F}$ to compute $\mathbf{F}^i$ we get

$$\mathbf{F}^i_{\text{wc}} = \begin{bmatrix} \gamma & 0 & \cdots & 0 \\ -\alpha & \ddots & \ddots & \vdots \\ \vdots & \ddots & \ddots & 0 \\ -\alpha & \cdots & -\alpha & \gamma \end{bmatrix}^i = \begin{bmatrix} \gamma^i & 0 & \cdots & 0 \\ S(1) & \ddots & \ddots & \vdots \\ \vdots & \ddots & \ddots & 0 \\ S(N-1) & \cdots & S(1) & \gamma^i \end{bmatrix} \tag{5.23}$$

where

$$S(m) = \sum_{p=1}^{\min(i,m)} (-1)^p \left({}^{i}C_p\right) \left({}^{m-1}C_{p-1}\right) \gamma^{i-p} \alpha^p. \tag{5.24}$$

The binomial coefficient ${}^mC_p$ is plotted for a constant m as a function of p in Fig. 5.1. It can be seen that the coefficient attains a very large value before converging to unity and if the convergence due to α and γ is slower than that of the initial divergence of ${}^mC_p$ then $S(m)$ would also follow the characteristic of ${}^mC_p$. This overshooting phenomenon will be illustrated in later examples. From the above results it is obvious that we need to guarantee monotonic convergence. Two conditions exist that guarantee monotonic convergence. The first one is time domain based while the other is frequency domain based and will be investigated later on. According to [119], if $|1 - \beta C \Gamma| < 1$ then the monotonic convergence is guaranteed if the following condition is satisfied

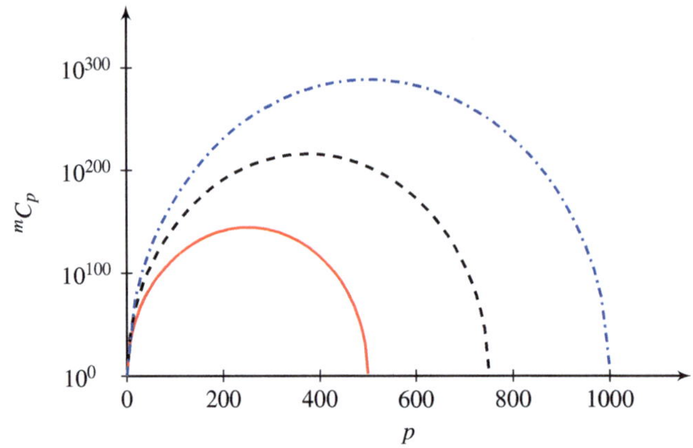

Fig. 5.1 Plot of ${}^mC_p$

$$|C\Gamma| > \sum_{j=2}^{N} |C\Phi^{j-1}\Gamma| \quad \forall \beta C\Gamma \in (0, 1)$$

$$|C\Gamma| < \frac{1}{|\beta|} - \sum_{j=2}^{N} |C\Phi^{j-1}\Gamma| \quad \forall \beta C\Gamma \in (1, 2). \tag{5.25}$$

Later it will be shown that with the proper selection of the learning function, $L(\mathrm{q})$, likelihood of satisfying the above condition is increased. Based on (5.24) and (5.25) it is worthwhile to ponder the effect of sampling period on the ILC convergence. Note that the parameter N in (5.23) and (5.25) represents the number of samples per iteration and that the lower the value of N (larger the sampling period) the lower the peak of the function ${}^mC_p$ as can be inferred from Fig. 5.1. Similarly, a lower value of N makes (5.25) more likely to be satisfied. However, this is true only for the ideal case of a stable LTI system with no disturbance. If the system is subjected to a repeatable disturbance then the sampling period must be selected such that as much as possible of the disturbance bandwidth is covered. Thus, a trade-off will exist between the selection of the sampling period and how much of the disturbance bandwidth we will need to cover.

5.3.2 D-Type and D²-Type ILC

In this section, two representative designs of the learning function, $L(\mathrm{q})$ will be considered and later on a detailed guideline will be presented from the frequency-domain perspective for the selection of appropriate learning functions. Consider the ILC

$$u_{i+1,k} = u_{i,k} + \beta L(\mathrm{q}) e_{i,k+1}. \tag{5.26}$$

If we are to consider the D-type ILC then $L(\mathrm{q})$ represents a first order difference and is given by, $L(\mathrm{q}) = 1 - \mathrm{q}^{-1}$. In the lifted form the learning function $L(\mathrm{q})$ is given by

$$\mathbf{L} = \begin{bmatrix} 1 & 0 & \cdots & \cdots & 0 \\ -1 & \ddots & \ddots & \ddots & \vdots \\ 0 & \ddots & \ddots & \ddots & \vdots \\ \vdots & \ddots & \ddots & \ddots & 0 \\ 0 & \cdots & 0 & -1 & 1 \end{bmatrix}. \tag{5.27}$$

Subtracting both sides of (5.26) from u_k^{d} and writing the result into the lifted form

$$\Delta\mathbf{u}_{i+1} = \Delta\mathbf{u}_i - \beta \begin{bmatrix} 1 & 0 & \cdots & \cdots & 0 \\ -1 & \ddots & \ddots & \ddots & \vdots \\ 0 & \ddots & \ddots & \ddots & \vdots \\ \vdots & \ddots & \ddots & \ddots & 0 \\ 0 & \cdots & 0 & -1 & 1 \end{bmatrix} \mathbf{e}_i. \tag{5.28}$$

Substituting (5.12) into (5.28) results in

$$\Delta\mathbf{u}_{i+1} = \begin{bmatrix} \gamma & 0 & \cdots & 0 \\ -\beta C\Phi^0(\Phi - I)\Gamma & \gamma & \ddots & \vdots \\ \vdots & \ddots & \ddots & 0 \\ -\beta C\Phi^{N-2}(\Phi - I)\Gamma & \cdots & -\beta C\Phi^0(\Phi - I)\Gamma & \gamma \end{bmatrix} \Delta\mathbf{u}_i \tag{5.29}$$

where $\gamma = (1 - \beta C\Gamma)$. A closer look at the matrix in (5.29) will reveal that the eigenvalues are similar to that of the matrix in (5.18). Thus, BIBO stability is guaranteed if

$$|(1 - \beta C\Gamma)| < 1. \tag{5.30}$$

Another point to note is that in (5.29) the term $(\Phi - I) \approx AT$ for small T. This indicates that in comparison to the matrix in (5.18) the non-diagonal elements in (5.29) are smaller with an order $O(T)$. If we revisit condition (5.25), for the D-type it is modified to

$$|C\Gamma| > \sum_{j=2}^{N} |C\Phi^{j-2}(I - \Phi)\Gamma|, \quad \forall \beta C\Gamma \in (0, 1)$$

$$|C\Gamma| < \frac{1}{|\beta|} - \sum_{j=2}^{N} |C\Phi^{j-2}(I - \Phi)\Gamma|, \quad \forall \beta C\Gamma \in (1, 2). \tag{5.31}$$

Since, $(\Phi - I) \approx AT$, the above conditions are more likely to be satisfied for the D-type as opposed to the P-type. If we proceed further and introduce the D^2-type where the learning function $L(\mathrm{q})$ represents a 2nd order difference given by

$$L(\mathrm{q}) = 1 - 2\mathrm{q}^{-1} + \mathrm{q}^{-2}. \tag{5.32}$$

In the lifted form the learning function $L(\mathrm{q})$ is given by

$$\mathbf{L} = \begin{bmatrix} 1 & 0 & \cdots & \cdots & \cdots & 0 \\ -2 & 1 & \ddots & \ddots & \ddots & \vdots \\ 1 & -2 & 1 & \ddots & \ddots & \vdots \\ 0 & \ddots & \ddots & \ddots & \ddots & \vdots \\ \vdots & \ddots & \ddots & \ddots & \ddots & 0 \\ 0 & \cdots & 0 & 1 & -2 & 1 \end{bmatrix}. \tag{5.33}$$

Following the same procedure as in the derivation of (5.29) we obtain

$$\Delta \mathbf{u}_{i+1} = \begin{bmatrix} \gamma & 0 & \cdots & & \cdots & 0 \\ -\beta C\Phi^0(2\Phi - I)\Gamma & \gamma & \ddots & & \ddots & \vdots \\ -\beta C\Phi^0(\Phi - I)^2\Gamma & \ddots & \ddots & & \ddots & \vdots \\ \vdots & \ddots & \ddots & & \ddots & 0 \\ -\beta C\Phi^{N-3}(\Phi - I)^2\Gamma & \cdots & -\beta C\Phi^0(\Phi - I)^2\Gamma & & -\beta C\Phi^0(2\Phi - I)\Gamma & \gamma \end{bmatrix} \Delta \mathbf{u}_i \tag{5.34}$$

where $\gamma = 1 - \beta C\Gamma$. Note that as in the case of the D-type ILC the eigenvalues are the same as that of the P-type ILC. However, most of the non-diagonal elements contain the term $(\Phi - I)^2$ which is significant since $(\Phi - I)^2 \approx (AT)^2 \approx O(T^2)$. Condition (5.25) for the D^2-type is modified to

$$|C\Gamma| < \frac{1}{|\beta|} - |C(I-2\Phi)\Gamma| - \sum_{j=3}^{N} |C\Phi^{j-3}(I-\Phi)^2\Gamma| \quad \forall \beta C\Gamma \in (1,2)$$

$$|C\Gamma| > |C(I-2\Phi)\Gamma| + \sum_{j=3}^{N} |C\Phi^{j-3}(I-\Phi)^2\Gamma| \quad \forall \beta C\Gamma \in (0,1). \tag{5.35}$$

Note that in condition (5.35) the term $(\Phi - I)^2$ is dominating. Since $(\Phi - I)^2 \approx O(T^2)$ it increases the likelihood of (5.35) being satisfied, thus, guaranteeing asymptotic convergence.

5.3.3 Effect of Time-Delay

Consider the following LTI system with control input time-delay

$$\begin{aligned} \dot{\mathbf{x}}_i(t) &= A\mathbf{x}_i(t) + Bu_i(t-\tau) \\ y_i(t) &= C\mathbf{x}_i(t) \end{aligned} \tag{5.36}$$

where τ is the time-delay. If system (5.36) is sampled with a sampling period T then it is possible to write

$$\tau = mT + T_{\mathrm{f}}$$

where $m \in Z^+$ and $0 \le T_{\mathrm{f}} < T$ such that

$$\begin{aligned} \mathbf{x}_{i,k+1} &= \Phi\mathbf{x}_{i,k} + \Gamma_1 u_{i,k-m} + \Gamma_2 u_{i,k-m-1} \\ y_{i,k} &= C\mathbf{x}_{i,k}. \end{aligned} \tag{5.37}$$

It can be easily shown that

$$\Gamma_1 = \int_{T_{\mathrm{f}}}^{T} e^{A\tau} B d\tau \quad \& \quad \Gamma_2 = \int_0^{T_{\mathrm{f}}} e^{A\tau} B d\tau. \tag{5.38}$$

Rewriting (5.37)

$$\begin{aligned} \mathbf{x}_{i,k+m+1} &= \Phi\mathbf{x}_{i,k+m} + \Gamma_1 u_{i,k} + \Gamma_2 u_{i,k-1} \\ y_{i,k+m} &= C\mathbf{x}_{i,k+m}. \end{aligned} \tag{5.39}$$

Define

$$\begin{aligned} \mathbf{x}^{\mathrm{d}}_{k+m+1} &= \Phi\mathbf{x}^{\mathrm{d}}_{k+m} + \Gamma_1 u^{\mathrm{d}}_k + \Gamma_2 u^{\mathrm{d}}_{k-1} \\ y^{\mathrm{d}}_{k+m} &= C\mathbf{x}^{\mathrm{d}}_{k+m} \end{aligned} \tag{5.40}$$

where $\mathbf{x}_k^{\rm d} \in \Re^n$, $y_k^{\rm d} \in \Re$ are the desired state and output trajectories, and $u_k^{\rm d} \in \Re$ is the control input required to achieve those trajectories. Subtracting (5.39) from (5.40) gives the following error dynamics

$$\begin{aligned}\Delta\mathbf{x}_{i,k+m+1} &= \Phi\Delta\mathbf{x}_{i,k+m} + \Gamma_1\Delta u_{i,k} + \Gamma_2\Delta u_{i,k-1}\\ e_{i,k+m} &= C\Delta\mathbf{x}_{i,k+m}.\end{aligned} \tag{5.41}$$

Similar to the previous case, define $\Delta\bar{\mathbf{x}}_i^\top = \left[\, \Delta\mathbf{x}_{i,m+1} \,\vdots\, \Delta\mathbf{x}_{i,m+2} \,\vdots \cdots \vdots\, \Delta\mathbf{x}_{i,N}\,\right]$, $\mathbf{e}_i^\top = \left[\, e_{i,m+1} \,\vdots\, e_{i,m+1} \,\vdots \cdots \vdots\, e_{i,N}\,\right]$ and $\Delta\mathbf{u}_i^\top = \left[\, \Delta u_{i,0} \,\vdots\, \Delta u_{i,1} \cdots \vdots\, \Delta u_{i,N-m-1}\,\right]$. Further, assume that the $i.i.c$ is satisfied ($\Delta\mathbf{x}_{i,m} = 0$ & $e_{i,m} = 0$), then (5.41) can be written as

$$\begin{aligned}\Delta\bar{\mathbf{x}}_i &= \begin{bmatrix} \Gamma_1 & \mathbf{0} & \cdots & \mathbf{0}\\ \Phi\Gamma_1+\Gamma_2 & \Gamma_1 & \cdots & \mathbf{0}\\ \vdots & \vdots & \ddots & \vdots\\ \Phi^{N-m-2}(\Phi\Gamma_1+\Gamma_2) & \Phi^{N-m-3}(\Phi\Gamma_1+\Gamma_2) & \cdots & \Gamma_1\end{bmatrix}\Delta\mathbf{u}_i\\ \mathbf{e}_i &= \begin{bmatrix} C\Gamma_1 & 0 & \cdots & 0\\ C\Phi\Gamma_1+C\Gamma_2 & C\Gamma_1 & \cdots & 0\\ \vdots & \vdots & \ddots & \vdots\\ C\Phi^{N-m-2}(\Phi\Gamma_1+\Gamma_2) & C\Phi^{N-m-3}(\Phi\Gamma_1+\Gamma_2) & \cdots & C\Gamma_1\end{bmatrix}\Delta\mathbf{u}_i.\end{aligned} \tag{5.42}$$

Consider the ILC law

$$u_{i+1,k} = u_{i,k} + \beta e_{i,k+m+1}. \tag{5.43}$$

Note that (5.43) has been modified for the time-delay. If both sides of (5.43) are subtracted from $u_k^{\rm d}$ and the result written in the lifted form similar to (5.12), then we get

$$\Delta\mathbf{u}_{i+1} = \Delta\mathbf{u}_i - \beta\mathbf{e}_i. \tag{5.44}$$

Substituting (5.42) into (5.44) leads to

$$\Delta\mathbf{u}_{i+1} = \begin{bmatrix} (1-\beta C\Gamma_1) & 0 & \cdots & 0\\ -\beta(C\Phi\Gamma_1+C\Gamma_2) & (1-\beta C\Gamma_1) & \cdots & 0\\ \vdots & \vdots & \ddots & \vdots\\ -\beta C\Phi^{N-m-2}(\Phi\Gamma_1+\Gamma_2) & -\beta C\Phi^{N-m-3}(\Phi\Gamma_1+\Gamma_2) & \cdots & (1-\beta C\Gamma_1)\end{bmatrix}\Delta\mathbf{u}_i. \tag{5.45}$$

Note that the matrix in (5.45) has $(1-\beta C\Gamma_1)$ as a repeated eigenvalue. BIBO stability of (5.45) requires that

$$|(1-\beta C\Gamma_1)| < 1. \tag{5.46}$$

Remark 5.4 From the above results it can be seen that for a special case of fractional time-delay, i.e. $m = 0$, (5.1) can still guarantee BIBO stability provided that condition (5.46) is satisfied. Also note that as T_{f} approaches T then Γ_1 will decrease and, therefore, the learning gain β must increase to satisfy (5.46).

5.4 General Iterative Learning Control: Frequency Domain

We have seen that based on the time-domain analysis it is only possible to guarantee BIBO stability. Based on this the system may not behave in a desirable manner. Thus, it is necessary to explore the possibility of guaranteeing monotonic convergence.

Consider the one-sided z-transformation of (5.4)

$$\begin{aligned} \Delta\mathbf{X}_i(z) &= (Iz - \Phi)^{-1}\Gamma\Delta U_i(z) \\ E_i(z) &= C\Delta\mathbf{X}_i(z) = \underbrace{C\,(Iz - \Phi)^{-1}\,\Gamma}_{P(z)}\,\Delta U_i(z). \end{aligned} \tag{5.47}$$

Obtain the z-transform of (5.1) and after subtracting both sides from $U_{\mathrm{m}}(z)$ we get

$$\Delta U_{i+1}(z) = Q(z)\left[\Delta U_i(z) - z\beta L(z)E_i(z)\right] + \left[1 - Q(z)\right]U_{\mathrm{m}}(z). \tag{5.48}$$

Substitution of (5.47) into (5.48) leads to

$$\Delta U_{i+1}(z) = \underbrace{Q(z)\left[1 - z\beta L(z)P(z)\right]}_{F(z)}\Delta U_i(z) + \left[1 - Q(z)\right]U_{\mathrm{m}}(z). \tag{5.49}$$

An important point to note is that if $F(z)$ is expanded as an infinite series

$$F(z) = f_0 + f_1z^{-1} + f_2z^{-2} + f_3z^{-3} + \cdots$$

then the first N coefficients of the series represent the first column of the Toeplitz matrix $\mathbf{F}$ in (5.15), in other words

$$\mathbf{F} = \begin{bmatrix} f_0 & 0 & \cdots & 0 \\ f_1 & \ddots & \ddots & \vdots \\ \vdots & \ddots & \ddots & 0 \\ f_{N-1} & \cdots & f_1 & f_0 \end{bmatrix}.$$

According to [127], if $F(z)$ in (5.49) is stable and causal then the condition

$$\sup_{\theta\in[-\pi,\pi]}\left|F(e^{j\theta})\right| = \sup_{\theta\in[-\pi,\pi]}\left|Q(e^{j\theta})\left[1 - e^{j\theta}\beta L(e^{j\theta})P(e^{j\theta})\right]\right| < 1 \tag{5.50}$$

where $\theta = \omega k T$, implies that the matrix norm of $\mathbf{F}$

$$\|\mathbf{F}\| < 1.$$

Remark 5.5 Condition (5.50) is more conservative than (5.19) as it implies that the norm $\|\mathbf{e}_i\|$ is monotonically decreasing and, thus, guarantees monotonic convergence.

Remark 5.6 Note that in many cases the system $P(z)$ may not be stable, however, a stable $P(z)$ is needed as a prerequisite to satisfying condition (5.50). This can be achieved using current-cycle iterative learning control.

5.4.1 Current-Cycle Iterative Learning

It was seen in the previous sections that along with condition (5.50) the system $P(z)$ must be stable in order to guarantee monotonic convergence of the ILC system. This can be achieved by including an inner closed-loop feedback to stabilize the system.

Consider once more the sampled-data system (5.2)

$$\begin{aligned} \mathbf{x}_{i,k+1} &= \Phi \mathbf{x}_{i,k} + \Gamma u_{i,k} \\ y_{i,k} &= C \mathbf{x}_{i,k} \end{aligned}$$

where it will be assumed that Φ has one or more unstable eigenvalues (or poles). The closed-loop control approach can be based on state feedback or output feedback depending on the availability of measured states. The state feedback approach is rather straight forward and will not be covered in details. Consider the closed-loop state feedback combined ILC law

$$u_{i,k} = -K \mathbf{x}_{i,k} + v_{i,k} \tag{5.51}$$

where v_i can be any of the ILC laws that were discussed up to this point. Note that the substitution of (5.51) in (5.2) results in

$$\mathbf{x}_{i,k+1} = (\Phi - \Gamma K)\mathbf{x}_{i,k} + \Gamma v_{i,k}. \tag{5.52}$$

Clearly the state feedback gain K can be designed such that the system has stable eigenvalues. From here on, all the results shown earlier apply.

Now consider if only the output measurement is available. In this case we will use an output feedback controller and the closed-loop output feedback combined ILC law is

$$u_{i,k} = G(\mathrm{q}) e_{i,k} + v_{i,k} \tag{5.53}$$

where $G(\mathrm{q})$ represents the controller function. For this analysis we will consider a general ILC. Thus, v_i is

$$v_{i,k} = Q(\mathrm{q})\left[v_{i-1}(k) + \beta L(\mathrm{q}) e_{i-1}(k+1)\right]. \tag{5.54}$$

The z-transformations of (5.53) and (5.54) are, [37],

$$U_i(z) = G(z)E_i(z) + V_i(z) \tag{5.55}$$

and

$$V_i(z) = Q(z)\left[V_{i-1}(z) + \beta z L(z) E_{i-1}(z)\right]. \tag{5.56}$$

The input-output relationship of the system in z domain is given by

$$Y_i(z) = \underbrace{C(Iz - \Phi)^{-1}\Gamma}_{P(z)} U_i(z). \tag{5.57}$$

Note that the tracking error is $e_{i,k} = y_k^{\mathrm{d}} - y_{i,k}$. Thus,

$$E_i(z) = Y_{\mathrm{m}}(z) - Y_i(z) = Y_{\mathrm{m}}(z) - P(z)U_i(z). \tag{5.58}$$

Substitution of (5.55) in (5.58) and simplifying the result we get

$$E_i(z) = \frac{1}{1+G(z)P(z)} Y_{\mathrm{m}}(z) - \frac{P(z)}{1+G(z)P(z)} V_i(z) \tag{5.59}$$

which can be rewritten as

$$-\frac{P(z)}{1+G(z)P(z)} V_i(z) = E_i(z) - \frac{1}{1+G(z)P(z)} Y_{\mathrm{m}}(z). \tag{5.60}$$

If both sides of (5.56) are multiplied by $-\frac{P(z)}{1+G(z)P(z)}$ we get

$$-\frac{P(z)}{1+G(z)P(z)} V_i(z) = -\frac{Q(z)P(z)}{1+G(z)P(z)} V_{i-1}(z) - \beta \frac{zQ(z)L(z)P(z)}{1+G(z)P(z)} E_{i-1}(z). \tag{5.61}$$

Substituting (5.60) and simplifying we get

$$E_i(z) = Q(z)\left(1 - \beta \frac{zL(z)P(z)}{1+G(z)P(z)}\right) E_{i-1}(z) + \frac{1-Q(z)}{1+G(z)P(z)} Y_{\mathrm{m}}(z). \tag{5.62}$$

Define $P_{\mathrm{CL}}(z)$ as the closed-loop transfer function

$$P_{\mathrm{CL}}(z) = \frac{P(z)}{1 + G(z)P(z)} \tag{5.63}$$

then (5.62) becomes

$$E_i(z) = Q(z)\left(1 - z\beta L(z)P_{\mathrm{CL}}(z)\right)E_{i-1}(z) + \frac{1 - Q(z)}{1 + G(z)P(z)}Y_{\mathrm{m}}(z). \tag{5.64}$$

From (5.64) we see that monotonic convergence requires that $P_{\mathrm{CL}}(z)$ be stable and the condition

$$\sup_{\theta\in[-\pi,\pi]} \left| Q(e^{j\theta})\left(1 - e^{j\theta}\beta L(e^{j\theta})P_{\mathrm{CL}}(e^{j\theta})\right)\right| < 1 \tag{5.65}$$

be satisfied. The poles of the transfer function $P_{\mathrm{CL}}(z)$ can be properly selected by designing $G(z)$ while condition (5.65) can be satisfied by the proper designs of $Q(z)$, $L(z)$ and β. However, note that the tracking error $E(z)$ is effected by $Y_{\mathrm{m}}(z)$ through $\frac{1-Q(z)}{1+G(z)P(z)}$. Thus, the designs of $G(z)$ and $Q(z)$ must also take into account that

$$\sup_{\theta\in[-\pi,\pi]} \left| \frac{1 - Q(e^{j\theta})}{1 + G(e^{j\theta})P(e^{j\theta})} \right| \ll 1. \tag{5.66}$$

5.4.2 Considerations for L(q) and Q(q) Selection

In this section we will discuss selection criteria for L(q) and Q(q) to achieve the desired monotone convergence of the ILC system.

Consider the ILC system given by (5.64). Without loss of generality we will assume that the $P(z)$ is stable and $G(z) \equiv 0$. Thus, without a feedback loop (5.64) becomes

$$E_i(z) = Q(z)\left[1 - \beta z L(z)P(z)\right]E_{i-1}(z) + \left[1 - Q(z)\right]Y_{\mathrm{m}}(z). \tag{5.67}$$

If it is assumed that $Q(z) = 1$ then, ideally, selecting $L(z) = \frac{1}{z\beta P(z)}$ would lead to the fastest possible convergence in the monotone sense. This, however, is impractical as it is not possible to identify $P(z)$ exactly for real systems. Consider the term $[1 - \beta z L(z)P(z)]$. The monotonic convergence requires that $[1 - \beta z L(z)P(z)]$ be within a unit circle centered at the origin of the complex plane. This can be restated as a requirement that $\beta z L(z)P(z)$ be within a unit circle centered at $(1, 0)$ on the complex plane as shown in Fig. 5.2. From this condition we observe that stability requires that

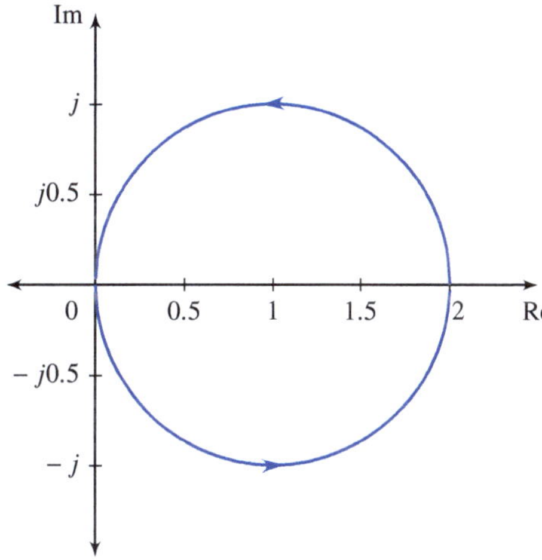

Fig. 5.2 Monotonic convergence region for $\beta z L(z) P(z)$

$$\angle(e^{j\theta} L(e^{j\theta}) P(e^{j\theta})) = \varphi \in (-\pi/2, \pi/2) \quad \forall\theta \in [-\pi, \pi]$$
$$\sup_{\theta\in[-\pi,\pi]} |\beta e^{j\theta(\varphi)} L(e^{j\theta(\varphi)}) P(e^{j\theta(\varphi)})| < 2\cos(\varphi). \tag{5.68}$$

An important fact to note is that $zL(z)$ should ensure that as $|\angle(e^{j\theta} L(e^{j\theta}) P(e^{j\theta}))| \rightarrow \pi/2$ then $|\beta e^{j\theta} L(e^{j\theta}) P(e^{j\theta})| \rightarrow 0$. On the other hand, the selection of $Q(z)$ must take into consideration that the term $[1 - Q(z)]$ be minimized and be as close as possible to zero at steady-state thereby preventing any steady state errors. Thus, $Q(z)$ is generally selected as a low pass filter. An advantage of using $Q(z)$ is that the stability region for certain frequencies can be increased if $Q(z)$ is a filter with a gain that is less than one. This can be seen from the condition shown below, [128],

$$\left|1 - \beta e^{j\theta} L(e^{j\theta}) P(e^{j\theta})\right| < \frac{1}{|Q(e^{j\theta})|} \quad \forall\theta \in [-\pi, \pi]. \tag{5.69}$$

Later on, some examples will be presented to highlight the above points.

5.4.3 *D-Type and D²-Type ILC*

Consider the ILC

$$u_{i+1,k} = Q(\mathrm{q}) \left[u_{i,k} + \beta \mathrm{q} L(\mathrm{q}) e_{i,k}\right] \tag{5.70}$$

substituting $L(\mathrm{q}) = 1 - \mathrm{q}^{-1}$ and performing the z-transform of (5.70) after subtracting both sides from u_m we obtain

$$\Delta U_{i+1}(z) = Q(z) [\Delta U_i(z) - (z-1)\beta E_i(z)] + [1 - Q(z)] U_\mathrm{m}(z). \tag{5.71}$$

Substitution of (5.47) into (5.71) leads to

$$\Delta U_{i+1}(z) = Q(z)\,[1-(z-1)\beta P(z)]\,\Delta U_i(z) + [1-Q(z)]\,U_{\mathrm{m}}(z). \tag{5.72}$$

According to the results with the P-type ILC, monotonic convergence is guaranteed if

$$\sup_{\theta\in[-\pi,\pi]} \left| Q(e^{j\theta})\left[1-\beta(e^{j\theta}-1)P(e^{j\theta})\right]\right| < 1. \tag{5.73}$$

Similarly for $L(\mathrm{q}) = 1 - 2\mathrm{q}^{-1} + \mathrm{q}^{-2}$, the stability condition would be

$$\sup_{\theta\in[-\pi,\pi]} \left| Q(e^{j\theta})\left[1-\beta(e^{j\theta}-2+e^{-j\theta})P(e^{j\theta})\right]\right| < 1. \tag{5.74}$$

In Fig. 5.3 the magnitude and phase diagrams of $zL(z)$ are plotted w.r.t to the frequency normalized by the sampling frequency $\omega_{\mathrm{s}} = \frac{2\pi}{T}$. The phase diagram indicates that for the D-type ILC at low frequency, the phase response is 90° which would violate the stability condition (5.68) if $P(z)$ has a phase greater than or equal to 0° at low frequencies. This learning function would work well if it is applied to a second order system with a single integrator or a third order system with a single integrator and a cut-off frequency near $\frac{\omega_{\mathrm{s}}}{2}$. Similarly for the D^2-type, the learning function would work ideally for a double integrator system or a third order system with a double integrator and cutoff frequency near $\frac{\omega_{\mathrm{s}}}{2}$.

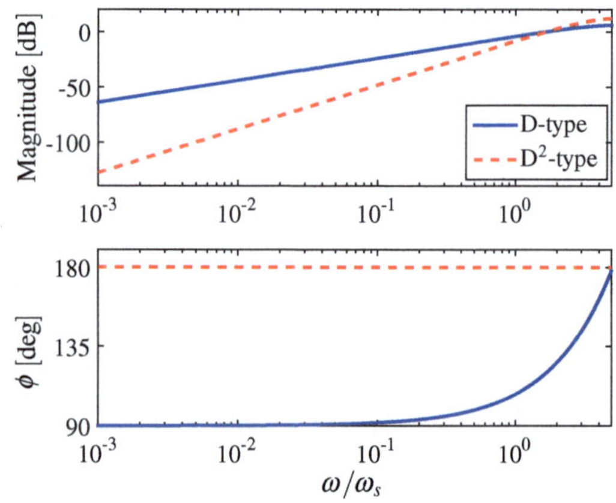

Fig. 5.3 Magnitude and Phase of $zL(z)$ for D-type and D^2-type ILC

5.5 Special Case: Combining ILC with Multirate Technique

A sampled-data system is considered a multirate system if the sampling at different locations occurs at different rates. Multirate control has been developed as an approach to achieve smoother control input and higher control bandwidth under lower sampling frequency (longer sampling periods). The basic idea is to update the control input signal at a higher frequency than the output sampling frequency [76].

Multirate control has been developed since the late 1950s. It was first developed as an analysis tool to calculate the inter-sample values of the state variables of continuous systems under discrete-time digital control. This technique was later used to design controllers for multi-variable systems with large variation of characteristic time constants in different parts, [46, 47].

5.5.1 Controller Design

This algorithm combines the multirate control technique with ILC. The multirate control technique is employed in the feedback loop so as to improve the efficiency of the measured data at relatively low sampling rates (large sampling periods). The ILC method can be applied in an off-line fashion so that both present and previous data, including tracking error and control signals, can be fully utilized. The basic idea of the design for such algorithm is to stabilize the feedback loop which contains multirate control technique and use the ILC method to make the tracking error converge in a few iterations. The controller design can be, therefore, divided into two structures: The multirate structure design and the ILC design.

5.5.2 Multirate Structure

In [46], a multirate digital controller was developed in time-domain with rigorous stability analysis. The design presented in this section follows the ideas presented in [46].

Suppose that the sampled-data representation of a system, with a sampling period of T_c is given by

$$\begin{aligned} \mathbf{x}_{k+1} &= \Phi_c \mathbf{x}_k + \gamma \Gamma_c u_k \\ y_k &= C\mathbf{x}_k \end{aligned} \tag{5.75}$$

and the same system at a different sampling period of T_s given by

$$\begin{aligned} \mathbf{x}_{k+1} &= \Phi_s \mathbf{x}_k + \gamma \Gamma_s u_k \\ y_k &= C\mathbf{x}_k \end{aligned} \tag{5.76}$$

where $\gamma \in \Re$ is the control gain and Φ_c, Γ_c, Φ_s, Γ_s and C are the parameters of the systems (5.75) and (5.76), respectively. Assume that all the parameter matrices have full rank. The two sampling periods have the relationship of $T_s = n_s T_c$, where $n_s > 1$ is the multirate index. Obviously, the frequency of the signals in system (5.76) is lower than that in (5.75). It is a general case that, in a computer controlled system, the data acquisition rate is incompatible with the computational speed. For instance, the computer can compute and operate, and thus generate control signals, at the sampling period of T_c but the data acquisition from the analog system is at the sampling period of T_s. How to synergize the different sampling rates yet improve the performance of the whole system? This is the primary goal of the multirate ILC design.

Define a new sample time index pair (k, l) to represent the time $t = kT_s + lT_c$ while $(k, 0)$ represents the time when signals from the continuous-time system are measured with $l = [0, \ldots, n_s - 1]$ being the index to indicate a finer sampling period T_c. The single index k indicates the sample instant at data acquisition rate, T_s. The state observer is implemented in open-loop to predict the state variables

$$\begin{aligned}
\hat{\mathbf{x}}_{k,l+1} &= \Phi_c \hat{\mathbf{x}}_{k,l} + \Gamma_c u_{k,l} \quad \forall l = [0, \ldots, n_s - 2] \\
\bar{\mathbf{x}}_{k+1,0} &= \Phi_c \hat{\mathbf{x}}_{k,n_s-1} + \Gamma_c u_{k,n_s-1} \\
\hat{\mathbf{x}}_{k+1,0} &= \bar{\mathbf{x}}_{k+1,0} + L_o \left[y_{k+1,0} - C\bar{\mathbf{x}}_{k+1,0} \right]
\end{aligned} \tag{5.77}$$

where, L_o is the observer gain. As the observer gain L_o in (5.77) works only at the time index $(k, 0)$, it can be obtained as if it were designed for the discrete-time state observer running at a sampling period of T_s. The observer designed in (5.77) splits the sampling period T_s into n portions and calculates the states of those n_s portions respectively. In this way, the system becomes observable at a faster sampling rate using the sampling period T_c. The multirate structure is required not to diverge within a finite time interval, i.e., the learning iteration, which is 'looser' than the stability shown in [46].

5.5.3 Iterative Learning Scheme

Generally speaking, an ILC control law can be expressed as

$$u_{i,k} = u_{i,k}^{\mathrm{fb}} + u_{i,k}^{\mathrm{ff}} \tag{5.78}$$

where, $u_{i,k}^{\mathrm{fb}}$ denotes the current cycle feedback portion, u_i^{ff} denotes the feedforward portion, and i denotes the iteration number. A previous and current cycle learning (PCCL) scheme is commonly applied to practical systems. The block diagram of the PCCL scheme is shown in Fig. 5.4 for easy understanding. In a PCCL scheme, a typical updating law may be designed as

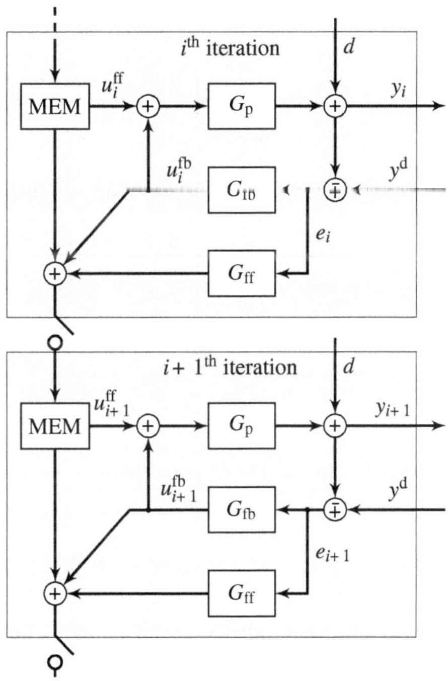

Fig. 5.4 Illustration of a previous and current cycle learning scheme (PCCL)

$$u^{\text{ff}}_{i+1,k} = u_{i,k} + G_{\text{ff}} e_{i,k} \tag{5.79}$$

$$u^{\text{fb}}_{i+1,k} = G_{\text{fb}} e_{i+1,k} \tag{5.80}$$

where (5.79) denotes the feedforward portion and (5.80) denotes the error feedback portion, $G_{\text{ff}}(z)$ is the feedforward portion gain which can be a compensator or a filter and $G_{\text{fb}}(z)$ is the error feedback portion gain which can be any controller, such as a PID or a lead compensator.

5.5.4 Convergence Condition

The controller design combines multirate technique and PCCL scheme. We need to obtain the open loop transfer function and the learning updating law before derive the convergence condition of the overall system. For a system sampled in T_{c} (as shown in (5.75)), it can be modified into

$$\begin{aligned} \mathbf{x}_{\text{e},k+1} &= \Phi_{\text{e}} \mathbf{x}_{\text{e},k} + \gamma \Gamma_{\text{e}} \mathbf{u}_{\text{e},k} + \gamma \eta_{\text{e}} \mathbf{u}_{\text{e},k+1} \\ y_k &= C_{\text{e}} \mathbf{x}_{\text{e},k} \end{aligned} \tag{5.81}$$

where $\mathbf{x}_{e,k}^{\top} = \left[\, \mathbf{x}_{k,0} \,\vdots\, \mathbf{x}_{k,1} \,\vdots\, \cdots \,\vdots\, \mathbf{x}_{k,n_s-1} \,\right]$, $\mathbf{u}_{e,k}^{\top} = \left[\, u_{k,0} \,\vdots\, u_{k,1} \,\vdots\, \cdots \,\vdots\, u_{k,n_s-1} \,\right]$ and

$$\Phi_e = \begin{bmatrix} \Phi_c^{n_s} & \mathbf{0} & \cdots & \mathbf{0} & \mathbf{0} \\ \mathbf{0} & \Phi_c^{n_s} & \cdots & \mathbf{0} & \mathbf{0} \\ \vdots & \vdots & \ddots & \vdots & \vdots \\ \mathbf{0} & \mathbf{0} & \cdots & \mathbf{0} & \Phi_c^{n_s} \end{bmatrix} \tag{5.82}$$

$$\Gamma_e = \begin{bmatrix} \Phi_c^{n_s-1}\Gamma_c & \Phi_c^{n_s-2}\Gamma_c & \cdots & \Phi_c\Gamma_c & \Gamma_c \\ \mathbf{0} & \Phi_c^{n_s-1}\Gamma_c & \cdots & \Phi_c^2\Gamma_c & \Phi_c\Gamma_c \\ \vdots & \vdots & \ddots & \vdots & \vdots \\ \mathbf{0} & \mathbf{0} & \cdots & \mathbf{0} & \Phi_c^{n_s-1}\Gamma_c \end{bmatrix} \tag{5.83}$$

$$\eta_e = \begin{bmatrix} \mathbf{0} & \mathbf{0} & \cdots & \mathbf{0} & \mathbf{0} \\ \Gamma_c & \mathbf{0} & \cdots & \mathbf{0} & \mathbf{0} \\ \vdots & \vdots & \ddots & \vdots & \vdots \\ \Phi_c^{n_s-2}\Gamma_c & \Phi_c^{n_s-3}\Gamma_c & \cdots & \Gamma_c & \mathbf{0} \end{bmatrix} \tag{5.84}$$

$$C_e = \left[\, C \; \mathbf{0} \cdots \mathbf{0} \; \mathbf{0} \,\right]. \tag{5.85}$$

Therefore, the modified system (5.81) can be considered being sampled in T_s. Similarly, the observer designed in (5.77) can be modified as

$$\hat{\mathbf{x}}_{e,k+1} = \hat{\Phi}_e \hat{\mathbf{x}}_{e,k} + \gamma \hat{\Gamma}_e \mathbf{u}_{e,k} + \gamma \eta_e \mathbf{u}_{e,k+1} + \hat{\zeta}_e \mathbf{y}_{e,k} \tag{5.86}$$

where, $\hat{\mathbf{x}}_{e,k}^{\top} = \left[\, \mathbf{x}_{k,0} \,\vdots\, \hat{\mathbf{x}}_{k,1} \,\vdots\, \cdots \,\vdots\, \hat{\mathbf{x}}_{k,n_s-1} \,\right]$, $\mathbf{y}_{e,k}^{\top} = \left[\, y_k \,\vdots\, \hat{y}_{k,1} \,\vdots\, \cdots \,\vdots\, \hat{y}_{k,n_s-1} \,\right]$ and $\hat{y}_{k,l} = C\hat{\mathbf{x}}_{n_s,l}$, for $l = [1, \ldots, n_s - 1]$, and

$$\hat{\Phi}_e = \begin{bmatrix} \Phi_c^{n_s} & \mathbf{0} & \cdots & \mathbf{0} \\ \mathbf{0} & \Phi_c \Theta \Phi_c^{n_s-2} & \cdots & \mathbf{0} \\ \vdots & \vdots & \ddots & \vdots \\ \mathbf{0} & \mathbf{0} & \cdots & \Phi_c^{n_s-1}\Theta \end{bmatrix} \tag{5.87}$$

$$\hat{\Gamma}_e = \begin{bmatrix} \Phi_c^{n_s-1}\Gamma_c & \Phi_c^{n_s-2}\Gamma_c & \cdots & \Gamma_c \\ \mathbf{0} & \Phi_c \Theta \Phi_c^{n_s-3}\Gamma_c & \cdots & \Phi_c \Omega \\ \vdots & \vdots & \ddots & \vdots \\ \mathbf{0} & \mathbf{0} & \cdots & \Phi_c^{n_s-1}\Omega \end{bmatrix} \tag{5.88}$$

$$\hat{\zeta}_e = \begin{bmatrix} L_o & \mathbf{0} & \cdots & \mathbf{0} \\ \Phi_c L_o & \mathbf{0} & \cdots & \mathbf{0} \\ \vdots & \vdots & \ddots & \vdots \\ \Phi_c^{n_s-1} L_o & \mathbf{0} & \cdots & \mathbf{0} \end{bmatrix} \tag{5.89}$$

where,

$$\Theta = \Phi_\mathrm{c} - L_\mathrm{o} C \Phi_\mathrm{c} \tag{5.90}$$

$$\Omega = \Gamma_\mathrm{c} - L_\mathrm{o} C \Gamma_\mathrm{c}. \tag{5.91}$$

According to the above definition, we have

$$\mathbf{y}_{\mathrm{e},k} = \hat{C}_\mathrm{e} \hat{\mathbf{x}}_{\mathrm{e},k} \tag{5.92}$$

where,

$$\hat{C}_\mathrm{e} = \begin{bmatrix} C & \cdots & \mathbf{0} \\ \vdots & \ddots & \vdots \\ \mathbf{0} & \cdots & C \end{bmatrix}. \tag{5.93}$$

Taking the z-transform of (5.81), (5.86) and (5.92), at the sampling rate of T_s,

$$\begin{aligned} \mathbf{x}_\mathrm{e}(z) &= (zI - \Phi_\mathrm{e})^{-1}(\gamma \Gamma_\mathrm{e} + \gamma \eta_\mathrm{e} z)\mathbf{u}_\mathrm{e}(z) \\ \hat{\mathbf{x}}_\mathrm{e}(z) &= (zI - \hat{\Phi}_\mathrm{e})^{-1}(\gamma \hat{\Gamma}_\mathrm{e} + \gamma \eta_\mathrm{e} z)\mathbf{u}_\mathrm{e}(z) + (zI - \hat{\Phi}_\mathrm{e})^{-1} \hat{\zeta}_\mathrm{e} \mathbf{y}_\mathrm{e}(z) \\ \mathbf{y}_\mathrm{e}(z) &= \hat{C}_\mathrm{e} \hat{\mathbf{x}}_\mathrm{e}(z) \end{aligned} \tag{5.94}$$

which implies

$$\begin{aligned} \mathbf{y}_\mathrm{e}(z) &= \hat{C}_\mathrm{e}(zI - \hat{\Phi}_\mathrm{e})^{-1}(\gamma \hat{\Gamma}_\mathrm{e} + \gamma \eta_\mathrm{e} z)\mathbf{u}_\mathrm{e}(z) + \hat{C}_\mathrm{e}(zI - \hat{\Phi}_\mathrm{e})^{-1} \hat{\zeta}_\mathrm{e} \mathbf{y}_\mathrm{e}(z) \\ &= (I - \hat{C}_\mathrm{e}(zI - \hat{\Phi}_\mathrm{e})^{-1} \hat{\zeta}_\mathrm{e})^{-1} \hat{C}_\mathrm{e}(zI - \hat{\Phi}_\mathrm{e})^{-1}(\gamma \hat{\Gamma}_\mathrm{e} + \gamma \eta_\mathrm{e} z)\mathbf{u}_\mathrm{e}(z) \\ &= G_p \mathbf{u}_\mathrm{e}(z) \end{aligned} \tag{5.95}$$

where the rank of $G_\mathrm{p} \in \Re^{n \times n}$ depends on the matrices $\hat{\Phi}_\mathrm{e}$, $\hat{\Gamma}_\mathrm{e}$, η_e, $\hat{\zeta}_\mathrm{e}$ and $\hat{C}_\mathrm{e}$.

If y^d is the target trajectory, define

$$\mathbf{y}^\mathrm{d}_{\mathrm{e},k} = \left[y^\mathrm{d}_{k,0} \,\vdots\, y^\mathrm{d}_{k,1} \,\vdots\, \cdots \,\vdots\, y^\mathrm{d}_{k,n_s-1} \right]^\top \tag{5.96}$$

$$\mathbf{e}_{\mathrm{e},k} = \left[e_k \,\vdots\, \hat{e}_{k,1} \,\vdots\, \cdots \,\vdots\, \hat{e}_{k,n_s-1} \right]^\top \tag{5.97}$$

where

$$e_k = y^\mathrm{d}_{k,0} - y_k \tag{5.98}$$

$$\hat{e}_{k,l} = y^\mathrm{d}_{k,l} - \hat{y}_{k,l} \quad \forall l = [1, \ldots, n_\mathrm{s} - 1]. \tag{5.99}$$

Let $\mathbf{u}_i = \mathbf{u}_\text{e}(z)$, $\mathbf{y}_i = \mathbf{y}_\text{e}(z)$, $\mathbf{e}_i$ and $\mathbf{y}_\text{d}$ be the z-transform (at sampling period T_s) of $\mathbf{u}_{\text{e},k}$, $\mathbf{y}_{\text{e},k}$, $\mathbf{e}_{\text{e},k}$ and $\mathbf{y}^\text{d}_{\text{e},k}$ at the ith iteration, respectively. We now derive the convergence condition of the designed controller. According to the updating law (5.79) and (5.80), in the modified system in T_s,

$$\mathbf{u}_{i+1} = \mathbf{u}_i + G_\text{ff}\mathbf{e}_i + G_\text{fb}\mathbf{e}_{i+1} \tag{5.100}$$

where G_ff and G_fb are appropriate feedforward and feedback learning gains. As shown in Fig. 5.4, recalling (5.95) and (5.100),

$$\mathbf{y}_{i+1} = G_\text{p}\mathbf{u}_{i+1} \tag{5.101}$$

$$\mathbf{e}_{i+1} = \mathbf{y}^\text{d} - \mathbf{y}_{i+1} \tag{5.102}$$

$$= \mathbf{y}^\text{d} - G_\text{p}[\mathbf{u}_i + G_\text{ff}\mathbf{e}_i + G_\text{fb}\mathbf{e}_{i+1}] \tag{5.103}$$

$$(I + G_\text{p}G_\text{fb})\mathbf{e}_{i+1} = (I - G_\text{p}G_\text{ff})\mathbf{e}_i \tag{5.104}$$

$$\Rightarrow \qquad \mathbf{e}_{i+1} = (I + G_\text{p}G_\text{fb})^{-1} \times (I - G_\text{p}G_\text{ff})\mathbf{e}_i. \tag{5.105}$$

Therefore, the convergence condition for the PCCL scheme is

$$\|G_\text{learn}\| = \|(I + G_\text{p}G_\text{fb})^{-1} \times (I - G_\text{p}G_\text{ff})\| \le \rho < 1. \tag{5.106}$$

A sufficient condition to satisfy (5.106), one may choose the estimator gain, the feedforward and feedback learning gains L, G_ff and G_fb such that all eigenvalues of the learning convergence matrix G_learn is strictly less than 1.

As shown in (5.79) and (5.80), both dimensions of G_ff and G_fb, are $n_\text{s} \times n_\text{s}$. There are n_s^2 degree of freedom (DOF) to design each gain matrix. Of course it is a tedious work. One simple and effective solution may be to choose

$$G_\text{ff} = g_\text{ff} I \quad \& \quad G_\text{fb} = g_\text{fb} I \tag{5.107}$$

where, g_ff and g_fb are scalars. Such G_ff and G_fb may not be the optimal solution. But appropriate g_ff and g_fb should exist to satisfy (5.106).

With this, we conclude the theoretical investigation of the various forms of ILC designs and proceed with a few numerical examples to highlight the implementation methodology of ILC.

5.6 Illustrative Example: Time Domain

5.6.1 P-Type ILC

Consider the second order system

$$\begin{aligned} \dot{\mathbf{x}}_i(t) &= A\mathbf{x}_i(t) + Bu_i(t) \\ y_i(t) &= C\mathbf{x}_i(t) \end{aligned} \tag{5.108}$$

where the system matrices are given by

$$A = \begin{bmatrix} 0 & 1 \\ 0 & -144 \end{bmatrix}, \quad B = \begin{bmatrix} 0 \\ 6 \end{bmatrix} \quad \& \quad C = \begin{bmatrix} 1 & 0 \end{bmatrix}.$$

This is the nominal model of a piezo-motor stage that will be used in the experimental application of the ILC laws. The sampled-data system representation of (5.108) is given by

$$\begin{aligned} \mathbf{x}_{i,k+1} &= \Phi\mathbf{x}_{i,k} + \Gamma u_{i,k} \\ y_{i,k} &= C\mathbf{x}_{i,k} \end{aligned} \tag{5.109}$$

where the system matrices are given by

$$\Phi = e^{AT} \quad \& \quad \Gamma = \int_0^T e^{A\tau} d\tau B \tag{5.110}$$

and T is the sampling period. If the sampling period is set to 1 ms, (5.110) becomes

$$\Phi = \begin{bmatrix} 1 & 9.313 \times 10^{-4} \\ 0 & 0.8659 \end{bmatrix} \quad \& \quad \Gamma = \begin{bmatrix} 2.861 \times 10^{-6} \\ 0.0056 \end{bmatrix}.$$

If the following ILC is used

$$\begin{aligned} u_{i+1,k} &= u_{i,k} + \beta e_{i,k+1} \\ e_{i,k} &= y_k^{\mathrm{d}} - y_{i,k} \end{aligned} \tag{5.111}$$

where the desired output trajectory is selected as $y_k^{\mathrm{d}} = 0.030 + 0.030 \sin\left(2\pi kT - \frac{\pi}{2}\right)$ and each iteration is 0.5 s in duration. The learning gain, β, is selected as 2×10^5 such that $1 - \beta C\Gamma = 0.4278$. The maximum error for each iteration is plotted in Fig. 5.5.

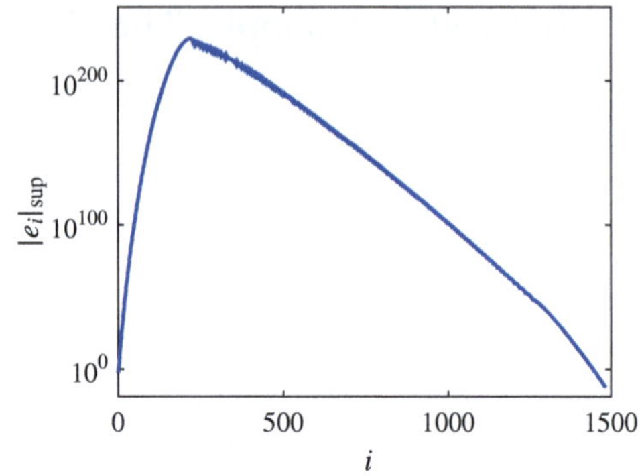

Fig. 5.5 Tracking error profile of the system using P-type ILC

5.6.2 D-Type and D^2-Type ILC

Now consider the D-type ILC

$$u_{i+1,k} = u_{i,k} + \beta(e_{i,k+1} - e_{i,k}) \tag{5.112}$$

where the learning gain, β, is selected as 2×10^5 similar to the P-type ILC case. This is because the eigenvalues of the system in the iteration domain are the same for both cases (Fig. 5.6). Thus, $1 - \beta C\Gamma = 0.4278$ for this example. The maximum error for each iteration is plotted in Fig. 5.7. In Fig. 5.8 the time-domain response is plotted. Note that the performance with the D-type ILC is similar to the P-type ILC. However, in the frequency domain analysis it will be shown that it is possible to select proper learning gain to achieve monotonic $\|\mathbf{e}_i\|$. It will also be shown that this is not possible for the P-type ILC.

Finally consider the D^2-type ILC

$$u_{i+1,k} = u_{i,k} + \beta\left(e_{i,k+1} - 2e_{i,k} + e_{i,k-1}\right) \tag{5.113}$$

where the learning gain, β, is selected as 2×10^5 similar to the P-type and D-type ILC cases (Fig. 5.9). The maximum error for each iteration is plotted in Fig. 5.10. Note that the performance with the D^2-type ILC is much better than the previous cases. Monotone convergence is achieved in this case as opposed to the P-type and D-type with the same learning gain. This is because in condition (5.35) the term $(\Phi - I)^2$ is rather small and, thus, the condition can be met easily.

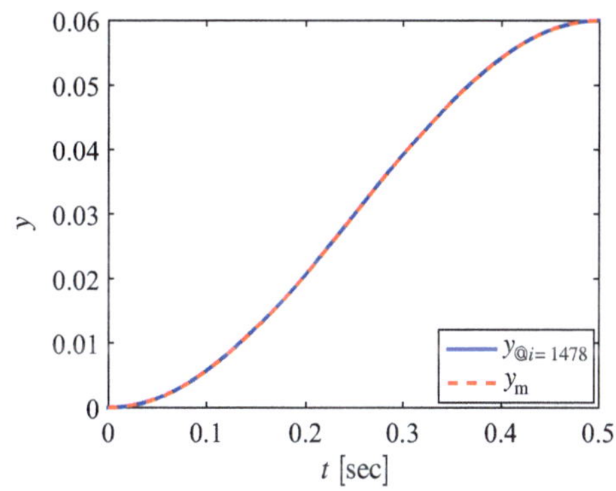

Fig. 5.6 Desired and actual output of the system using P-type ILC

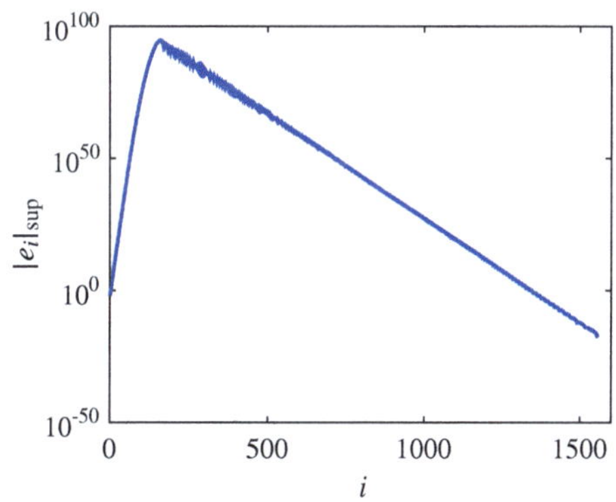

Fig. 5.7 Tracking error profile of the system using D-type ILC

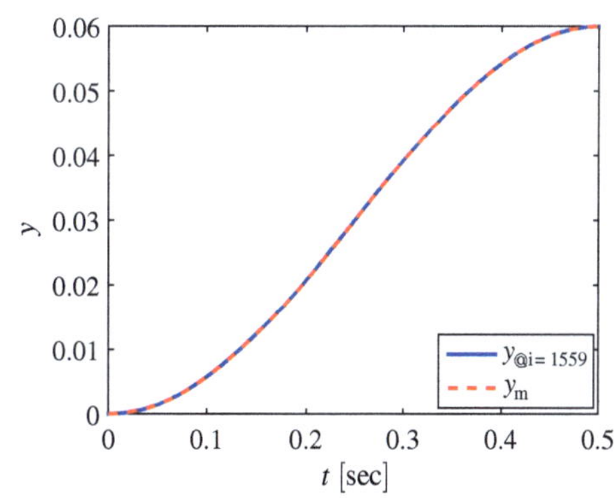

Fig. 5.8 Desired and actual output of the system using D-type ILC

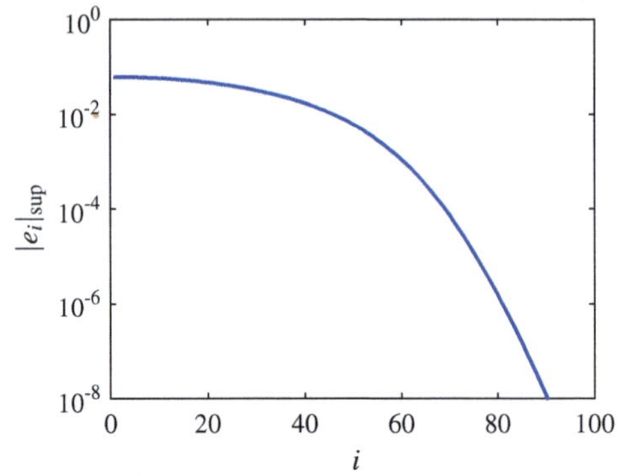

Fig. 5.9 Tracking error profile of the system using D^2-type ILC

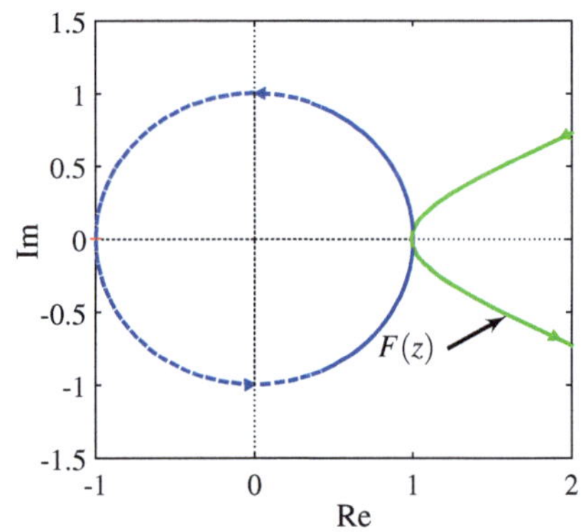

Fig. 5.10 Nyquist plot of $F(z)$ for P-type ILC

5.7 Illustrative Example: Frequency Domain

5.7.1 P-Type ILC

In order to have more insight into the example considered in the Time domain analysis, again consider the system (5.109), the input-output relationship is given by

$$Y(z) = P(z)U(z). \tag{5.114}$$

Using the same parameters as in the time-domain example, the transfer function of the system, $P(z)$, is

$$P(z) = C(Iz - \Phi)^{-1}\Gamma = \frac{2.861 \times 10^{-6}z + 2.727 \times 10^{-6}}{z^2 - 1.866z + 0.866}. \tag{5.115}$$

The filter $Q(z)$ and learning function $L(z)$ are set to unity. Thus, the Nyquist diagram of $F(z)$ is constructed in Fig. 5.10. From the figure we see that $|F(e^{j\theta})|$ does not lie inside the unit circle for any frequencies and as $\theta \to 0$ then $|F(e^{j\theta})| \to \infty$. Thus, condition (5.50) is not satisfied and monotonically decreasing $\|\mathbf{e}_i\|$ is not guaranteed.

5.7.2 D-Type and D^2-Type ILC

Consider now if the Nyquist diagram of $Q(z)\,[1 - \beta z L(z) P(z)]$ is constructed using the same parameters as the P-type ILC example while using $L(z) = 1 - z^{-1}$. From Fig. 5.11 we see that the Nyquist diagram of $Q(e^{j\theta})\left[1 - \beta(e^{j\theta} - 1)P(e^{j\theta})\right]$ lies outside the unit circle but there is a possibility of selecting learning gains β that would allow it to stay inside the unit circle. For example choosing the value of β at around 4.75×10^4 or below will lead to a Nyquist plot inside the unit circle as shown in Fig. 5.12. Now, if we go back to the time-domain analysis and use $\beta = 4.75 \times 10^4$ then the repeated eigenvalue of (5.29) is $1 - \beta C\Gamma = 0.8641 < 1$. The maximum tracking error for each iteration is plotted in Fig. 5.13 which shoes a monotonic convergence of the error in the iteration domain.

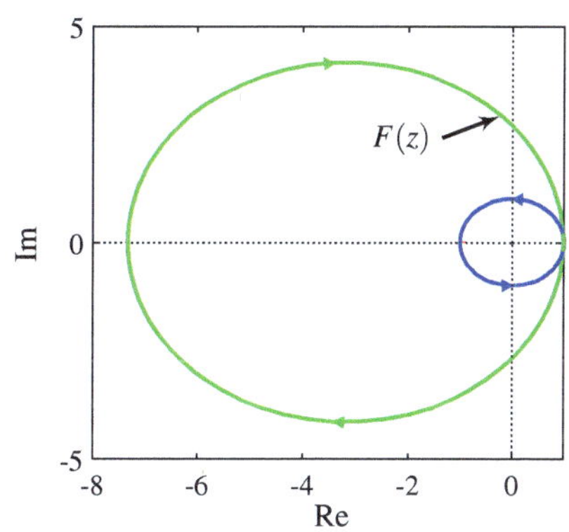

Fig. 5.11 Nyquist plot for D-type ILC example with $\beta = 2 \times 10^5$

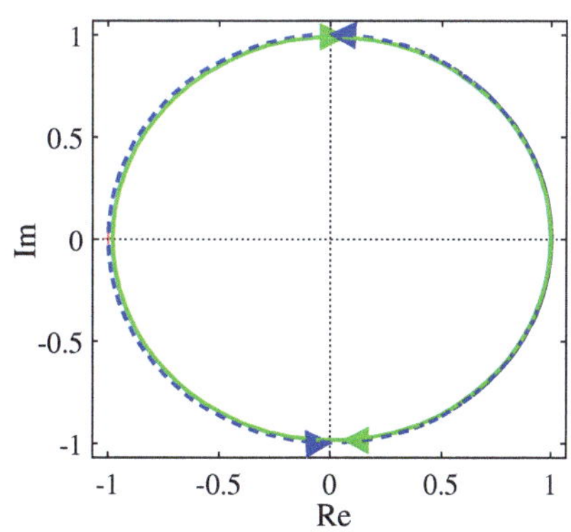

Fig. 5.12 Nyquist plot for D-type ILC example with $\beta = 4.75 \times 10^4$

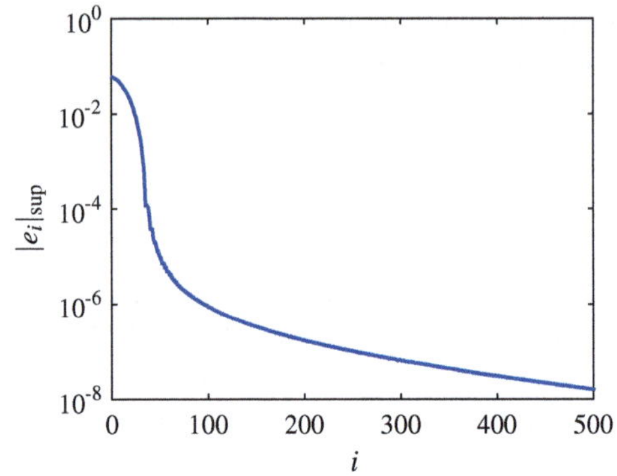

Fig. 5.13 Tracking error profile of the system using D-type ILC and $\beta = 4.75 \times 10^4$

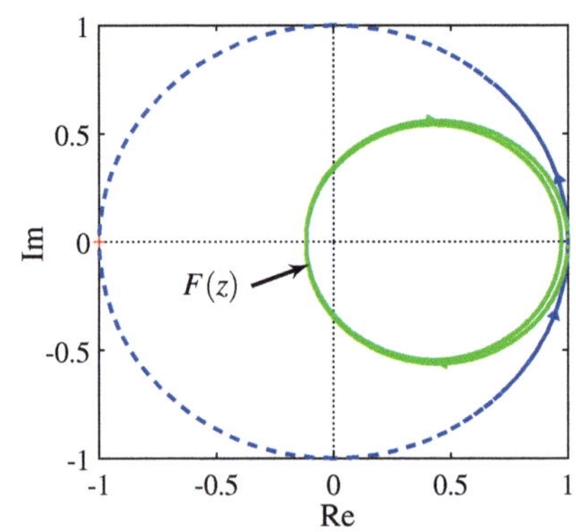

Fig. 5.14 Nyquist plot for D^2-type ILC example with $\beta = 2 \times 10^5$

If instead the D^2-type or $L(z) = 1 - 2z^{-1} + z^{-2}$ is used with the same initial learning gain for the D-type, we see that from Fig. 5.14 the Nyquist plot of $Q(z)\,[1 - \beta z L(z) P(z)]$ is within the unit circle but takes a value close to 1 at very low frequencies. This is because $(z-1)^2 \to 0$ as $\omega \to 0$. If the system was a double integrator type then this problem would not exist.

5.7.3 Current-Cycle Iterative Learning Control

We had seen in the previous P-type ILC example that monotonic convergence was not possible due to the presence of an integrator in $P(z)$ which would lead to $|zP(z)| \to \infty$ as $\omega \to 0$. Thus, we will try to eliminate the undesired pole by employing closed-loop feedback. As a start, we will check the possibility of achieving a stable closed-loop with simple proportional feedback. Thus, $G(z)$ is simply

$$G(z) = K_{\mathrm{p}}$$

where K_{p} is the proportional gain. For this we plot the root locus of $P(z)$ shown by Figs. 5.15 and 5.16. We select the proportional gain as, $K_{\mathrm{p}} = 834$ (no overshoot and

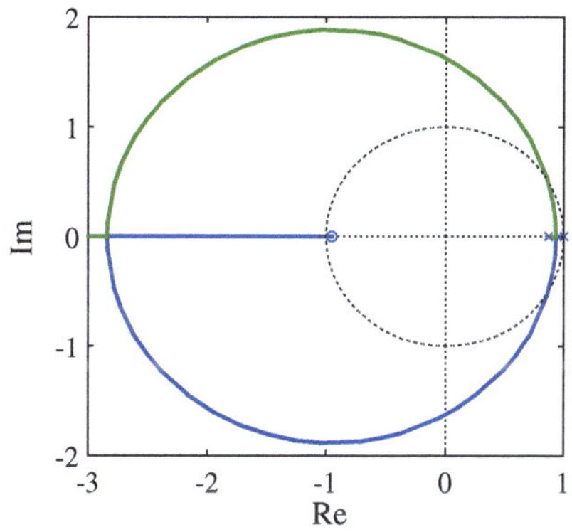

Fig. 5.15 Root locus plot for $P(z)$

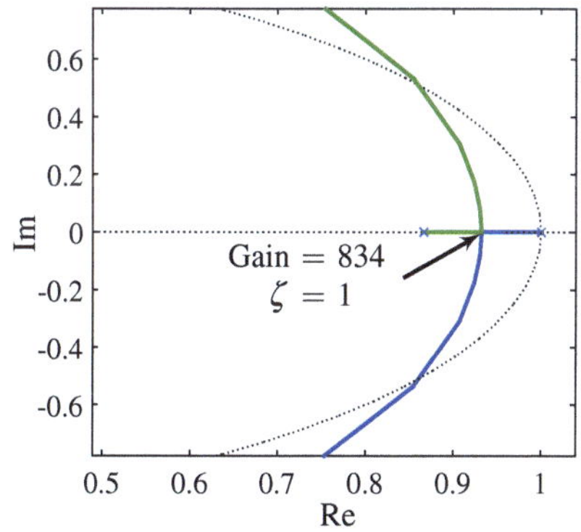

Fig. 5.16 Root locus plot for $P(z)$ (close up)

damping ratio equal to 1), and proceed to plot the Nyquist diagram of $|1 - \beta z P_{\mathrm{CL}}(z)|$ shown by Fig. 5.17. We see that the Nyquist plot is so close to the edge of the unit circle and escapes it for frequencies larger than 77 rad/s. Thus, we decide to include filtering as well in the ILC and select the learning function $L(z) = 1$ while the filter $Q(z)$ is

$$Q(z) = \frac{0.4337z^2 + 0.8675z + 0.4337}{z^2 + 0.5159z + 0.219}$$

which is a 2nd order Butterworth with a cut-off frequency of 200 rad/s. The reason for this selection is to have as simple as possible filter design while at the same time achieving a minimum $[1 - Q(z)]$ for as wide as possible range of frequencies. Now we plot the Nyquist diagram for $Q(z)[1 - \beta z L(z) P_{\mathrm{CL}}(z)]$ shown in Fig. 5.18. We see now that the Nyquist diagram is within the unit circle for all frequencies. The maximum tracking error for the system at every iteration is shown in Fig. 5.19. From Fig. 5.19 it can be seen that monotonic convergence of the tracking error is achieved. The time responses are shown in Fig. 5.20 for the system at $i = 0$ and $i = 500$ respectively. The ILC achieves better performance than with simple proportional control.

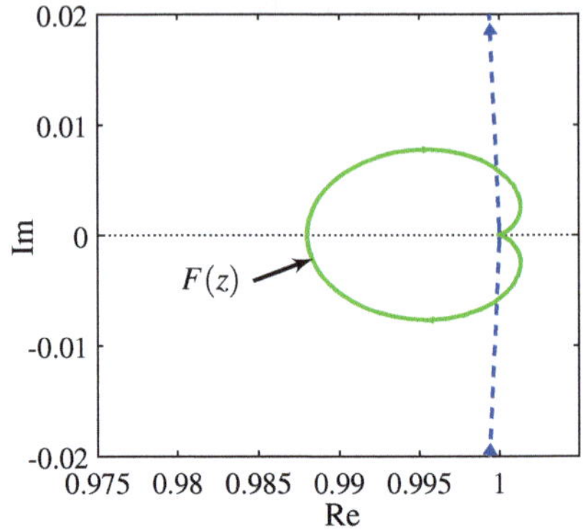

Fig. 5.17 Nyquist plot for P-type ILC with closed-loop P-control

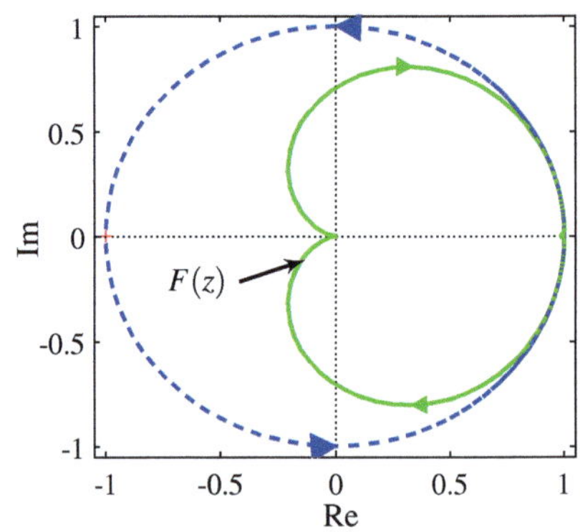

Fig. 5.18 Nyquist plot for P-type ILC with closed-loop P-control and Filtering

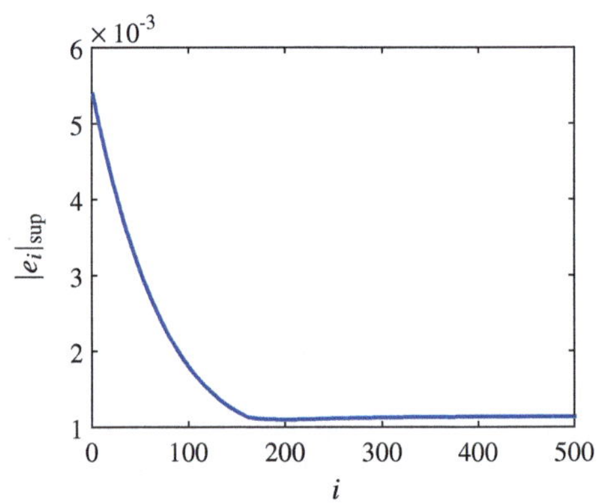

Fig. 5.19 Tracking error profile of the system using P-type ILC with closed-loop P-control and filtering

5.7.4 $L(q)$ Selection

In the previous cases the learning function, $L(\mathrm{q})$, was either selected as unity or the special case of D-type and D^2-type. In this example we will select the learning function, $L(\mathrm{q})$, in order to obtain the best possible performance of the ILC system.

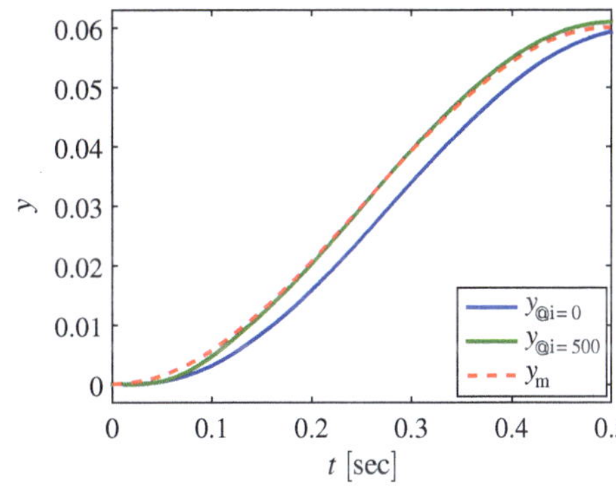

Fig. 5.20 Desired and actual output of the system using P-type ILC with closed-loop P-control and filtering

Consider the system (5.115) which is stable and has the magnitude and phase diagrams shown in Fig. 5.21. We see that the phase varies from 0° to −270°. We also note that roughly at 0.1 rad/s the phase changes to −90° and at 200 rad/s it changes from −90° to −270°. So in order to keep the overall phase between 90° and −90°, we select our learning function as a combination of two lead compensators as follows

$$L(z) = \left(\frac{5001z + 5000}{z}\right)\left(\frac{14.17z - 12.5}{z + 0.6667}\right). \tag{5.116}$$

The inverse of $L(z)$ has very similar phase characteristics as that of the system $P(z)$ as seen from Fig. 5.22. Plotting $zL(z)P(z)$ in Fig. 5.23 we see that the combination $zL(z)P(z)$ has a phase within the stability range and magnitude also in the stability range. If we plot the Nyquist plot in Fig. 5.24 for $[1 - zL(z)P(z)]$ we see that it is well within the unit circle and, thus, the condition for monotone convergence is satisfied. Figure 5.25 shows the maximum error at each iteration.

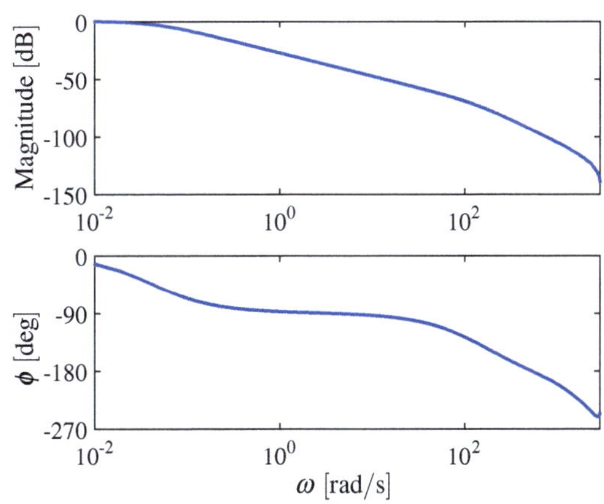

Fig. 5.21 Bode plot of $P(z)$ in (5.115)

Fig. 5.22 Bode plot of $L^{-1}(z)$

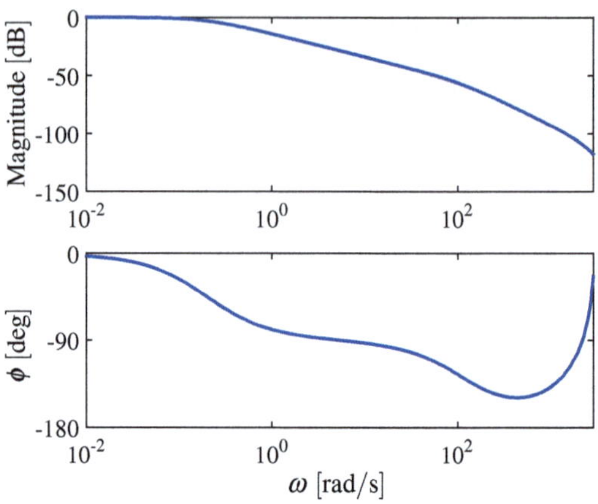

Fig. 5.23 Bode plot of $zL(z)P(z)$

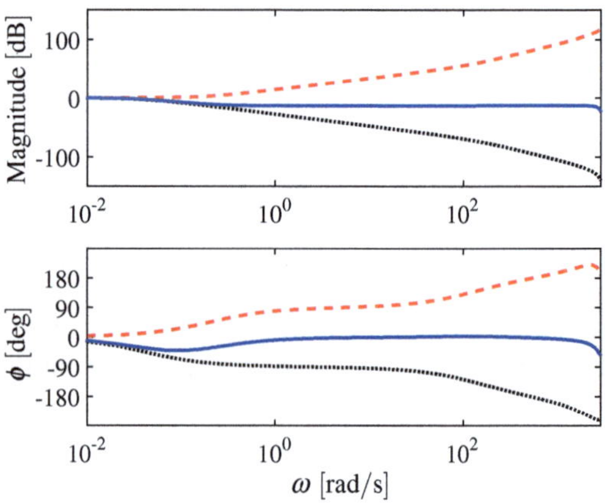

5.7.5 Sampling Period Selection

Consider the system $P(z)$ defined by (5.115) at sampling period $T = 1$ ms. From Fig. 5.21 we see that the phase response crosses $-90°$ at nearly 2π rad/s. If we also look at the phase diagram of z we see that it is linearly increasing from 0° to 180° as a function of frequency (Fig. 5.26). From here it seems obvious that if we select a larger sampling period such that the phase response of $P(z)$ slightly crosses the $(-90°, 90°)$ stability bound, then combined with z the overall phase response, φ, would be within $(-90°, 90°)$. Thus, we select sampling period $T = 10$ ms and draw the magnitude and phase of $zP(z)$ in Fig. 5.27. We see from Fig. 5.27 that the overall phase response of $zP(z)$ still crosses the stability bound $(-90°, 90°)$, hence, we will increase the sampling period to $T = 15$ ms and redraw the magnitude and phase of $zP(z)$ in Fig. 5.27. We see from Fig. 5.27 that with the new sampling period, the phase response of $zP(z)$ is now within the stability bound $(-90°, 90°)$ and since for all the cases the magnitude of $zP(z)$ was within the stability bound, the system

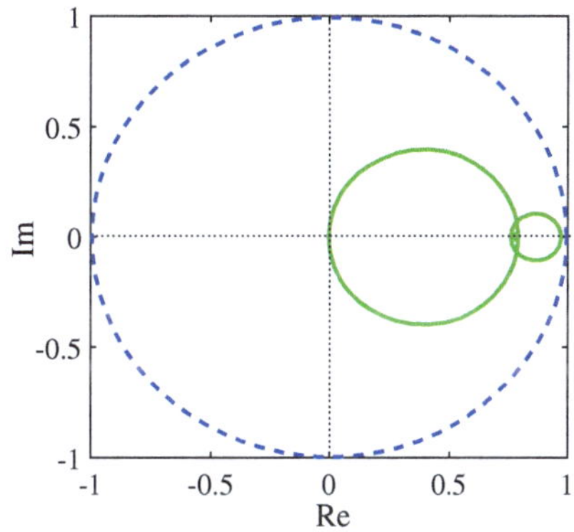

Fig. 5.24 Nyquist plot of $1 - zL(z)P(z)$

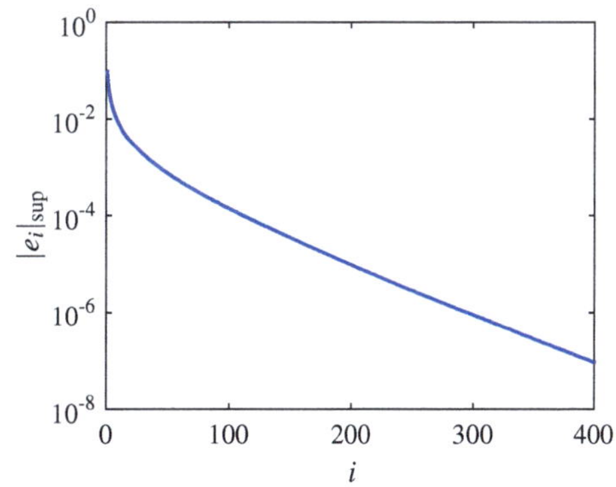

Fig. 5.25 Tracking error profile of the system using P-type ILC with closed-loop P-control and filtering

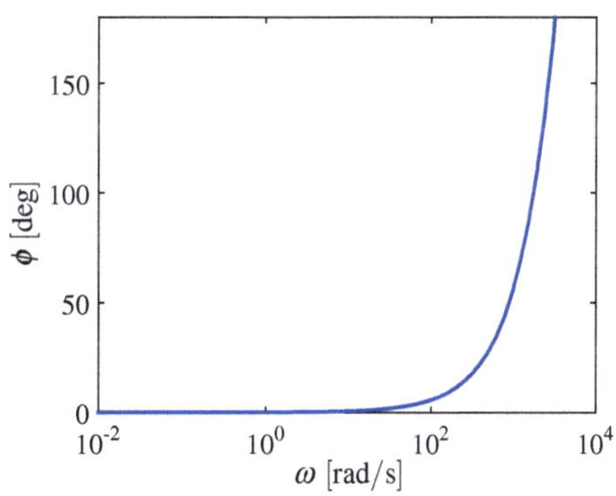

Fig. 5.26 Phase diagram of z

can now achieve monotone convergence of the tracking error. To conclude we can tabulate all the results in the form of a guideline to help with the ILC design. This can be seen in Table 5.1.

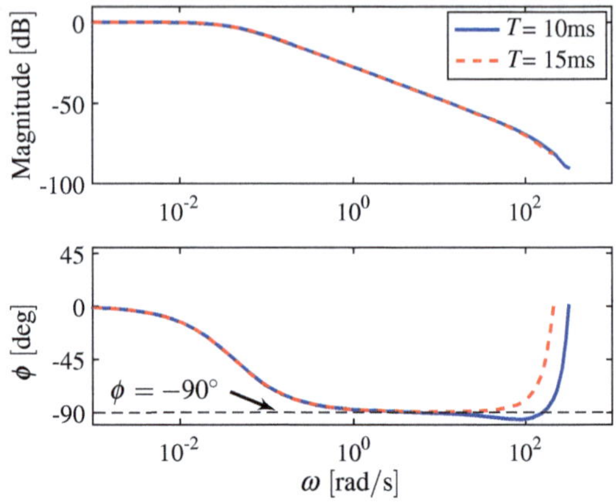

Fig. 5.27 Bode plot of $zP(z)$

Table 5.1 Guideline for ILC design

Design factor		Design considerations
Sampling Period T		Selection of a larger sampling period increases the chances of achieving monotone convergence, however, the trade-off would be that the system bandwidth does not cover the whole range of disturbances and uncertainties that may exist and, therefore, incur large tracking errors
Q(z)		The filter $Q(z)$ increases the stability bound, however, it would create steady state errors and, therefore, should only be used if stability cannot be achieved by $L(z)$ alone. $Q(z)$ is typically selected as a low-pass filter
$L(z)$	P-type	P-type is suitable if $P(z)$ has a phase within $(-90°, 90°)$
	D-type	D-type is suitable if $P(z)$ is 2nd order with a single integrator
	D^2-type	D^2-type is suitable for either a 2nd order or a 3rd order $P(z)$ with at most two integrators
	Filter	For cases where $P(z)$ is of high order or does not satisfy the above conditions then $L(z)$ can be designed as a combination of lead compensators depending on the order of $P(z)$

5.8 Conclusion

This work summarizes the theoretical results of ILC for sampled-data SISO systems in the time and frequency domain. Stability and convergence criterias are shown as well as design procedures with numerous examples. Finally, the discussed design procedure is applied to a real system with promising results.

Chapter 6
Discrete-Time Fuzzy PID Control

Abstract In this study, a parallel structure of fuzzy PID control systems is presented. It is associated with a new tuning method which, based on gain margin and phase margin specifications, determines the parameters of the fuzzy PID controller. In comparison with conventional PID controllers, the presented fuzzy PID controller shows higher control gains when system states are away from equilibrium and, at the same time, retains lower profile of control signals. Consequently better control performance is achieved. With the presented formula, the weighting factors of a fuzzy logic controller can be systematically selected according to the plant under control. By virtue of using the simplest structure of fuzzy logic control, the stability of the nonlinear control system is able to be analyzed and a sufficient BIBO stability condition is given. The superior performance of the controller is demonstrated through both numerical and experimental examples.

6.1 Introduction

In recent years, fuzzy logic controllers (FLC), especially fuzzy proportional-integral-derivative (PID) controllers have been widely used for industrial processes owing to their heuristic nature associated with simplicity and effectiveness for both linear and nonlinear systems. In fact, for single-input single-output systems, most of fuzzy logic controllers are essentially the fuzzy PD type, fuzzy PI type or fuzzy PID type associated with nonlinear gains. Because of the nonlinear property of control gains, fuzzy PID controllers possess the potential to improve and achieve better system performance over the conventional PID controller if the nonlinearity can be suitably utilized. On the other hand, due to the existence of nonlinearity, it is usually difficult to conduct theoretical analyses to explain why fuzzy PID can achieve better performance. Consequently it is important, from both the theoretical and practical point of view, to explore the essential nonlinear control properties of fuzzy PID and find out appropriate design methods which will assist the control engineers to confidently utilize the nonlinearity of the fuzzy PID controllers to improve the closed-loop performance.

K. Abidi and J.-X. Xu, *Advanced Discrete-Time Control*, Studies in Systems, Decision and Control 23, DOI 10.1007/978-981-287-478-8_6

To fulfill the above target, we need to answer following three fundamental questions: (1) what is the suitable structure for fuzzy PID controllers; (2) how to systematically tune the fuzzy PID controller parameters; and (3) how to analyze and evaluate the designed fuzzy PID controllers.

To answer the first question, let us investigate existing fuzzy PID controllers. Although various types of fuzzy PID (including PI and PD) controllers have been proposed, [44, 108, 137, 162, 176, 180], they can be classified into two major categories according to the way of construction. One category of fuzzy PID controllers is composed of the conventional PID control system in conjunction with a set of fuzzy rules (knowledge base) and a fuzzy reasoning mechanism. The PID gains are tuned on-line in terms of the knowledge base and fuzzy inference, and then the PID controller generates the control signal. The other category of fuzzy PID controllers is a typical fuzzy logic controller constructed as a set of heuristic control rules, hence the control signal is directly deduced from the knowledge base and the fuzzy inference. They are referred as fuzzy PID controllers because, from the viewpoint of input-output relationship, their structure is analogous to that of the conventional PID controller. Once the structure is fixed, the nonlinear property of the fuzzy PID controller is uniquely determined.

In this chapter we propose a new type of the fuzzy PID controller of the second category. It has the simplest structure: Only two fuzzy labels are used for each fuzzy input variable and three fuzzy labels for the fuzzy control output variable. The considerations behind our selection are as follows. First, from the practical point of view, it seems that the heuristic knowledge of the second category is more analogous to that of human operator or expert, therefore is easier to be acquired. Second, owing to the similarity of the input-output relationship between the fuzzy and conventional PID controllers, it is possible for us to borrow conventional PID tuning methods to decide the fuzzy PID controller parameters. Third, with the simplest structure the dynamics of the fuzzy PID, [44, 108, 176], it is convenient for us to conduct more theoretical analysis and evaluation.

It should be pointed out that, for fuzzy PID controllers, normally a 3-D rule base is required. It is difficult to obtain since 3-D information is usually beyond the sensing capability of a human expert. To obtain all of proportional, integral and derivative control action, it is intuitive and convenient to combine PI and PD type fuzzy logic controllers together to form a fuzzy PID controller. Therefore, in the proposed fuzzy control system there are only four control rules for the fuzzy PI and fuzzy PD control channels respectively and the two channels are combined in parallel.

After determining the structure, we are ready to answer the second and the third questions. There are two ways of determining fuzzy logic controller parameters, depending on how much is known about the process under control. Without knowing the process dynamics or its approximation, the FLC parameters can only be tuned through trial and error. On the other hand, it is well known that for most industrial control problems, the effective tuning of conventional PID controllers is based on estimating and approximating the process dynamics, whether linear or nonlinear, by a simple linear first or second order system with time-delay. There exist many tuning or auto-tuning algorithms such as Ziegler-Nichols tuning formula, internal model control tuning method, [15, 19, 75], etc. Because a fuzzy controller of the second category is essentially a PD type, PI type or PID type controller with nonlinear gains, it is possible to borrow the standard estimation method, for instance the relay test, and tuning methods of a conventional PID controller to design the fuzzy PID controller. Gain margin and phase margin are important measures of the closed-loop system characteristics and they offer a convenient means for designing control systems. The auto-tuning method for PI/PID controllers to satisfy a pair of specified gain margin and phase margin has been proven to be effective, [75, 80]. In this chapter, we introduce a tuning method, which is similarly based on gain margin and phase margin, to determine the parameters of the proposed fuzzy PID controller. The auto-tuning formula is applied here to decide the fuzzy PID parameters (the weighting coefficients for error, the change of error and controller output) with respect to a selected equivalent gain/phase margin contour on which both fuzzy and conventional PID controllers possess the same gain and phase margins. This ensures the required gain and phase margins, which give the local stability, of the fuzzy PID controller on the selected contour. The fine-tuning of the error coefficient, which does not affect the given gain and phase margins, is based on heuristic knowledge.

As for the analysis and evaluation of the fuzzy PID controllers, we will focus on an important issue: the stability problem. Although fuzzy logic controllers have been adopted in many engineering applications, their performance is not guaranteed since there is lack of stability analysis. Notice that the concept of local stability based on the gain and phase margins is essentially from the linear control systems. Therefore more general stability analysis methods which can incorporate the nonlinear nature of the fuzzy PID controller are preferred. In this chapter, the well known small gain theorem is employed to evaluate the bounded-input/bounded-output stability condition of the proposed fuzzy PID control system. Through analysis we will show another important property of the new fuzzy PID controller: It possesses higher control gains but yields less control efforts than the conventional PID controllers.

6.2 Design of Fuzzy PID Control System

6.2.1 Fuzzy PID Controller with Parallel Structure

Usually a fuzzy controller is either a PD or a PI type depending on the output of fuzzy control rules. A fuzzy PID controller may be constructed by introducing the third information besides error and change in error, which is either rate of change in error or sum of error, with a 3-D rule base. Such a fuzzy PID controller with a 3-D rule base is difficult to construct because: (1) for the case of using rate of change in error, a human expert can hardly sense the third dimension of information, for instance, the acceleration besides position and velocity in a motion control system, and thus it is difficult to obtain the control rules; (2) for the case of using sum of error, it is difficult to quantize its linguistic value since a different system needs different integral gain

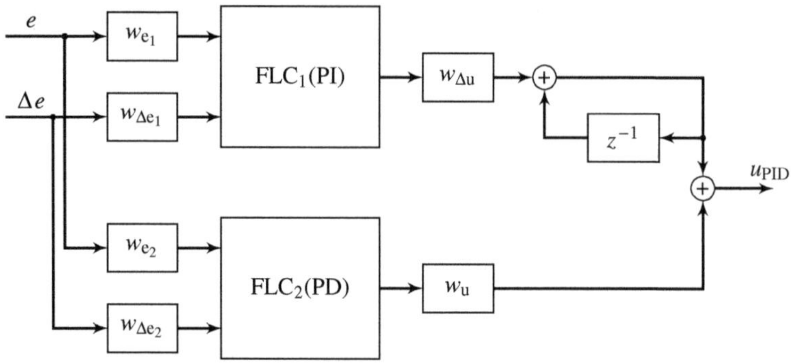

Fig. 6.1 The overall structure of fuzzy PID controller

and steady state value of sum of error; (3) a 3-D rule base can be very complex when the number of quantizations of each dimension increases; in this situation, the control rule number increases cubically with the number of quantizations.

In this chapter, we propose a parallel combination of a fuzzy PI controller and a fuzzy PD controller to achieve a fuzzy PID controller. The overall structure is shown in Fig. 6.1. Simplest structures are used in each FLC. There are only two fuzzy labels (Positive and Negative) used for the fuzzy input variables and three fuzzy labels (Positive, Zero and Negative) for the fuzzy output variable. There are two main reasons which motivate us to choose this type of FLCs: (1) theoretical analysis is possible owing to the simplicity and (2) the nonlinearity of the simplest fuzzy controller is the strongest, [38]. Therefore, we can expect better control performance from this simplest structure controller as long as we can correctly use its nonlinear property.

First of all, the error and the change of error are defined as:

$$\begin{aligned} e_k &= r_k - y_k \\ \Delta e_k &= e_k - e_{k-1}. \end{aligned} \tag{6.1}$$

The inputs of the fuzzy controller are normalized error ($w_{e*}e_k$) and normalized change of error ($w_{\Delta e*}\Delta e_k$) where w_{e*} and $w_{\Delta e*}$ are weighting factors. The notation $*$ ($* = \{1, 2\}$) denotes different types of FLCs. The membership functions $\mu(\bullet)$ of fuzzified inputs are defined in Fig. 6.2. According to this kind of triangular shape membership functions, there are four fuzzy labels P_e, $P_{\Delta e}$, N_e and $N_{\Delta e}$ for the two fuzzy input variables and the corresponding membership functions are described as:

$$\mu_{P_{e*}} = \begin{cases} 0 & \text{if } w_{e*} \cdot e < -1 \\ \frac{1}{2} + \frac{1}{2} w_{e*} \cdot e & \text{if } -1 \le w_{e*} \cdot e < 1 \\ 1 & \text{if } w_{e*} \cdot e \ge 1 \end{cases} \tag{6.2}$$

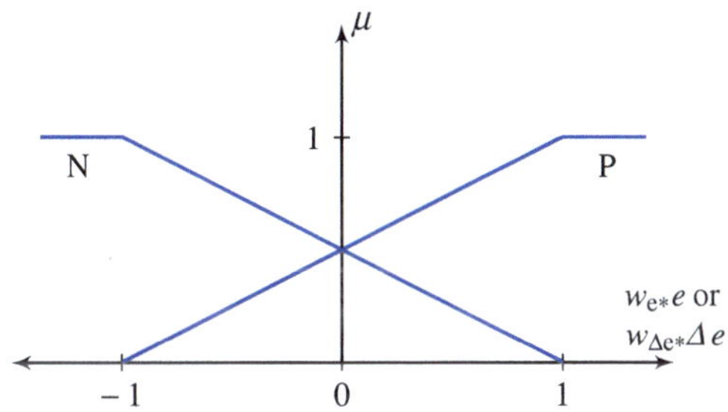

Fig. 6.2 Membership functions of e_k and Δe_k

$$\mu_{N_{e*}} = \begin{cases} 1 & \text{if } w_{e*} \cdot e < -1 \\ \frac{1}{2} - \frac{1}{2} w_{e*} \cdot e & \text{if } -1 \le w_{e*} \cdot e < 1 \\ 0 & \text{if } w_{e*} \cdot e \ge 1 \end{cases} \tag{6.3}$$

$$\mu_{P_{\Delta e*}} = \begin{cases} 0 & \text{if } w_{\Delta e*} \cdot \Delta e < -1 \\ \frac{1}{2} + \frac{1}{2} w_{\Delta e*} \cdot \Delta e & \text{if } -1 \le w_{\Delta e*} \cdot \Delta e < 1 \\ 1 & \text{if } w_{\Delta e*} \cdot \Delta e \le 1 \end{cases} \tag{6.4}$$

$$\mu_{N_{\Delta e*}} = \begin{cases} 1 & \text{if } w_{\Delta e*} \cdot \Delta e < -1 \\ \frac{1}{2} - \frac{1}{2} w_{\Delta e*} \cdot \Delta e & \text{if } -1 \le w_{\Delta e*} \cdot \Delta e < 1 \\ 0 & \text{if } w_{\Delta e*} \cdot \Delta e \ge 1. \end{cases} \tag{6.5}$$

Consequently there are only four simple fuzzy control rules used in each FLC (see Table 6.1). The fuzzy labels of control outputs are singletons defined as $P = 1, Z = 0$ and $N = -1$. By using Larsen's product inference method with Zadeh fuzzy logic AND and Lukasiewicz fuzzy logic OR, using the center of gravity defuzzification method, and for simplicity choosing $w_{e_1} = w_{e_2} = w_e$ and $w_{\Delta e_1} = w_{\Delta e_2} = w_{\Delta e}$, the control output of each FLC can be obtained, inside the universe of discourse, [176], as

$$\Delta u_1^{(F)} = \frac{w_{\Delta u_1}}{4 - 2\max\left(w_{e_1}|e|, w_{\Delta e_1}|\Delta e|\right)} \left(w_{e_1} e + w_{\Delta e_1} \Delta e\right) = \frac{w_{\Delta u}}{4 - 2\alpha} \left(w_e e + w_{\Delta e} \Delta e\right) \tag{6.6}$$

$$u_2^{(F)} = \frac{w_{u_2}}{4 - 2\max\left(w_{e_2}|e|, w_{\Delta e_2}|\Delta e|\right)} \left(w_{e_2} e + w_{\Delta e_2} \Delta e\right) = \frac{w_u}{4 - 2\alpha} \left(w_e e + w_{\Delta e} \Delta e\right) \tag{6.7}$$

where

$$\alpha = \max(w_{e_1}|e|, w_{\Delta e_1}|\Delta e|) = \max(w_{e_2}|e|, w_{\Delta e_2}|\Delta e|) = \max(w_e|e|, w_{\Delta e}|\Delta e|).$$

Table 6.1 Fuzzy control rules (FLC_1 and FLC_2)

PI part FLC_1	Rule 1	If error is N and change of error is N, change in control action is N
	Rule 2	If error is N and change of error is P, change in control action is Z
	Rule 3	If error is P and change of error is N, change in control action is Z
	Rule 4	If error is P and change of error is P, change in control action is P
PD part FLC_2	Rule 1	If error is N and change of error is N, control action is N
	Rule 2	If error is N and change of error is P, control action is Z
	Rule 3	If error is P and change of error is N, control action is Z
	Rule 4	If error is P and change of error is P, control action is P

N negative; P positive; Z zero

The overall fuzzy control output will be given as

$$
\begin{aligned}
u_{\mathrm{PID}}^{(\mathrm{F})} &= \sum_{0}^{k} \Delta u_1^{(\mathrm{F})} + u_2^{(\mathrm{F})} \\
&= \sum_{0}^{k} \frac{w_{\Delta\mathrm{u}} w_{\Delta\mathrm{e}}}{4-2\alpha}\left(\Delta e + \frac{\Delta t}{\frac{w_{\Delta\mathrm{e}}}{w_{\mathrm{e}}}\Delta t} e\right) + \frac{w_{\mathrm{u}} w_{\mathrm{e}}}{4-2\alpha}\left(e + \frac{w_{\Delta\mathrm{e}}\Delta t}{w_{\mathrm{e}}}\frac{\Delta e}{\Delta t}\right). \qquad (6.8)
\end{aligned}
$$

If we choose

$$
\begin{cases}
K_{\mathrm{c}}^{(\mathrm{F})} &= \dfrac{w_{\Delta\mathrm{u}} w_{\Delta\mathrm{e}}}{4-2\alpha} \\
T_{\mathrm{i}}^{(\mathrm{F})} &= \dfrac{w_{\Delta\mathrm{e}}}{w_{\mathrm{e}}}\Delta t \\
K_{\mathrm{c}}^{(\mathrm{F})} \dfrac{T_d^{(\mathrm{F})}}{T_{\mathrm{i}}^{(\mathrm{F})}} &= \dfrac{w_{\mathrm{u}} w_{\mathrm{e}}}{4-2\alpha}
\end{cases} \qquad (6.9)
$$

then the fuzzy control output in (6.8) can be rewritten as

$$
u_{\mathrm{PID}}^{(\mathrm{F})} = \sum_{0}^{k} K_{\mathrm{c}}^{(\mathrm{F})}\left(\Delta e + \frac{\Delta t}{T_{\mathrm{i}}^{(\mathrm{F})}} e\right) + K_{\mathrm{c}}^{(\mathrm{F})} \frac{T_{\mathrm{d}}^{(\mathrm{F})}}{T_{\mathrm{i}}^{(\mathrm{F})}}\left(e + T_{\mathrm{i}}^{(\mathrm{F})}\frac{\Delta e}{\Delta t}\right). \qquad (6.10)
$$

Now assume that the time constants of the system are sufficiently large compared with the sampling period, which is common and reasonable in process control, such that

$$
\dot{e} \approx \frac{\Delta e}{\Delta t}
$$

then the overall control output can be approximated as

$$
\begin{aligned}
u_{\mathrm{PID}}^{(\mathrm{F})} &\approx \int_0^{k\cdot\Delta t} K_{\mathrm{c}}^{(\mathrm{F})}\left(de + \frac{e}{T_{\mathrm{i}}^{(\mathrm{F})}}dt\right) + K_{\mathrm{c}}^{(\mathrm{F})}\frac{T_{\mathrm{d}}^{(\mathrm{F})}}{T_{\mathrm{i}}^{(\mathrm{F})}}\left(e + T_{\mathrm{i}}^{(\mathrm{F})}\frac{de}{dt}\right) \\
&= \int_0^{k\cdot\Delta t} K_{\mathrm{c}}^{(\mathrm{F})}\dot{e}dt + \int_0^{k\cdot\Delta t} \frac{K_{\mathrm{c}}^{(\mathrm{F})}}{T_{\mathrm{i}}^{(\mathrm{F})}}edt + \frac{K_{\mathrm{c}}^{(\mathrm{F})}T_{\mathrm{d}}^{(\mathrm{F})}}{T_{\mathrm{i}}^{(\mathrm{F})}}\left(e + T_{\mathrm{i}}^{(\mathrm{F})}\dot{e}\right). \quad (6.11)
\end{aligned}
$$

Notice that the linear PID controller in series form is

$$
G_{\mathrm{c}}(s) = \frac{K_{\mathrm{c}}}{T_{\mathrm{i}}s}\left(1 + sT_{\mathrm{i}}\right)\left(1 + sT_{\mathrm{d}}\right)
$$

or

$$
u = \int_0^t K_{\mathrm{c}}\dot{e}dt + \int_0^t \frac{K_{\mathrm{c}}}{T_{\mathrm{i}}}edt + \frac{K_{\mathrm{c}}T_{\mathrm{d}}}{T_{\mathrm{i}}}\left(e + T_{\mathrm{i}}\dot{e}\right)
$$

in time domain. Comparing (6.11) with above formula we can conclude that the fuzzy PID controller (6.8) is a nonlinear PID controller with variable proportional gain.

Remark 6.1 By adopting the simplest FLC in Fig. 6.1, its nonlinear structure and the inherent relationships between its components and their functioning can be made transparent to the designer. With the formulas (6.6), (6.7) and (6.11), the FLC is in essence a *nonlinear* PID type controller because its structure is analogous to that of a common linear PID controller. Moreover, the equivalent proportional control gain $K_{\mathrm{c}}^{(\mathrm{F})}$, integral time $T_{\mathrm{i}}^{(\mathrm{F})}$ and derivative time $T_{\mathrm{d}}^{(\mathrm{F})}$ are composed of FLC parameters w_{e}, $w_{\Delta\mathrm{e}}$, w_{u}, and $w_{\Delta\mathrm{u}}$ as shown in Eq. (6.9).

This greatly facilitate the property analysis and setting of control parameters, as will be shown subsequently.

Remark 6.2 It is worthwhile pointing out that the fuzzy PID control system has six control parameters free for design, whereas the conventional PID only has three. In this chapter we choose $w_{\mathrm{e}_1} = w_{\mathrm{e}_2} = w_{\mathrm{e}}$ and $w_{\Delta\mathrm{e}_1} = w_{\Delta\mathrm{e}_2} = w_{\Delta\mathrm{e}}$ to reduce the undermined fuzzy PID control parameters. The purpose is to facilitate the application of conventional PID tuning algorithms to the fuzzy PID controller. It is clear that if all the control parameters are used, we actually have more degrees of freedom in designing fuzzy PID to achieve multiple control targets such as robustness or adaptation. However, in this chapter we will not pursue any further discussions in this direction.

Remark 6.3 Notice here that the series form of conventional PID controller is used in the comparison. This structure is also implemented in many commercial controllers and can be easily transformed to the parallel structure, [15, 19]. Although this kind of PID controllers cannot introduce complex zeros, it is sufficient for the purpose of process control. Moreover it is sometimes claimed that the series form is easier to tune, [19].

6.2.2 Tuning of the Fuzzy PID Controller

Suppose that a process can be modeled by a second order system with time-delay which has the transfer function of

$$G(s) = \frac{K_{\mathrm{p}} e^{-sL}}{(1 + s\tau_1)(1 + s\tau_2)} \tag{6.12}$$

where $\tau_1 \geq \tau_2$ and a pair of gain margin and phase margin $(A_{\mathrm{m}}, \Phi_{\mathrm{m}})$ is given as the closed-loop performance specification. The tuning formula of a conventional PID controller can be obtained as, [80],

$$\begin{cases} \omega_{\mathrm{p}} = \dfrac{A_{\mathrm{m}}\Phi_{\mathrm{m}} + \frac{1}{2}\pi A_{\mathrm{m}}(A_{\mathrm{m}} - 1)}{(A_{\mathrm{m}}^2 - 1)L} \\ K_{\mathrm{c}} = \dfrac{\omega_{\mathrm{p}}\tau_1}{A_{\mathrm{m}} K_{\mathrm{p}}} \\ T_{\mathrm{i}} = \dfrac{1}{2\omega_{\mathrm{p}} - \dfrac{4\omega_{\mathrm{p}}^2 L}{\pi} + \dfrac{1}{\tau_1}} \\ T_{\mathrm{d}} = \tau_2 \end{cases} \tag{6.13}$$

where ω_{p} is the resultant phase crossover frequency.

Remark 6.4 In many practical control problems, ranging from the level control in chemical industry to the servo tracking control in disk drive industry, it is sufficient and some times also necessary to use such a simple model as (6.12) to design a controller. It is sufficient because a PI/PID controller based on the linearized model can work well even though the original process has high nonlinearities and uncertainties, [13, 15, 19, 75, 81, 93, 108, 136, 142]. It is also necessary because other advanced control methods may either be too complicated to be implemented, or require too many measurements which are not available. For instance, consider the level control of a coupled-tank system, the complete model takes a nonlinear cascaded form. Since the liquid level of the 1st tank is usually not available, many advanced control methods such as backstepping approach based robust adaptive control methods, which are dedicated to this kind of nonlinear systems, are not applicable. Whereas PID, as simple as it is, gives satisfactory control performance.

Comparing (6.13) with (6.9), let

$$K_{\mathrm{c}}^{(\mathrm{F})} = K_{\mathrm{c}}; \quad T_{\mathrm{i}}^{(\mathrm{F})} = T_{\mathrm{i}}; \quad T_{\mathrm{d}}^{(\mathrm{F})} = T_{\mathrm{d}}.$$

It is easy to derive

$$\begin{cases} w_{\Delta e} = \dfrac{w_e}{\left(2\omega_p - \dfrac{4\omega_p^2 L}{\pi} + \dfrac{1}{\tau_1}\right)\Delta t} \\ w_{\Delta u} = \dfrac{\omega_p \tau_1 \left(2\omega_p - \dfrac{4\omega_p^2 L}{\pi} + \dfrac{1}{\tau_1}\right)}{A_m K_p w_e}(4 - 2\alpha)\Delta t \\ w_u \;\; = \dfrac{\omega_p \tau_1 \tau_2 \left(2\omega_p - \dfrac{4\omega_p^2 L}{\pi} + \dfrac{1}{\tau_1}\right)}{A_m K_p w_e}(4 - 2\alpha). \end{cases} \tag{6.14}$$

Thus we have three independent equations with four undermined control parameters. Usually the system output has a working range which is highly related to the changing range of set-point. If the working range is large, w_e should be relative small and vice versa. Such a suitable w_e ensures the normalized error fitted into the interval of $[-1, 1]$. This is reasonable because the quantization values of the fuzzy linguistic variables e or Δe are dependent on what range the system is working in. Thus the normalizing factor of error should be proportional to the reciprocal of the working range, or specifically in our study, the set-point changing range, i.e.

$$w_e = \frac{\chi}{r_0}. \tag{6.15}$$

Based on extensive numerical studies, we choose $\chi = 0.2$ to make possible compromise among rise-time, overshoot and settling time, where r_0 is the set-point change. For a fuzzy PI controller, this w_e can be used to approximately minimize the *ITAE* to set-point response, [162]. With Eqs. (6.14) and (6.15), the coefficients of the fuzzy PID controller (6.8) can be uniquely determined with respect to any systems in the form of (6.12).

In the tuning algorithm, α can be interpreted as an equivalent gain/phase contour in the sense of the gain and phase margins. When determining the fuzzy PID parameters in terms of gain and phase margins, we need to assign a fixed value to the quantity α. Let $\alpha = \alpha_0$ where $\alpha_0 \in [0, 1]$. The coefficient α_0 actually specifies a particular contour on the normalized $e/\Delta e$ plane such that, on this contour the gain and phase margins (which are measures of the local stability of the closed-loop system) satisfy the specifications (Fig. 6.3). According to the tuning formula (6.14) and (6.15), when a particular α_0 is selected, the weighting factors w_e, $w_{\Delta e}$ $w_{\Delta u}$ and w_u will be fixed. Since we have

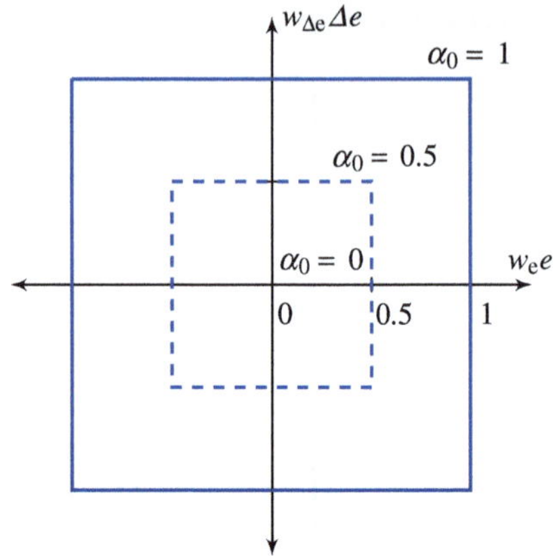

Fig. 6.3 Equivalent gain/phase margin contours of different α_0

$$\begin{cases} T_i^{(F)}(e, \Delta e) = \dfrac{w_{\Delta e}\Delta t}{w_e} \\ T_d^{(F)}(e, \Delta e) = \dfrac{w_u}{w_{\Delta u}}\Delta t \\ K_c^{(F)}(e, \Delta e) = \dfrac{w_{\Delta u} w_{\Delta e}}{4 - 2\alpha} \\ \qquad = \dfrac{4 - 2\alpha_0}{4 - 2\max(w_e|e|, w_{\Delta e}|\Delta e|)} \cdot \dfrac{1}{4 - 2\alpha_0} w_{\Delta u} w_{\Delta e} \\ \qquad = \gamma \cdot \dfrac{1}{4 - 2\alpha_0} w_{\Delta u} w_{\Delta e} \\ \qquad = \gamma \cdot K_{c\alpha_0}^{(F)}, \end{cases} \tag{6.16}$$

where $K_{c\alpha_0}$ is the gain of FLC when the system is at its α_0 contour. Clearly, the fuzzy PID controller has the property that its $T_i^{(F)}$ and $T_d^{(F)}$ are fixed and $K_c^{(F)}$ is variable in terms of different e and Δe. Because we will have

$$\gamma(e, \Delta e) = \frac{4 - 2\alpha_0}{4 - 2\max(w_e|e|, w_{\Delta e}|\Delta e|)} \tag{6.17}$$

which means $\gamma < 1$, $\gamma = 1$ or $\gamma > 1$ when normalized system states ($w_e e$ and $w_{\Delta e}\Delta e$) are inside, on, or outside the α_0 contour, respectively. From the second equation of (6.13), replacing K_c by $K_c^{(F)}$, we obtain the closed-loop gain margin for the fuzzy PID as:

$$A_m^{(F)} = \frac{A_{m\alpha_0}}{\gamma} \tag{6.18}$$

where

$$A_{m\alpha_0} = \frac{\omega_p \tau_1}{K_{c\alpha_0}^{(F)} K_p}.$$

Equation (6.18) shows that the gain margin is always reciprocal to γ. Therefore from (6.17) we can conclude that no matter what α_0 is used, the further away from outside the equivalent gain/phase margin contour, the larger is the γ and consequently the smaller is the gain margin. Similarly, the closer to the steady state, the smaller is the γ and then the larger is the gain margin. In short, $A_{\mathrm{m}}^{(\mathrm{F})}$ increases when $|e|$ or $|\Delta e|$ decreases, which means the loop gain $K_{\mathrm{c}}^{(\mathrm{F})} K_{\mathrm{p}}$ decreases and the system 'safety factor' increases. Note that this is a desirable property, as this ensures that the system is less sensitive to measurement noise nearby the steady state and is quick in response when off the steady state.

The property of α_0 can be used to allocate the equivalent gain/phase margin contour. For example, if $\alpha_0 = 0$ then the system has the same local stability property as the one controlled by the conventional PID controller around the steady state, and the equivalent gain/phase margin contour shrinks to a single point located at the center of the $e/\Delta e$ plane. In this situation, the controller gain will reach its minimum only when the system is at its steady state. In this study, we set $\alpha_0 = 0$ to ensure that the fuzzy PID controller has the same local stability as the conventional PID controller around the steady state and higher gain property off the steady state.

6.3 Stability and Performance Analysis

In the previous section, the properties of gain scheduling and local stability have been discussed in the sense of gain and phase margins. Since the concepts of gain and phase margins are essentially for linear control systems, the above discussions are qualitative and approximate ones. In order to explore the quantitative relationship between the fuzzy and conventional PID controllers and evaluate the global stability, we need more strict and more general analysis methods which can be applied to both nonlinear processes and nonlinear controllers. The small gain theorem is an appropriate tool for this purpose. It should be noted that, the new fuzzy PID is tuned based on a simple model of second order system with time-delay. This implies that less information is available in the stage of controller design. Hence the stability analysis is imperative, especially when the controlled process is of general nonlinear uncertain classes such as BIBO types.

6.3.1 BIBO Stability Condition of the Fuzzy PID Control System

In this Section, we will analyze the bounded-input-bounded-output (BIBO) stability of the fuzzy PID control system. By using the small gain theorem, [57], we will find the generalized sufficient BIBO stability condition of the proposed fuzzy PID control system. Consider a general case where the process under control, which is denoted

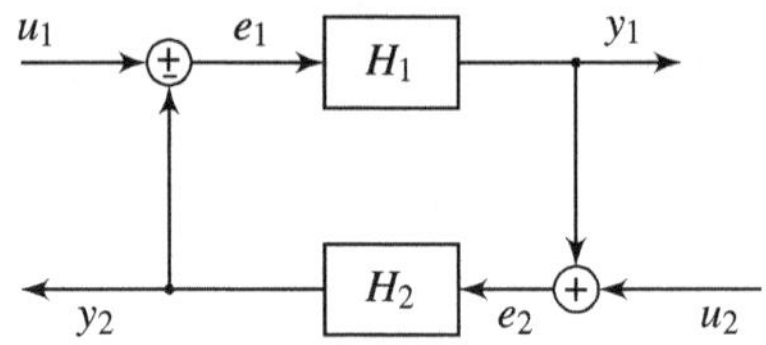

Fig. 6.4 An equivalent closed-loop fuzzy PID control system

by $g(\bullet)$, is nonlinear and the reference input is r_k. By using the control law (6.10), we have:

$$\begin{aligned} u^{(F)}_{\mathrm{PID},k} &= \Delta u^{(F)}_{1,k} + u^{(F)}_{2,k} + \sum_{i=0}^{k-1} \Delta u^{(F)}_{1,i} \\ &= \Delta u^{(F)}_{\mathrm{PID},k} + u^{(F)}_{\mathrm{PID},k-1}, \end{aligned} \tag{6.19}$$

where

$$\Delta u^{(F)}_{\mathrm{PID},k} = \Delta u^{(F)}_{1,k} + u^{(F)}_{2,k} - u^{(F)}_{2,k-1}. \tag{6.20}$$

By denoting

$$\begin{cases} \Delta u^{(F)}_{\mathrm{PID},k} = f\left(e_k\right) \\ y_k = g\left(\Delta u^{(F)}_{\mathrm{PID},k}\right) \end{cases}$$

it is easy to obtain an equivalent closed-loop control system as shown in Fig. 6.4.

Theorem 6.1 *A sufficient condition for the nonlinear fuzzy PID control system to be BIBO stable is that the given nonlinear process has a bounded norm (gain) as* $\|g\| < \infty$ *and the parameters of the fuzzy PID controller,* w_{e}, $w_{\Delta\mathrm{e}}$, w_{u} *and* $w_{\Delta\mathrm{u}}$, *(or* $K^{(F)}_{\mathrm{c},k}$, $T^{(F)}_{\mathrm{i}}$ *and* $T^{(F)}_{\mathrm{d}}$ *in (6.17)), satisfy:*

$$K^{(F)}_{\mathrm{c},k}\left(1 + \frac{T^{(F)}_{\mathrm{d}}}{T^{(F)}_{\mathrm{i}}} + \frac{\Delta t}{T^{(F)}_{\mathrm{i}}} + \frac{T^{(F)}_{\mathrm{d}}}{\Delta t}\right) \cdot \|g\| < 1 \tag{6.21}$$

where $\|g\|$ *is the operator norm of the given* $g(\bullet)$, *or the gain of the given nonlinear system, usually defined as, [50, 57],*

$$\|g\| = \sup_{v_1 \neq v_{2,k} \geq 0} \frac{|g(v_{1,k}) - g(v_{2,k})|}{|v_{1,k} - v_{2,k}|}. \tag{6.22}$$

Proof The fuzzy PID controller can be considered as a self-tuning adaptive nonlinear PID controller since the gains of this kind of controllers vary in terms of the combination of the inputs $(e, \Delta e)$ of the fuzzy controller. The combination of $(e, \Delta e)$ can be divided into nine regions as shown in Fig. 6.5. When the inputs of $(e_k, \Delta e_k)$ are in region I, the control law is given by (6.20), therefore from (6.6), (6.7) and (6.9):

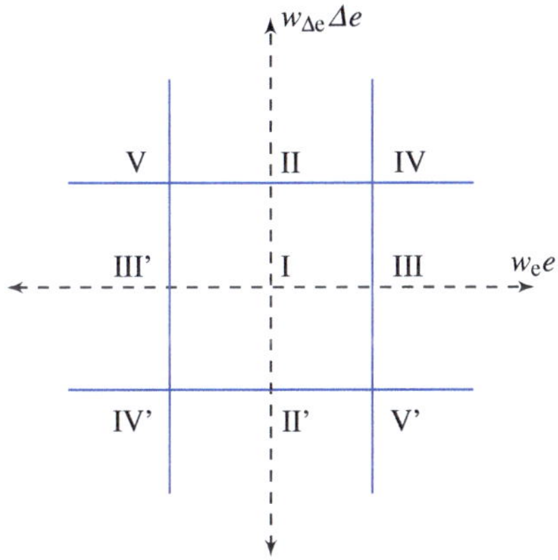

Fig. 6.5 Different regions of fuzzy PID controller's inputs combinations

$$\begin{aligned}
\Delta u_{\mathrm{PID},k}^{(\mathrm{F})} &= \Delta u_{1,k}^{(\mathrm{F})} + u_{2,k}^{(\mathrm{F})} - u_{2,k-1}^{(\mathrm{F})} \\
&= K_{\mathrm{c},k}^{(\mathrm{F})}\left[\Delta e + \frac{\Delta t}{T_{\mathrm{i}}^{(\mathrm{F})}} e_k\right] + K_{\mathrm{c},k}^{(\mathrm{F})}\frac{T_{\mathrm{d}}^{(\mathrm{F})}}{T_{\mathrm{i}}^{(\mathrm{F})}}\left[e_k + \frac{T_{\mathrm{i}}^{(\mathrm{F})}}{\Delta t}\Delta e_k\right] \\
&\quad - K_{c,k-1}^{(\mathrm{F})}\frac{T_{\mathrm{d}}^{(\mathrm{F})}}{T_{\mathrm{i}}^{(\mathrm{F})}}\left[e_{k-1} + \frac{T_{\mathrm{i}}^{(\mathrm{F})}}{\Delta t}\Delta e_{k-1}\right] \\
&= K_{\mathrm{c},k}^{(\mathrm{F})}\left(1 + \frac{T_{\mathrm{d}}^{(\mathrm{F})}}{T_{\mathrm{i}}^{(\mathrm{F})}} + \frac{\Delta t}{T_{\mathrm{i}}^{(\mathrm{F})}} + \frac{T_{\mathrm{d}}^{(\mathrm{F})}}{\Delta t}\right)e_k - K_{\mathrm{c},k}^{(\mathrm{F})}\left(1 + \frac{T_{\mathrm{d}}^{(\mathrm{F})}}{\Delta t}\right)e_{k-1} \\
&\quad - K_{\mathrm{c},k-1}^{(\mathrm{F})}\left(\frac{T_{\mathrm{d}}^{(\mathrm{F})}}{T_{\mathrm{i}}^{(\mathrm{F})}} + \frac{T_{\mathrm{d}}^{(\mathrm{F})}}{\Delta t}\right)e_{k-1} + K_{\mathrm{c},k-1}^{(\mathrm{F})}\frac{T_{\mathrm{d}}^{(\mathrm{F})}}{\Delta t}e_{k-2}
\end{aligned} \tag{6.23}$$

and, thus,

$$\|f(e_k)\| \le K_{\mathrm{c},k}^{(\mathrm{F})}\left|1 + \frac{T_{\mathrm{d}}^{(\mathrm{F})}}{T_{\mathrm{i}}^{(\mathrm{F})}} + \frac{\Delta t}{T_{\mathrm{i}}^{(\mathrm{F})}} + \frac{T_{\mathrm{d}}^{(\mathrm{F})}}{\Delta t}\right| \cdot |e_k| + \nu_1 e_{\max} \tag{6.24}$$

where

$$\begin{aligned}
\nu_1 &= \max(K_{\mathrm{c}}^{(\mathrm{F})}) \cdot \left(1 + \frac{T_{\mathrm{d}}^{(\mathrm{F})}}{T_{\mathrm{i}}^{(\mathrm{F})}} + 3\frac{T_{\mathrm{d}}^{(\mathrm{F})}}{\Delta t}\right) \le 2K_{\mathrm{c}\alpha_0}^{(\mathrm{F})}\left(1 + \frac{T_{\mathrm{d}}^{(\mathrm{F})}}{T_{\mathrm{i}}^{(\mathrm{F})}} + 3\frac{T_{\mathrm{d}}^{(\mathrm{F})}}{\Delta t}\right) \\
e_{\max} &= \max\left(|e_0|, |e_1|, \ldots, |e_{k-1}|\right).
\end{aligned}$$

On the other hand

$$\|g(u_{\mathrm{PID},k}^{(\mathrm{F})})\| \le \|g\| \cdot |u_{\mathrm{PID},k}^{(\mathrm{F})}|. \tag{6.25}$$

Applying the small gain theorem, we can obtain the sufficient condition for the BIBO stability given by the theorem:

$$K_c^{(F)} \left| 1 + \frac{T_d^{(F)}}{T_i^{(F)}} + \frac{\Delta t}{T_i^{(F)}} + \frac{T_d^{(F)}}{\Delta t} \right| \cdot \|g\| < 1.$$

Similarly when the system state $(e_k, \Delta e_k)$ is in the regions of II and II', in which the term $w_{\Delta e} \Delta e$ is outside the interval of $[-1, 1]$ and becomes a constant due to the saturation. Hence, we can obtain the sufficient BIBO stability condition as

$$K_c^{(F)} \left| \frac{T_d^{(F)}}{T_i^{(F)}} + \frac{\Delta t}{T_i^{(F)}} \right| \cdot \|g\| < 1.$$

When $(e_k, \Delta e_k)$ is in the regions of III and III', in which $w_e e$ is outside the interval of $[-1, 1]$, the sufficient BIBO stability conditions are found to be

$$K_c^{(F)} \left| 1 + \frac{T_d^{(F)}}{\Delta t} \right| \cdot \|g\| < 1.$$

Finally, when $(e_k, \Delta e_k)$ is in the regions of IV, IV', V and V', since the control effort is constant, the sufficient BIBO stability is that $\|g\|$ is bounded.

By combining all the above conditions together, and noticing that $K_c^{(F)} > 0$, $T_i^{(F)} > 0$, $T_d^{(F)} > 0$ and $\Delta t > 0$, the result for the stability of the fuzzy PID control system will be given as

$$K_c^{(F)} \left(1 + \frac{T_d^{(F)}}{T_i^{(F)}} + \frac{\Delta t}{T_i^{(F)}} + \frac{T_d^{(F)}}{\Delta t} \right) \cdot \|g\| < 1.$$

Note that in (6.21), if we eliminate the superscript (F), we will arrive at the sufficient BIBO stability condition for a linear PID controlled closed-loop system (see Appendix). Moreover, if same gain and phase margin specifications are used, the tuning formula (6.14) and (6.15) result in a fuzzy PID controller that has same gains on its α_0 contour as a conventional PID controller. In other words, the proposed fuzzy PID controller will retain at least the same stability property as its conventional counterpart on and inside its α_0 contour. This result is summarized as follows:

Corollary 6.1 *For a nonlinear process controlled stably by a conventional PID controller with gain K_c, integral time constant T_i and derivative time constant T_d, if the PID controller is replaced by the proposed fuzzy PID controller whose control parameters $w_{\Delta e}$, $w_{\Delta u}$ and w_u are given by the tuning formula (6.15) with $\alpha = \alpha_0 \in [0, 1]$, the resulting fuzzy PID control system will ensure at least the same (local) stability on and within the α_0 contour.*

Remark 6.5 The significance of above conclusion is that, one can always replace a conventional PID controller by the proposed nonlinear fuzzy PID controller without losing the stability margin around the equilibrium. In particular, if we take $\alpha_0 = 0$, then in steady state $e_k = \Delta e_k = 0$, we have $K_c^{(F)} = K_c$, the conventional PID and the fuzzy PID control systems have exactly the same stability.

Remark 6.6 From Eqs. (6.15), (6.17) and (6.21), it is easy to derive that the stability condition (6.21) is independent of the error weighting factor w_e. Therefore we can use this extra degree of freedom in the design to improve system responses, at discussed in previous section, and at the same time maintaining the system stability.

6.3.2 Control Efforts Between Fuzzy and Conventional PID Controllers

In the previous section we have shown that, by choosing $\alpha_0 = 0$, the fuzzy PID control gain is higher than that of the conventional PID except for the equilibrium in which both are the same. This property ensures that the load disturbance rejection of the fuzzy PID will be at least as good as the conventional one. Here we will explore another novel property of the new fuzzy PID controller: in set-point control the proposed fuzzy PID control system will generate lower control signal profiles compared with the conventional PID controllers while maintaining the higher control gain. This property will effectively reduce the overshoot phenomenon in set-point control. Note that in the PI or PID control, the initial value of the control signal plays an important role because of the integral action. By limiting the initial control effort at low level, the overall control profile will be kept lower. In the remainder of the Section we will show that, the proposed fuzzy PID control system does generate a lower initial control signal compared to the conventional PID control system.

To eliminate the derivative kick in the implementation of PID control a modified derivative term is used as follows, [19],

$$D = -K_c T_d \frac{dy}{dt}.$$

Similarly, the input of Δe to the fuzzy PID controller is replaced by $-\Delta y$. By transforming the series form of conventional PID controller to the parallel form, and subtracting u_{k-1} from u_k it is obtained that

$$\Delta u_{\text{PID},k} = K_c \frac{T_i + T_d}{T_i}[e_k - e_{k-1}] + \frac{K_c}{T_i} e_{k-1} \Delta t - K_c T_d \frac{y_k - 2y_{k-1} + y_{k-2}}{\Delta t}. \tag{6.26}$$

Similarly, for the fuzzy PID controller we have

$$\Delta u_{\text{PID},k}^{(F)} = \Delta u_{1,k} + u_{2,k} - u_{2,k-1}. \tag{6.27}$$

From (6.23) we have

$$\Delta u_{\mathrm{PID},k}^{(\mathrm{F})} = K_{\mathrm{c},k}^{(\mathrm{F})}\left[y_{k-1} - y_k + \frac{T_{\mathrm{d}}^{(\mathrm{F})}}{T_{\mathrm{i}}^{(\mathrm{F})}}(e_k - e_{k-1})\right] + \frac{K_{\mathrm{c},k}^{(\mathrm{F})}}{T_{\mathrm{i}}^{(\mathrm{F})}} e_k \Delta t$$
$$-T_{\mathrm{d}}^{(\mathrm{F})} \frac{K_{\mathrm{c},k}^{(\mathrm{F})}\left[y_k - y_{k-1}\right] - K_{\mathrm{c},k-1}^{(\mathrm{F})}\left[y_{k-1} - y_{k-2}\right]}{\Delta t}.$$

When a set-point change occurs, we have $k = 0$, $e_0 = r$, $e_{-1} = 0$ and $y_0 = y_{-1} = y_{-2}$. Therefore, from (6.26) and (6.28), we have

$$\begin{aligned}\delta u_0 &= \Delta u_{\mathrm{PID},0}^{(\mathrm{F})} - \Delta u_{\mathrm{PID},0}\\ &= \left(K_{\mathrm{c},0}^{(\mathrm{F})} \frac{T_{\mathrm{d}}^{(\mathrm{F})} + \Delta t}{T_{\mathrm{i}}^{(\mathrm{F})}} - K_{\mathrm{c}} \frac{T_{\mathrm{i}} + T_{\mathrm{d}}}{T_{\mathrm{i}}}\right) \cdot r\\ &= \left(K_{\mathrm{c},0}^{(\mathrm{F})} \frac{T_{\mathrm{d}} + \Delta t}{T_{\mathrm{i}}} - K_{\mathrm{c}} \frac{T_{\mathrm{i}} + T_{\mathrm{d}}}{T_{\mathrm{i}}}\right) \cdot r.\end{aligned}$$

From the tuning formula (6.15) and choosing $\alpha_0 = 0$, we have

$$K_{\mathrm{c},0}^{(\mathrm{F})} = \frac{4 - 2\alpha_0}{4 - 2\max(w_{\mathrm{e}}|e_0|, w_{\Delta\mathrm{e}}|\Delta e_0|)} K_{\mathrm{c}} = \frac{10}{9} K_{\mathrm{c}}, \tag{6.28}$$

and, thus,

$$\delta u_0 = \frac{K_{\mathrm{c}}}{9T_{\mathrm{i}}}(T_{\mathrm{d}} + 10\Delta t - 9T_{\mathrm{i}}) \cdot r. \tag{6.29}$$

From (6.15) we can obtain, when the specified gain and phase margin pair is $(3, 45°)$, Δt is sufficient small and $\frac{L}{\tau_1} \geq 0.023$, that $\delta u_0 < 0$ (noticing that $T_{\mathrm{d}} = \tau_2 \leq \tau_1$). Moreover, if $\frac{L}{\tau_1} \geq 0.1$, we will have $T_{\mathrm{i}} \in (0.35\tau_1, \tau_1)$, and we will arrive at the result that

$$\delta u_0 \approx -K_{\mathrm{c}} r < 0. \tag{6.30}$$

Consequently,

$$\Delta u_{\mathrm{PID},0}^{(\mathrm{F})} < \Delta u_{\mathrm{PID},0} \tag{6.31}$$

namely, the initial control effort of fuzzy controller is smaller than that of the conventional PID controller. Moreover, if the original Δe is used in derivative control action for both controllers, the fuzzy PID controller will still give a smaller initial control effort. This is because the saturation of the controller output occurs when there is a set-point change, which leads to the normalized change of error be outside the universe of discourse. This smaller initial control effort of fuzzy controller prevents it from injecting large amount of energy, which may cause large overshoot, to the system. In simulation examples we will further demonstrate the low control profile property.

6.4 Illustrative Example

Numerical simulation examples of system responses to step set-point change and load disturbance are used to demonstrate the effectiveness of the proposed fuzzy PID controller associated with the tuning formula. A linear PID controller with tuning formula (6.15), which shows better control performance than other PID tuning algorithms, [80], is used for comparison. Both fuzzy and linear PID controllers are tuned by same gain and phase margin specifications of $3/45°$, which give good system response to both set-point change and load disturbance. The simulations are carried out during a time period from 0 to T. At the instant of 0, there is a step set-point change to the system and at the instant of $\frac{T}{2}$, there is a step load disturbance with negative magnitude. A unique sampling period of 10 ms is used in all simulations. In the simulations, the derivative action is applied only to the process output, as common in practice, to prevent any drastic change in Derivative control mode. In the figures, the solid and dotted lines are for fuzzy and conventional PID controllers respectively, if it is not otherwise mentioned.

In the first case, the controllers are designed to control a lag dominant system. The magnitudes of set-point change and load disturbance are 1 and -4 respectively. From the simulation results (Fig. 6.6), it is found that the overshoot of the system response to set point change is very large when the system is controlled by the conventional PID controller. On the other hand, the well designed fuzzy PID controller eliminates system overshoot without losing much of response speed. This means that the nonlinearity introduced by fuzzy inference improves the closed-loop system performance by suppressing the overshoot to set-point change. The control efforts of different controllers are shown below.

From the control effort profiles in Fig. 6.7, we find that control output of the fuzzy PID controller is smaller than that of the conventional PID controller when the set-point change occurs, as discussed in previous section. In this case, $r = 1$ and $K_c = 4.9$, $T_i = 0.35$, $T_d = 0.5$. From Eq. (6.30) the initial discrepancy of control efforts is $\delta u_0 \approx 3.98$, which is consistent with the simulation result shown in Fig. 6.7.

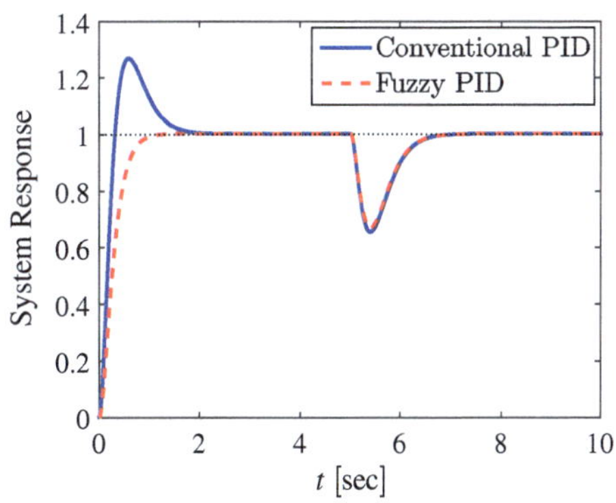

Fig. 6.6 Fuzzy and conventional PID for $\frac{e^{-0.1s}}{(1+s)(1+0.5s)}$

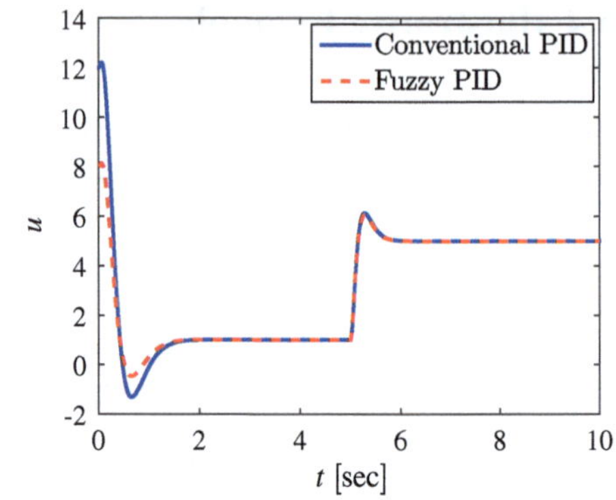

Fig. 6.7 Control efforts of Fuzzy and conventional PID

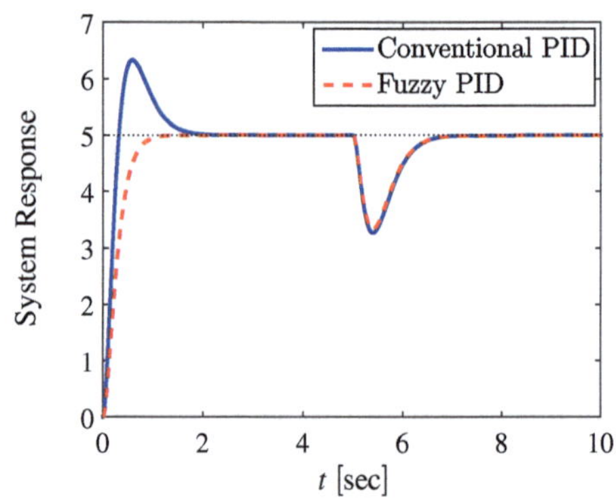

Fig. 6.8 Fuzzy and conventional PID for $\frac{e^{-0.1s}}{(1+s)(1+0.5s)}$ (large set-point change)

When the system has time-delay, the initial feedback control action may not behave correctly and it tends to inject too much energy into the system. For this reason it is intuitive to control the system slowly with care at the initial stage of control. On the other hand, reasonable initial energy injection associated with higher gains will ensure the system response in a suitable speed and easy to be stopped at the set-point. The system performance, when the fuzzy logic controller is used, gives us a clear view on this property.

The second example is the closed-loop system response with a large set-point change with the magnitude of 5 and a large load disturbance with the magnitude of -20. From the simulation results in Fig. 6.8, we can find that both conventional and fuzzy PID controllers give similar closed-loop performance as that in the previous case. Although the magnitudes of both set-point change and load disturbance are quite different from those in previous case, the selected w_e, which is reciprocal to the magnitude of set-point change, ensures the normalized error and change of error inside the universe of discourse and thus, gives similar profiles of system responses.

In the third example a relatively small set-point change of 1 and a relatively large load disturbance of -10 are considered. From the simulation results shown in Fig. 6.9, we find that the advantage of the fuzzy PID controller to reject load

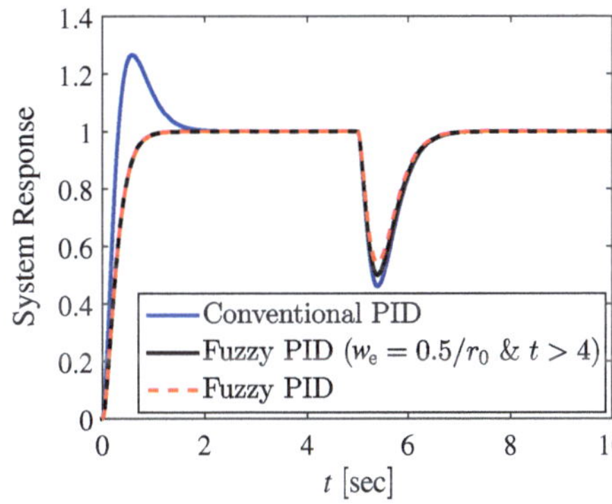

Fig. 6.9 Fuzzy and conventional PID for $\frac{e^{-0.1s}}{(1+s)(1+0.5s)}$

disturbance is more obvious than the previous examples. From the figure we can observe that, by increasing χ in (6.15) from 0.2 to 0.5, the fuzzy PID controller gives a better system response to load disturbance.

From the simulation results we can conclude that the tuning formula based on gain and phase margin specifications is valid to determine the weighting coefficients of a fuzzy PID controller. The resulting controller performs better than the conventional PID controller as the fuzzy logic controller yields less overshoot to set-point change and leads to faster convergence (shorter settling time), while the system responses to load disturbance are similar for both conventional and fuzzy PID controllers.

6.5 Conclusion

In this chapter, a new structure of fuzzy PID controller is presented. The parallel combination of fuzzy PI and PD controllers shows its simplicity in determining the control rules and controller parameters. A tuning formula based on gain and phase margins is introduced by which the weighting factors of a fuzzy PID controller can be selected with respect to second order systems with time-delay. The validity of the proposed fuzzy PID controller and gain and phase margin based tuning formula are confirmed through theoretical analysis and numerical simulations. The theoretical results show that the fuzzy PID controller has the nonlinear properties of: (1) higher control gains when the system is away of its steady states; and (2) lower control profile when set-point changes occur. As a result, these nonlinear properties provide the fuzzy PID control system with a superior performance over the conventional PID control system.

Chapter 7
Benchmark Precision Control of a Piezo-Motor Driven Linear Stage

Abstract In this study, practical application of ISM control and ILC is investigated using a piezo-motor driven stage as the platform. Theoretical developments have shown ISM control and ILC to be highly suitable for high-precision motion control problems. The experimental results show that ISM control and ILC can indeed produce superior performance to conventional control methods.

7.1 Introduction

In this chapter, the discrete-time ISM control, TSM control and ILC are applied to a linear piezo-motor driven stage which has many promising applications in industries. The piezo-motors are characterized by low speed and high torque, which are in contrast to the high speed and low torque properties of the conventional electromagnetic motors. Moreover, piezo-motors are compact, light, operates quietly, and robust to external magnetic or radioactive fields. Piezo-electric motors are mainly applied to high precision control problems as it can easily reach the precision scale of micro-meters or even nano-meters. This gives rise to extra difficulty in establishing an accurate mathematical model for piezo-motors: any tiny factors, nonlinear and uncertain, will severely affect their characteristics and control performance.

It is well known that sliding mode (SM) control is a very popular robust control method owing to its ease of design and robustness to "matched" disturbances, hence was widely adopted in various industrial applications, [3, 62]. Computer implementation of control algorithms presents a great convenience and has, hence, caused the research in the area of discrete-time control to intensify. This also necessitated a rework in the sliding mode control strategy for sampled-data systems. Most of the discrete-time sliding mode approaches are based on full state information, [1, 3, 62].

On the other hand, this chapter considers the output tracking of the piezo-motor driven stage. To accomplish the task of arbitrary output reference tracking in the presence of disturbances, an output feedback controller with a state observer and a disturbance observer are designed. The objective is to drive the output tracking error to a certain neighborhood of the origin. For this purpose discrete-time integral sliding surfaces are proposed for the controller and observers.

K. Abidi and J.-X. Xu, *Advanced Discrete-Time Control*, Studies in Systems, Decision and Control 23, DOI 10.1007/978-981-287-478-8_7

Delays in the state or disturbance estimation in sampled-data systems is an inevitable phenomenon and must be studied carefully. In [1] it was shown that in the case of delayed disturbance estimation a worst case accuracy of $O(T)$ can be guaranteed for deadbeat sliding mode control design and a worst case accuracy of $O(T^2)$ for integral sliding mode control. While deadbeat control is a desired phenomenon, it is undesirable in practical implementation due to the over large control action required. In [1] the integral sliding mode design avoided the deadbeat response by eliminating the poles at zero. In this chapter, we extend the integral sliding mode design to output tracking.

In output feedback based sliding mode control, [3, 146], there are mainly two design approaches: design using only the output measurement, [3, 55], and design based on observers to construct the missing states, [67, 146]. The purely output based design imposed extra stability requirements that are not practical in most cases. Hence, in this chapter we adopt a discrete-time state observer.

As it was shown in the classical SM and ISM approaches the system state converges to the equilibrium asymptotically with infinite settling time. It was shown that further improvement to the dynamic response of the closed-loop system is possible by introducing nonlinear sliding surfaces such as TSM. TSM can ensure the finite-time convergence of the system during the sliding mode stage. The finite-time control of dynamical systems is of interest because the systems with finite-time convergence demonstrate some nice features such as faster convergence rate, and better robustness and disturbance rejection properties. These properties are especially desirable in high-precision applications that involve piezo-motors.

Finally, for high-precision repetitive tasks it would be desirable to implement a control approach that uses as less as possible system information while retaining high robustness and disturbance rejection properties. In this case we implement ILC which has been shown to require very little system information while retaining high robustness and ability to reject repeatible disturbances. Piezo-motors are generally difficult to model and demonstrate highly nonlinear behavior. In systems where moving parts are involved, friction is also of concern. It can be shown that the nonlinear behavior of a piezo-motor and friction will have an almost repeatible characteristic for repeatible tasks. Thus, using ILC the undesirable effects due to the nonlinearity and friction can be 'learned' and rejected to produce very high-precision motion.

7.2 Model of the Piezo-Motor Driven Linear Motion Stage

In this section we first discuss the continuous-time piezo-motor stage model, then the friction model. Next the discretized model and disturbance properties are presented.

7.2.1 Overall Model in Continuous-Time

A major objective of this study is to design a controller based on the simplest possible model of the piezo-motor stage. Therefore, we consider the following second-order continuous-time model of the piezo-motor stage

$$
\begin{aligned}
\dot{x}_1(t) &= x_2(t) \\
\dot{x}_2(t) &= -\frac{k_{\mathrm{fv}}}{m}x_2(t) + \frac{k_{\mathrm{f}}}{m}u(t) - \frac{1}{m}(f(\mathbf{x},t) + g(t)) \\
y(t) &= x_1(t)
\end{aligned}
\tag{7.1}
$$

where x_1 is the linear displacement which is measurable, x_2 is the linear velocity which is not available, u is the voltage input, $f(\mathbf{x}, t)$ represents the friction, $g(t)$ represents the effect of process perturbations and is assumed smooth and bounded. The constants m, k_{fv}, and k_{f} are the nominal mass, damping, and force constants respectively. This model closely represents the dynamics of the system as shown in Fig. 7.1. As can be seen for a wide spectrum of frequencies the dynamics of the real system is indeed that of a second-order system.

7.2.2 Friction Models

Through experiments, it is found there exists a large friction in the piezo-motor stage, which is discontinuous when the velocity across zero. In this chapter, we treat the friction as an uncertain disturbance and use a disturbance observer to estimate.

In order to understand the behavior of the piezo-motor stage under friction $f(\mathbf{x}, t)$, hence facilitate the performance analysis on the controller and disturbance observer, we consider three widely accepted friction models, in the sequel determine the most appropriate model.

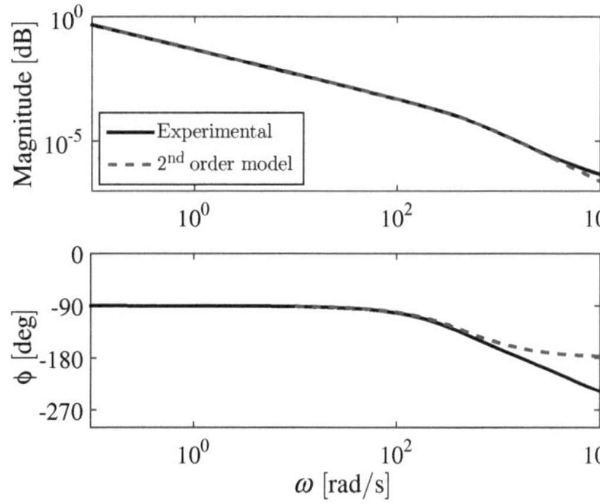

Fig. 7.1 Frequency responses of the piezo-motor stage

7.2.2.1 Static Friction Model

Here the friction is modelled as a bounded piece-wise continuous function and the discontinuity occurs only when x_2 changes sign. In details $f(\mathbf{x}, t)$ can be represented as follows, [61],

$$f(\mathbf{x}, t) = \begin{cases} k_{\mathrm{f}} a_{\max} & x_2(t) > 0 \\ k_{\mathrm{f}} \mathrm{sat}[a(t)] & x_2(t) = 0 \\ k_{\mathrm{f}} a_{\min} & x_2(t) < 0 \end{cases} \tag{7.2}$$

where $\mathrm{sat}[a(t)]$ is a saturation function given by

$$\mathrm{sat}[a(t)] = \begin{cases} a_{\max} & u(t) \geq a_{\max} \\ a(t) & a_{\min} < a(t) < a_{\max} \\ a_{\min} & a(t) < a_{\min} \end{cases} \tag{7.3}$$

where $a_{\min}$ and $a_{\max}$ are uncertain constant coefficients of the static friction.

7.2.2.2 Gaussian Friction Model

This model [12] considers three kinds of frictions—the static friction, viscous friction, and kinetic friction

$$f(x, t) = -\frac{1}{m}\left(\left(f_{\mathrm{c}} + (f_{\mathrm{s}} - f_{\mathrm{c}})e^{-(\frac{x_2}{v_{\mathrm{s}}})^{\delta}}\right)\mathrm{sgn}(x_2) + f_{\mathrm{v}} x_2\right) \tag{7.4}$$

where f_{c} is the minimum level of kinetic friction, f_{s} is the level of static friction, f_{v} is the level of viscous friction, $v_{\mathrm{s}} > 0$ and $\delta > 0$ are empirical parameters. The signum function from static friction represents a discontinuity crossing the zero velocity.

7.2.2.3 Lugre Friction

One motivation behind the LuGre model is to offer a regularised static friction model with stiction. The model captures several friction characteristics, such as increased friction force at lower velocities, [61]. It is a first order dynamic model and the most commonly used form is

$$\begin{aligned} \dot{z} &= x_2 - \rho_0 \frac{|x_2|}{g(x_2)} z \\ g(x_2) &= \alpha_0 + \alpha_1 e^{-\left(\frac{x_2}{x_{2,\mathrm{s}}}\right)^2} \\ f &= \rho_0 z + \rho_1 \dot{z} \end{aligned} \tag{7.5}$$

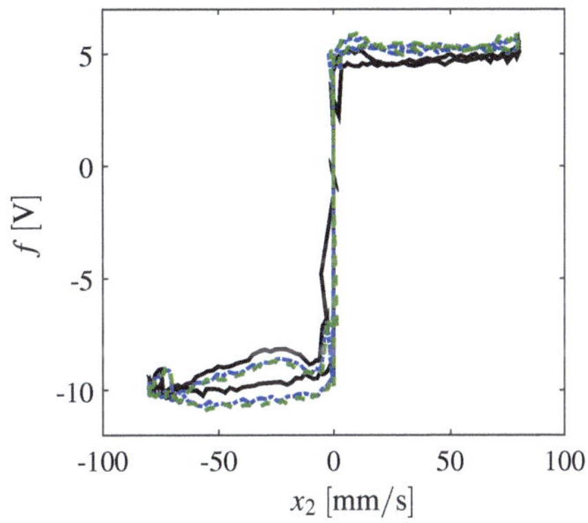

Fig. 7.2 Experimentally obtained friction f w.r.t velocity x_2

where α_0, α_1, $x_{2,s}$, ρ_0 and ρ_1 are positive parameters. Since the state z cannot be measured, it is necessary to use an observer to get an estimate of the friction based on this model.

The three models presented allow different degrees of accuracy. The first model is the simplest, the second model is more generic while the third model is dynamic. However, it is in general difficult to determine the model parameters. A number of experimental tests were conducted and the results of three trials were shown in Fig. 7.2. In the experiment, a slow sinusoidal input was injected into the to generate a low speed motion with very low acceleration. In this way the control input injected is solely to overcome the friction of the piezo-motor stage. Thus the force-velocity relationship in Fig. 7.2 can be obtained. It can be seen that none of the three friction models can perfectly capture the behaviors of the piezo-motor stage. Comparatively the static friction model can better fit the experimental results by choosing $k_f a_{max} = 5\,\text{V}$ and $k_f a_{min} = -10\,\text{V}$. Thus we can use the static friction model in the performance analysis.

The modeling mismatching can be considered as some uncertain disturbance, due to the presence of many uncertain factors such as unmodelled electrical dynamics, the hysteresis, measurement errors, system and sensor noise, as well as other random purterbations which cannot be modelled. We will introduce disturbance observer to estimate and compensate it ultimately. Moreover, by virtue of the robustness in sliding mode control, we may not need a perfect plant model.

7.2.3 Overall Model in Discrete-Time

The sampled-data counterpart of (7.1) can be given by

$$\begin{aligned} \mathbf{x}_{k+1} &= \Phi \mathbf{x}_k + \boldsymbol{\gamma} u_k + \mathbf{d}_k \\ y_k &= \mathbf{c}\mathbf{x}_k = x_{1,k}, \quad y_0 = y(0) \end{aligned} \tag{7.6}$$

where

$$\Phi(T) = \begin{bmatrix} \phi_{11} & \phi_{12} \\ \phi_{21} & \phi_{22} \end{bmatrix} = \exp\left(\begin{bmatrix} 0 & 1 \\ 0 & -\frac{k_{fv}}{m} \end{bmatrix} T\right),$$

$$\gamma = \begin{bmatrix} \gamma_1 \\ \gamma_2 \end{bmatrix} = \int_0^T \Phi(\tau)d\tau \begin{bmatrix} 0 \\ \frac{k_f}{m} \end{bmatrix} \quad \& \quad \mathbf{c} = \begin{bmatrix} c_1 \vdots c_2 \end{bmatrix} = \begin{bmatrix} 1 \; 0 \end{bmatrix}.$$

The equivalent disturbance term $\mathbf{d}_k = \begin{bmatrix} d_1 \vdots d_2 \end{bmatrix}^\top$ can be calculated for the three scenarios given in (7.2)

$$\mathbf{d}_k = \mathbf{h}_k - \begin{cases} \gamma a_{\max} & x_{2,k} > 0 \\ \gamma \text{sat}[a(t)] & x_{2,k} = 0 \\ \gamma a_{\min} & x_{2,k} < 0 \end{cases} \tag{7.7}$$

where $\mathbf{h}_k$ is given by

$$\mathbf{h}_k = -\int_0^T \Phi(\tau) \begin{bmatrix} 0 \\ \frac{k_f}{m} \end{bmatrix} g((k+1)T - \tau)d\tau,$$

and T is the sampling period. Here the disturbance $\mathbf{h}_k$ represents the influence accumulated from kT to $(k+1)T$.

From *Lemma* 2.1 in Chap. 2 the following useful properties of the disturbance $\mathbf{d}_k$ when the motor speed is not zero are known:

Property 7.1 $\mathbf{d}_k = \gamma(g_k + a_m) + \frac{1}{2}\gamma v_k T + O(T^3)$.

Property 7.2 $\mathbf{d}_k = O(T)$.

Property 7.3 $\mathbf{d}_k - \mathbf{d}_{k-1} = O(T^2)$.

Property 7.4 $\mathbf{d}_k - 2\mathbf{d}_{k-1} + \mathbf{d}_{k-2} = O(T^3)$.

Property 7.5 *For stable dynamics* $x_{k+1} = \lambda x_k + \delta_k$, $|\lambda| < 1$ *and* $\delta_k = O(T^n)$, *then* $|x_k| = O(T^{n-1})$ *when* $k \to \infty$.

where a_m is either $a_{\min}$ or $a_{\max}$, $v_k = v(kT)$ and $v(t) = \frac{d}{dt}g(t)$. Note that the magnitude of the mismatched part in the disturbance $\mathbf{d}_k$ is of the order $O(T^3)$.

7.3 Discrete-Time Output ISM Control

In this section we will discuss the design of the output tracking controller for the piezo-motor stage. The controller will be designed based on an appropriate integral sliding-surface. Further, the stability conditions of the closed-loop system will be analyzed. Appropriate observers for the disturbance and the uncertain state x_2 will be derived and this section will conclude with a discussion on the tracking-error bound.

7.3.1 Controller Design and Stability Analysis

Consider the discrete-time integral sliding-surface below,

$$\begin{aligned}\sigma_k &= e_k - e_0 + \varepsilon_k \\ \varepsilon_k &= \varepsilon_{k-1} + \beta e_{k-1}\end{aligned} \tag{7.8}$$

where $e_k = r_k - y_k$ is the output tracking error, e_0 is the initial tracking error, r_k is an arbitrary time-varying reference, σ_k, ε_k are the sliding function and integral of the tracking error, and β is a design constant. The output tracking problem is to force $y_k \to r_k$.

Let us first derive the discrete-time ISM control law by using to the concept of equivalent control.

Theorem 7.1 *The new ISM control law proposed is*

$$u_k = \gamma_1^{-1}[r_{k+1} - \lambda e_k - \phi_{11}x_{1,k} - \phi_{12}\hat{x}_{2,k} + \sigma_k] - \hat{\eta}_{k-1} \tag{7.9}$$

where $\hat{\eta}_{k-1}$ and $\hat{x}_{2,k}$ are estimates of the disturbance observer and state observer respectively as will be shown later, $\lambda = 1 - \beta$.

Further, the controller (7.9) drives the sliding variable to

$$\sigma_{k+1} = \gamma_1\hat{\eta}_{k-1} - d_{1,k} - \phi_{12}\tilde{x}_{2,k}$$

and results in the output error dynamics

$$e_{k+1} - \lambda e_k + \delta_k,$$

where $\tilde{x}_{2,k} = x_{2,k} - \hat{x}_{2,k}$ is state estimation error, and

$$\delta_k = -(d_{1,k} - \gamma_1\hat{\eta}_{k-1} - d_{1,k-1} + \gamma_1\hat{\eta}_{k-2}) + \phi_{12}(\tilde{x}_{2,k} - \tilde{x}_{2,k-1}), \tag{7.10}$$

which consists disturbance and state estimation errors.

Proof The control law (7.9) can be derived using the design method based on equivalent control. To proceed, consider a forward expression of (7.8)

$$\begin{aligned}\sigma_{k+1} &= e_{k+1} - e_0 + \varepsilon_{k+1} \\ \varepsilon_{k+1} &= \varepsilon_k + \beta e_k.\end{aligned} \tag{7.11}$$

The objective of a sliding mode controller is to achieve $\sigma_{k+1} = 0$, therefore, we need to derive an explicit expression in terms of the sliding surface and system dynamics. For this substitute ε_{k+1} and the expression $\varepsilon_k - e_0 = \sigma_k - e_k$ into the expression of the sliding surface in (7.11) in order to eliminate the term ε_k from the resulting expression. Equating the resulting expression of σ_{k+1} to zero we obtain

$$\sigma_{k+1} = e_{k+1} + \beta e_k - e_0 + \varepsilon_k = e_{k+1} - (1-\beta)e_k + \sigma_k = 0. \tag{7.12}$$

Note that (7.6) can be rewritten as

$$\begin{aligned} x_{1,k+1} &= \phi_{11}x_{1,k} + \phi_{12}x_{2,k} + \gamma_1 u_k + d_{1,k} \\ x_{2,k+1} &= \phi_{21}x_{1,k} + \phi_{22}x_{2,k} + \gamma_2 u_k + d_{2,k}. \end{aligned} \tag{7.13}$$

Using the relation $e_{k+1} = r_{k+1} - y_{k+1} = r_{k+1} - x_{1,k+1}$, and substituting y_{k+1} or $x_{1,k+1}$ dynamics into (7.12) and solve for the equivalent control u_k^{eq}, we have

$$u_k^{\mathrm{eq}} = \gamma_1^{-1}\left[r_{k+1} - \lambda e_k - \phi_{11}x_{1,k} - \phi_{12}x_{2,k} - d_{1,k} + \sigma_k.\right] \tag{7.14}$$

Note that the control (7.14) is based on the state $x_{2,k}$ as well as the current value of the disturbance $d_{1,k}$ which are uncertain and therefore cannot be implemented in this current form. To overcome this, we will introduce the state estimate and disturbance estimate. Therefore, the final controller structure is given by (7.9) which is to replace $x_{2,k}$ and $d_{1,k}$ in the equivalent control (7.14) by the state estimate $\hat{x}_{2,k}$ and disturbance estimate $\hat{d}_{1,k-1} = \gamma_1\hat{\eta}_{k-1}$.

In order to verify the second part of Theorem 7.1 with regard to the losed-loop stability, first derive the closed-loop state dynamics. Substitute u_k in (7.9) and $\hat{x}_{2,k}$ into (7.6), we obtain

$$\begin{aligned} \mathbf{x}_{k+1} = &\left[\Phi - \boldsymbol{\gamma}\gamma_1^{-1}\left([\phi_{11}\ \phi_{12}] - \lambda\mathbf{c}\right)\right]\mathbf{x}_k + \mathbf{d}_k - \boldsymbol{\gamma}\hat{\eta}_{k-1} + \boldsymbol{\gamma}\gamma_1^{-1}\phi_{12}\tilde{x}_{2,k} \\ &+ \boldsymbol{\gamma}\gamma_1^{-1}(r_{k+1} - \lambda r_k) + \boldsymbol{\gamma}\gamma_1^{-1}\sigma_k \end{aligned} \tag{7.15}$$

where $\tilde{x}_{2,k} = x_{2,k} - \hat{x}_{2,k}$ is state estimation error.

Now rewrite (7.12) as follows

$$\sigma_{k+1} = r_{k+1} - \mathbf{c}\mathbf{x}_{k+1} - \lambda(r_k - \mathbf{c}\mathbf{x}_k) + \sigma_k = 0 \tag{7.16}$$

and substitute (7.15) into (7.16), which yields the closed-loop sliding dynamics

$$\sigma_{k+1} = \gamma_1\hat{\eta}_{k-1} - d_{1,k} - \phi_{12}\tilde{x}_{2,k}. \tag{7.17}$$

As expected, due to the fact that the estimates $\hat{x}_2$ and $\hat{d}_1$ are used in the control law, the sliding function σ_k no longer converges to the origin as desired but convergences to a region around the origin. The size of this region depends on the performance of the state and disturbance estimation, and will be shown to be of $O(T^2)$.

Returning to the stability issue of (7.15). Since the system being studied is of 2nd order, it is easy to compute the closed-poles $z_1 = \lambda$ and $z_2 = \frac{k_{\mathrm{fv}}Te^{-k_{\mathrm{fv}}T} + e^{-k_{\mathrm{fv}}T} - 1}{k_{\mathrm{fv}}T + e^{-k_{\mathrm{fv}}T} - 1}$. The first pole is a function of the integral constant β while the second pole is the open-loop zero and is stable for $T > 0$. Thus, the system is stable as long as β is properly selected.

Finally, since it is desired to achieve proper performance characteristics for the output tracking error, we will derive the tracking error dynamics in terms of the design parameter λ. Substitution of (7.15) into $y_{k+1} = \mathbf{c}\mathbf{x}_{k+1}$ yields the dynamics

$$y_{k+1} = -\lambda e_k + r_{k+1} + d_{1,k} - \gamma_1 \hat{\eta}_{k-1} + \phi_{12}\tilde{x}_{2,k} + \sigma_k. \tag{7.18}$$

Substituting the result $\sigma_k = \gamma_1 \hat{\eta}_{k-2} - d_{1,k-1} - \phi_{12}\tilde{x}_{2,k-1}$ obtained from (7.17) into (7.18)

$$y_{k+1} = -\lambda e_k + r_{k+1} + d_{1,k} - \gamma_1 \hat{\eta}_{k-1} - d_{1,k-1} + \gamma_1 \hat{\eta}_{k-2} + \phi_{12}(\tilde{x}_{2,k} - \tilde{x}_{2,k-1}) \tag{7.19}$$

which yields the tracking error dynamics

$$e_{k+1} = \lambda e_k + \delta_k \tag{7.20}$$

where δ_k is given by (7.10) as a sum of state and disturbance estimation errors.

Remark 7.1 It will be shown in subsequent subsections that under smoothness and boundedness conditions for the disturbance, the disturbance estimate $\hat{\eta}_k$ and the state estimate $\hat{x}_{2,k}$ converge to their actual values.

7.3.2 Disturbance Observer Design

In order to design the observer we need to first note that according to *Property* 7.1 the disturbance can be written as

$$\mathbf{d}_k = \boldsymbol{\gamma}\eta_k + O(T^3) \tag{7.21}$$

where the magnitude of $O(T^3)$ term is in proportion to T^3. Define the observer

$$\begin{aligned} \mathbf{x}_{d,k} &= \Phi \mathbf{x}_{d,k-1} + \boldsymbol{\gamma} u_{k-1} + \boldsymbol{\gamma}\hat{\eta}_{k-1} \\ y_{d,k-1} &= \mathbf{c}\mathbf{x}_{d,k-1} \end{aligned} \tag{7.22}$$

where $\mathbf{x}_d$ is the observer state vector, y_d is the observer output vector, $\hat{\eta}_k$ is the disturbance estimate and will act as the 'control input' to the observer, therefore the estimate $\hat{\mathbf{d}}_{k-1} = \boldsymbol{\gamma}\hat{\eta}_{k-1}$ and $\hat{d}_{1,k-1} = \gamma_1 \hat{\eta}_{k-1}$. Since the disturbance estimate will be used in the final control signal it must not be overly large, therefore, it is wise to avoid a deadbeat design. For this reason we will use an observer based on an integral sliding surface

$$\begin{aligned} \sigma_{d,k} &= e_{d,k} - e_{d,0} + \varepsilon_{d,k} \\ \varepsilon_{d,k} &= \varepsilon_{d,k-1} + \beta_d e_{d,k-1} \end{aligned} \tag{7.23}$$

where $e_{d,k} = y_k - y_{d,k}$ is the output estimation error, $e_{d,0}$ is the initial estimation error, $\sigma_{d,k}$, $\varepsilon_{d,k}$ are the sliding function and integral vectors, and β_d is an integral gain matrix.

Since the sliding surface (7.23) is the same as (7.8), the set $(y_k, \mathbf{x}_{d,k}, u_k + \hat{\eta}_k, y_{d,k}, \sigma_{d,k})$ has duality with the set $(r_k, \mathbf{x}_k, u_k, y_k, \sigma_k)$, therefore, $\hat{\eta}_k$ is given by

$$\hat{\eta}_{k-1} = \gamma_1^{-1}\left(y_k - \lambda_d e_{d,k-1} - [\phi_{11}\ \phi_{12}]\mathbf{x}_{d,k-1} + \sigma_{d,k-1}\right) - u_{k-1} \quad (7.24)$$

where $\lambda_d = 1 - \beta_d$. Expression (7.24) is the required disturbance estimate and is similar in form to (7.14).

The stability and convergence property of the observer is summarized in the following theorem.

Theorem 7.2 *The state dynamics (7.22) is stable when closing the loop with the disturbance estimate (7.24). The disturbance estimate $\hat{\eta}_{k-1}$ from (7.24) converges to an $O(T)$ bound around the actual disturbance η_{k-1} asymptotically.*

Proof To analyze the stability of the observer, substitute (7.24) into (7.22) and follow the same steps of the derivation of (7.15) to obtain

$$\mathbf{x}_{d,k} = \left[\Phi - \boldsymbol{\gamma}\gamma_1^{-1}\left([\phi_{11}\ \phi_{12}] - \lambda_d\mathbf{c}\right)\right]\mathbf{x}_{d,k-1} + \boldsymbol{\gamma}\gamma_1^{-1}[y_k - \lambda_d y_{k-1}] + \boldsymbol{\gamma}\gamma_1^{-1}\sigma_{d,k-1}. \quad (7.25)$$

By substituting (7.25) into $\sigma_{d,k+1}$, the sliding dynamics becomes $\sigma_{d,k} = 0$. Therefore,

$$\mathbf{x}_{d,k} = \left[\Phi - \boldsymbol{\gamma}\gamma_1^{-1}\left([\phi_{11}\ \phi_{12}] - \lambda_d\mathbf{c}\right)\right]\mathbf{x}_{d,k-1} + \boldsymbol{\gamma}\gamma_1^{-1}[y_k - \lambda_d y_{k-1}]. \quad (7.26)$$

Subtracting (7.26) from a delayed form of the system (7.6) and substituting $\mathbf{d}_{k-1} = \boldsymbol{\gamma}\eta_{k-1} + O(T^3)$ we obtain

$$\Delta\mathbf{x}_{d,k} = \left[\Phi - \boldsymbol{\gamma}\gamma_1^{-1}\left([\phi_{11}\ \phi_{12}] - \lambda_d\mathbf{c}\right)\right]\Delta\mathbf{x}_{d,k-1} + O(T^3)$$

where $\Delta\mathbf{x}_{d,k} = \mathbf{x}_k - \mathbf{x}_{d,k}$. From (7.27) we see that the convergence of the disturbance observer states, $\mathbf{x}_{d,k}$, to the actual system states $\mathbf{x}_k$, depends only on the matrix $\left[\Phi - \boldsymbol{\gamma}\gamma_1^{-1}\left([\phi_{11}\ \phi_{12}] - \lambda_d\mathbf{c}\right)\right]$ whose stability is dependent on the selection of the constant λ_d. Also note that premultiplication of (7.27) with $\mathbf{c}$ yields the tracking error dynamics

$$e_{d,k} = \lambda_d e_{d,k-1}. \quad (7.27)$$

To prove the second part of theorem, subtract (7.22) from a delayed (7.6) to obtain

$$\Delta\mathbf{x}_{d,k} = \Phi\Delta\mathbf{x}_{d,k-1} + \boldsymbol{\gamma}(\eta_{k-1} - \hat{\eta}_{k-1}) + O(T^3). \quad (7.28)$$

To obtain the relationship between η_k and $\hat{\eta}_k$, premultiplying both sides of (7.28) with $\mathbf{c}$ and substituting (7.27) yield

$$\hat{\eta}_{k-1} = \gamma_1^{-1}\left([\phi_{11}\ \phi_{12}] - \lambda_{\mathrm{d}}\mathbf{c}\right)\Delta\mathbf{x}_{\mathrm{d},k-1} + \eta_{k-1}. \tag{7.29}$$

Substituting (7.27) recursively we have

$$\Delta\mathbf{x}_{\mathrm{d},k-1} = \left[\boldsymbol{\Phi} - \boldsymbol{\gamma}\gamma_1^{-1}\left([\phi_{11}\ \phi_{12}] - \lambda_{\mathrm{d}}\mathbf{c}\right)\right]^{k-1}\Delta\mathbf{x}_{\mathrm{d},0} + O(T^2).$$

Substituting (7.30) into (7.29) we obtain

$$\begin{aligned}\hat{\eta}_{k-1} = {} & \gamma_1^{-1}\left([\phi_{11}\ \phi_{12}] - \lambda_{\mathrm{d}}\mathbf{c}\right)\left[\boldsymbol{\Phi} - \boldsymbol{\gamma}\gamma_1^{-1}\left([\phi_{11}\ \phi_{12}] - \lambda_{\mathrm{d}}\mathbf{c}\right)\right]^{k-1}\Delta\mathbf{x}_{\mathrm{d},0} \\ & + \gamma_1^{-1}\left([\phi_{11}\ \phi_{12}] - \lambda_{\mathrm{d}}\mathbf{c}\right)O(T^2) + \eta_{k-1}.\end{aligned} \tag{7.30}$$

For this particular system it can be shown that $\phi_{11} = 1$, $\phi_{12} = \frac{1-e^{-k_{\mathrm{fv}}T}}{k_{\mathrm{fv}}} = O(T)$, $\gamma_1 = O(T^2)$ and that a reasonable choice of the controller pole is $\lambda \approx 1 - O(T)$. From these it can be found that $([\phi_{11}\ \phi_{12}] - \lambda_{\mathrm{d}}\mathbf{c}) = O(T)$ and, since, the matrix $\left[\boldsymbol{\Phi} - \boldsymbol{\gamma}\gamma_1^{-1}\left([\phi_{11}\ \phi_{12}] - \lambda_{\mathrm{d}}\mathbf{c}\right)\right]$ is stable, then

$$\lim_{k\to\infty}\left[\boldsymbol{\Phi} - \boldsymbol{\gamma}\gamma_1^{-1}\left([\phi_{11}\ \phi_{12}] - \lambda_{\mathrm{d}}\mathbf{c}\right)\right]^{k-1} = 0$$

and the disturbance estimate will converge to a worst case of $O(T)$ around the actual disturbance.

Remark 7.2 It should be noted that the sliding dynamics (7.26), output error dynamics (7.27), hence the disturbance estimation error, are independent of the control input u_k and the state estimation error. This decoupling property is highly desirable for any control system combined with observers.

7.3.3 State Observer Design

State estimation is accomplished with the following state observer

$$\hat{\mathbf{x}}_{k+1} = \boldsymbol{\Phi}\hat{\mathbf{x}}_k + \boldsymbol{\gamma}u_k + \mathbf{l}(y_k - \hat{y}_k) + \hat{\mathbf{d}}_{k-1} \tag{7.31}$$

where $\hat{\mathbf{x}}_k$, $\hat{y}_k$ are the state and output estimates and $\mathbf{l}$ is a vector valued observer gain. Notice that in observer (7.31), the term $\hat{\mathbf{d}}_{k-1}$ has been added to compensate for the disturbance. Since only the delayed disturbance is available it is necessary to investigate the effect it may have on the state estimation. Subtracting (7.31) from (7.6) yields

$$\tilde{\mathbf{x}}_{k+1} = [\Phi - \mathbf{lc}]\tilde{\mathbf{x}}_k + \mathbf{d}_k - \hat{\mathbf{d}}_{k-1} \tag{7.32}$$

where $\tilde{\mathbf{x}}_k = \mathbf{x}_k - \hat{\mathbf{x}}_k$ is the state estimation error. The solution of (7.32) is given by

$$\tilde{\mathbf{x}}_k = [\Phi - \mathbf{lc}]^k \tilde{\mathbf{x}}_0 + \sum_{i=0}^{k-1} \left([\Phi - \mathbf{lc}]^{k-1-i} (\mathbf{d}_i - \hat{\mathbf{d}}_{i-1}) \right). \tag{7.33}$$

The state estimation error $\tilde{x}_{2,k} = x_{2,k} - \hat{x}_{2,k}$ is given by

$$\tilde{x}_{2,k} = [0\ 1]\,[\Phi - \mathbf{lc}]^k\, \tilde{\mathbf{x}}_0 + \sum_{i=0}^{k-1} [0\ 1] \left([\Phi - \mathbf{lc}]^{k-1-i} (\mathbf{d}_i - \hat{\mathbf{d}}_{i-1}) \right). \tag{7.34}$$

Using *Property* 7.1, $\mathbf{d}_k - \hat{\mathbf{d}}_{k-1} = O(T^2)$ when no discontinuity occurs. From (7.32) and *Property* 7.2 we know that the ultimate bound on $\tilde{\mathbf{x}}_k$, hence $\tilde{x}_{2,k}$ is $O(T)$.

7.3.4 Ultimate Tracking Error Bound

Now we are in a position to derive the ultimate tracking error bound of the piezo-motor stage (7.1) when the proposed discrete-time ISM control is applied.

Theorem 7.3 *Using the discrete-time ISM control law (7.9), the disturbance observer (7.24) and (7.22), the state observer (7.31), the ultimate bound of output tracking error is $O(T^2)$.*

Proof In order to calculate the output tracking error bound we must find the bound of δ_k in (7.20). From (7.34) we can derive the difference $\tilde{x}_{2,k} - \tilde{x}_{2,k-1}$ as

$$\begin{aligned}\tilde{x}_{2,k} - \tilde{x}_{2,k-1} = {} & [0\ 1][I - (\Phi - \mathbf{lc})](\Phi - \mathbf{lc})^{k-1}\tilde{\mathbf{x}}_0 - \sum_{i=0}^{k-1} [0\ 1] \left([\Phi - \mathbf{lc}]^{k-1-i} (\mathbf{d}_i - \hat{\mathbf{d}}_{i-1}) \right) \\ & + \sum_{i=0}^{k-2} [0\ 1] \left([\Phi - \mathbf{lc}]^{k-1-i} (\mathbf{d}_i - \hat{\mathbf{d}}_{i-1}) \right) \end{aligned} \tag{7.35}$$

where I is a unity matrix. (7.35) can be simplified to

$$\tilde{x}_{2,k} - \tilde{x}_{2,k-1} = \begin{bmatrix} 0 & 1 \end{bmatrix} [I - (\Phi - \mathbf{lc})]\,(\Phi - \mathbf{lc})^{k-1}\,\tilde{\mathbf{x}}_0 - \left(d_{2,k} - \gamma_2 \hat{\eta}_{k-1}\right). \tag{7.36}$$

Since $(\Phi - \mathbf{lc})^k \to 0$ ultimately, we have

$$\tilde{x}_{2,k} - \tilde{x}_{2,k-1} = -(d_{2,k} - \gamma_2 \hat{\eta}_{k-1}) \tag{7.37}$$

as $k \to \infty$. Substituting (7.37) into (7.10) yields

$$\delta_k = -(d_{1,k} - \gamma_1 \hat{\eta}_{k-1} - d_{1,k-1} + \gamma_1 \hat{\eta}_{k-2}) - \phi_{12}(d_{2,k} - \gamma_2 \hat{\eta}_{k-1}). \tag{7.38}$$

Next by substitution of the relations $d_{1,k} = \gamma_1 \eta_k + O(T^3)$, $d_{2,k} = \gamma_2 \eta_k + O(T^3)$, and into (7.38), we obtain

$$\delta_k = -\gamma_1(\eta_k - \hat{\eta}_{k-1} - \eta_{k-1} + \hat{\eta}_{k-2}) - \phi_{12}\gamma_2(\eta_k - \hat{\eta}_{k-1}) + O(T^3). \tag{7.39}$$

Since we are trying to calculate the steady state error bound, using the fact that at steady state $\hat{\eta}_k = \eta_k + O(T)$ and substituting it in (7.39)

$$\delta_k = -\gamma_1(\eta_k - 2\eta_{k-1} + \eta_{k-2} + O(T)) - \phi_{12}\gamma_2(\eta_{k-1} - \eta_{k-2} + O(T)) + O(T^3). \tag{7.40}$$

For sampled-data system (7.6), $\boldsymbol{\gamma} = O(T)$ and $\phi_{12} = O(T)$. If η_k is smooth and bounded, then from *Property* 7.1

$$\delta_k = O(T^2) \cdot O(T) + O(T) \cdot O(T) \cdot O(T) + O(T^3) = O(T^3). \tag{7.41}$$

In order to derive the output tracking error bound, look into output tracking error dynamics derived in Theorem 7.1, $e_{k+1} = \lambda e_k + \delta_k$, whose solution is

$$e_k = \lambda^k e_0 + \sum_{i=0}^{k-1} \lambda^i \delta_{k-i-1}. \tag{7.42}$$

According to *Property* 7.2, the ultimate error bound of e_k will be one order higher than the bound of δ_k, therefore, since the bound of δ_k is $O(T^3)$ the ultimate bound of e_k is $O(T^2)$, i.e.,

$$|e_k| = O(T^2). \tag{7.43}$$

We have computed the tracking error in the case when the disturbance is smooth and bounded. Now, we look at what happens to the tracking error when there is a discontinuity in the disturbance, i.e., when there is a change in the sign of x_2. Consider the disturbance term associated with the closed-loop system (7.38). It can be reasonably assumed that the discontinuity occurs rarely, therefore, if we assume that the discontinuity occurs at the kth sampling point, then $\delta_k = O(T^2)$ rather than $O(T^3)$ as the difference $d_{1,k} - 2d_{1,k-1} + d_{1,k-2}$ will no longer be $O(T^3)$ but of the order of $d_{1,k}$ which is $O(T)$. If the discontinuity occurs at a time instance k', then $\delta_k = O(T)$ at $k = k', k'+1, k'+2$, and return to $\delta_k = O(T^3)$ for subsequent sampling instances. Therefore the solution of (7.42) would lead to the worst case error bound

$$|e_k| = O(T) \tag{7.44}$$

for certain time interval but $O(T^2)$ ultimately.

We have computed the tracking error in the case when the disturbance is smooth and bounded. Now, we look at what happens to the tracking error when there is a discontinuity in the disturbance, i.e., when there is a change in the sign of x_2. Consider the disturbance term associated with the closed-loop system (7.38). It can be reasonably assumed that the discontinuity occurs rarely, therefore, if we assume that the discontinuity occurs at the kth sampling point, then $\delta_k = O(T^2)$ rather than $O(T^3)$ as the difference $d_{1,k} - 2d_{1,k-1} + d_{1,k-2}$ will no longer be $O(T^3)$ but of the order of $d_{1,k}$ which is $O(T)$. If the discontinuity occurs at a time instance k', then $\delta_k = O(T)$ at $k = k', k'+1, k'+2$, and return to $\delta_k = O(T^3)$ for subsequent sampling instances. Therefore the solution of (7.42) would lead to the worst case error bound

$$|e_k| = O(T) \tag{7.45}$$

for certain time interval but $O(T^2)$ ultimately.

7.3.5 Experimental Investigation

The experimental system used is shown in Fig. 7.3. The nominal parameters of the system are $m = 1\,\text{kg}$, $k_{\text{fv}} = 144\,\text{N}$ and $k_{\text{f}} = 6\,\text{N/V}$. This simple linear model does not contain any nonlinear and uncertain effects such as the frictional force in the mechanical part, high-order electrical dynamics of the driver, loading condition, etc., which are hard to model in practice. In general, producing a high precision model will require more efforts than performing a control task with the same level of precision.

7.3.5.1 Determination of Controller Parameters

In order to select an appropriate sampling period T, the open-loop zero of the system is plotted in Fig. 7.4 as a function of sampling period. We see from Fig. 7.4 that a sampling period below 10^{-4} second would produce a less stable open-loop zero.

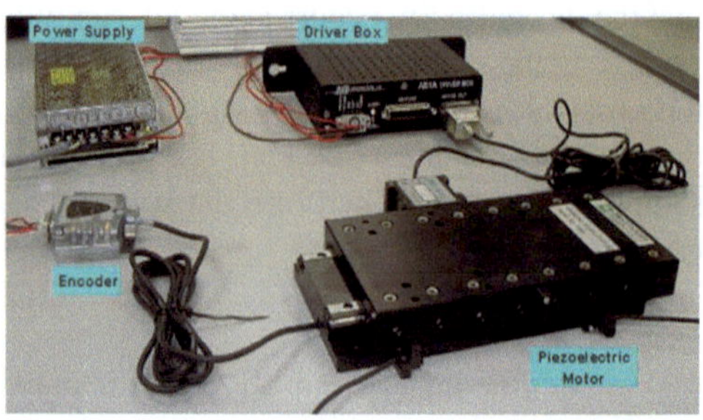

Fig. 7.3 The piezo motor driven linear motion stage

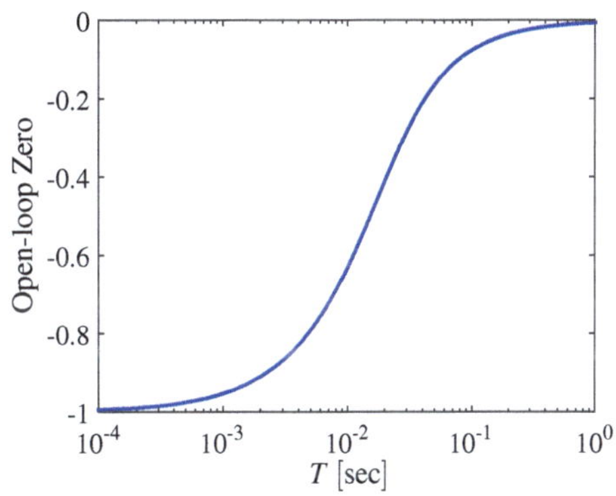

Fig. 7.4 Open-loop zero with respect to sampling period

On the other hand, an over large sampling period will degrade the feedback effect. In the experimental tests we select two sampling periods of 1ms and 10ms respectively.

To proceed with the implementation, three parameters need to be designed: the state observer gain $\mathbf{l}$ (computed such that $\Phi - \mathbf{lc}$ has the desired poles), the disturbance observer integral gain matrix β_{d}, and the controller integral gain β. The state observer gain is selected such that the observer poles are (0.4, 0.4). This selection is arbitrary, but, the poles are selected to ensure quick convergence. Next, the constant β_{d} is designed. To ensure the quick convergence of the disturbance observer, β_{d} is selected such that the observer pole at 1ms sampling is $\lambda_{\mathrm{d}} = 0.9$ and at 10 ms sampling is $\lambda_{\mathrm{d}} = 0.6$. Since the remaining pole of the observer is the non-zero open-loop zero (-0.958 at 1 ms and -0.683 at 10 ms), it is the dominant pole. Finally, the controller pole is selected as $\lambda = 0.958$ at 1 ms sampling and $\lambda = 0.683$ at 10 ms sampling which are found to be the best possible after some trials. Thus, the design parameters are as follows:

$$\begin{aligned}
\mathbf{l}_{@T=10\mathrm{ms}} &= \begin{bmatrix} 0.4269 & 5.3260 \end{bmatrix}^\top \\
\mathbf{l}_{@T=1\mathrm{ms}} &= \begin{bmatrix} 1.059 & 231.048 \end{bmatrix}^\top \\
\beta_{\mathrm{d},@T=10\mathrm{ms}} &= 1 - \lambda_{\mathrm{d}} = 0.4 \\
\beta_{\mathrm{d},@T=1\mathrm{ms}} &= 1 - \lambda_{\mathrm{d}} = 0.1 \\
\beta_{@T=10\mathrm{ms}} &= 1 - \lambda = 0.317 \\
\beta_{@T=1\mathrm{ms}} &= 1 - \lambda = 0.042.
\end{aligned}$$

The reference trajectory r_k is shown in Fig. 7.5 and as it can be seen the initial conditions are $e_0 = 0$ and $e_{\mathrm{d},0} = 0$. For comparison purposes, PI control is also applied to piezo-motor and PI gains were optimized through intensive tests. The PI gains are at $k_{\mathrm{p}} = 1.5$ and $k_{\mathrm{i}} = 55$ at the sampling period of 1ms and $k_{\mathrm{p}} = 0.6$ and $k_{\mathrm{i}} = 6$ at the sampling period of 10ms. To verify that the PI gains used are optimally tuned, PI gains are made to vary from their optimal values by $\pm 20\,\%$. The optimally tuned PI gains can be determined when other PI values either produce larger tracking errors or lead to oscillatory responses (Figs. 7.6, 7.7 and 7.8).

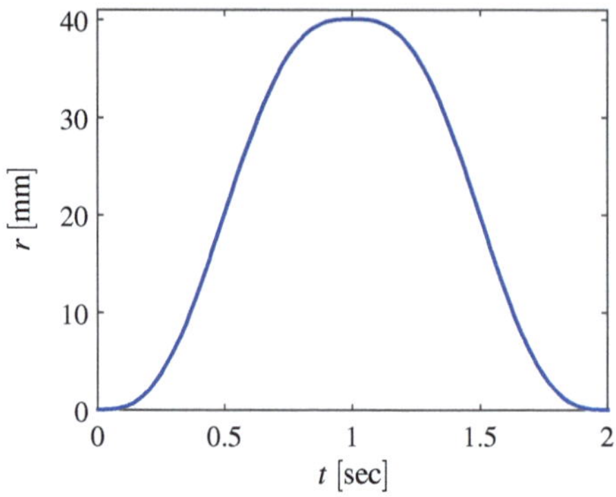

Fig. 7.5 The reference trajectory

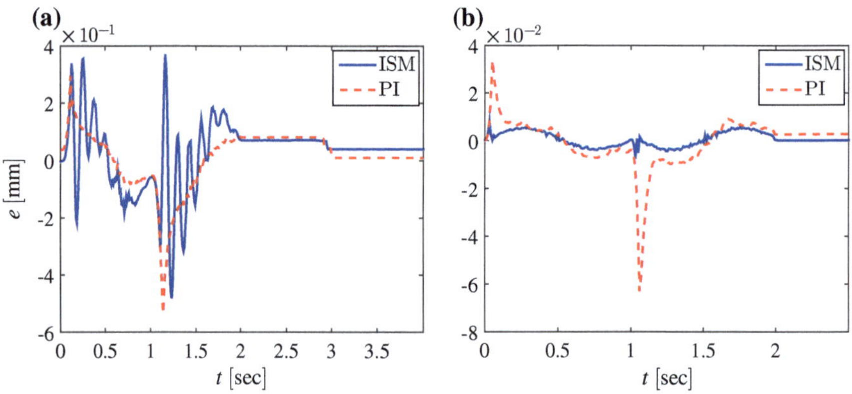

Fig. 7.6 Tracking error of ISM control and PI control at **a** 10 ms sampling period and **b** 1 ms sampling period

7.3.5.2 Experimental Results and Discussions

ISM control is applied with both sampling periods of 10ms and 1ms. For comparison the PI control is also applied. The tracking errors of both controllers are shown in Fig. 7.12. It can be seen that at 10ms the performance of ISM control and PI controller are comparable whereas at 1ms the performance of the ISM control is far better. Figure 7.13 shows the control signals of ISM control and PI. It can be seen that ISM control profile at 1ms is smoother comparing with at 10ms. In Fig. 7.14 the reference velocity and the estimated velocity under the state observer is plotted. It is clearly seen that the smaller sampling period of 1ms produces a better estimate, $\hat{x}_2$, in comparison with 10ms sampling period. Figure 7.9 demonstrates estimation result of the disturbance observer. It can be observed that the sliding mode control will produce some chattering due to the limited sampling frequency. It is well known that sliding mode control requires a fast switching frequency in order to maintain

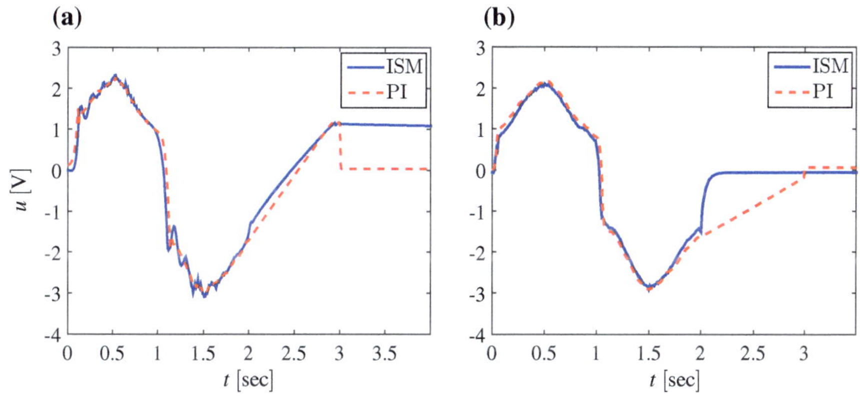

Fig. 7.7 Comparison of the control inputs of ISM and PI controllers at **a** 10 ms sampling period and **b** 1 ms sampling period

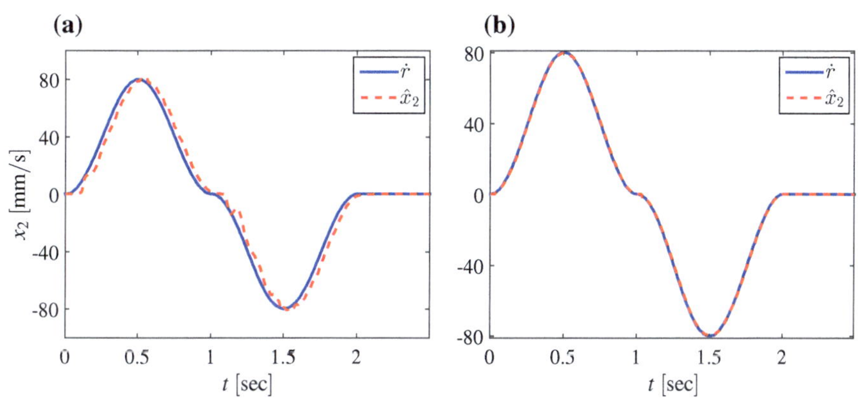

Fig. 7.8 Estimated state $\hat{x}_2$ and reference velocity $\dot{r}$ at **a** 10 ms sampling period and **b** 1 ms sampling period

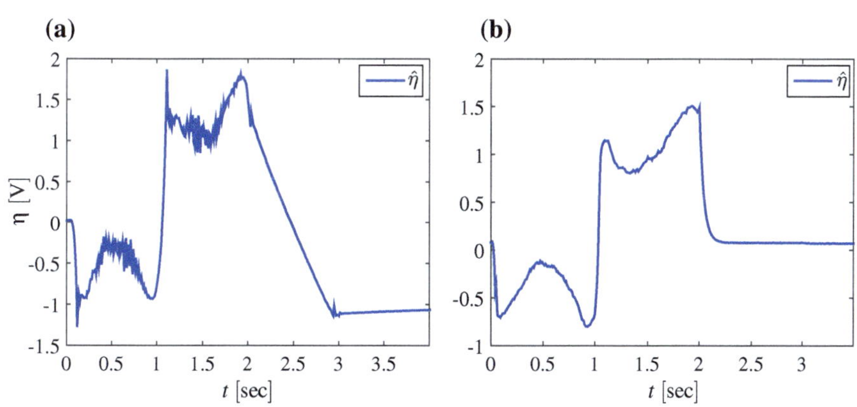

Fig. 7.9 Disturbance observer response at **a** 10 ms sampling period **b** 1 ms sampling period

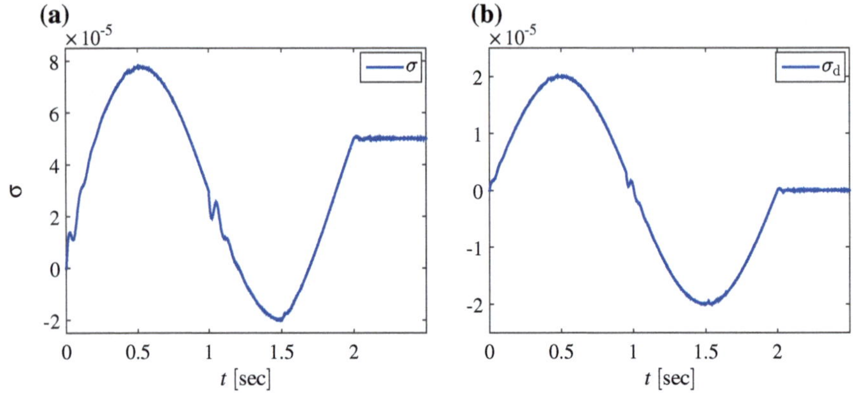

Fig. 7.10 Sliding function **a** σ and **b** $\sigma_{\rm d}$ at 1ms sampling period

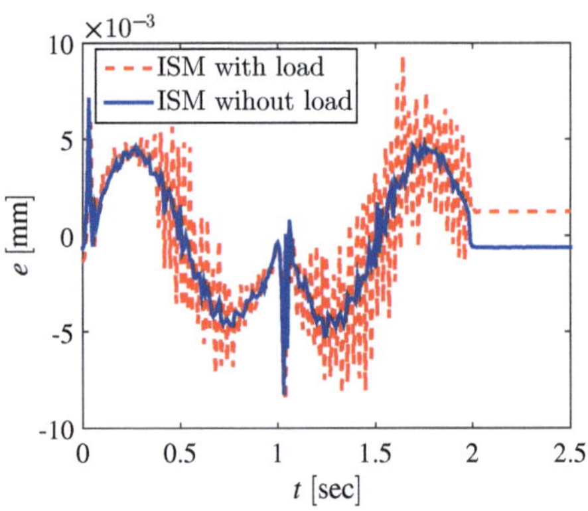

Fig. 7.11 Tracking errors of ISM control with and without the 2.5 kg load at 1ms sampling period

the sliding motion. In this chapter, the output ISM control is designed in discrete-time, but the real plant is analog in nature. Moreover, through analysis we have shown the tracking error bound is proportional to the size of the sampling period T. Therefore, we can expect a smoother control response and lower tracking error when the sampling period is reduced to 1ms. Nonetheless, the magnitude of the tracking error is at the scale of 5×10^{-4}, confirming the theoretical error bound of $O(T^2) = O(0.01^2) = O(10^{-4})$. It is interesting to note that, when reducing the sampling period by 10 times, the tracking error bound is about 100 times less. This result is consistent the theoretical analysis, because the magnitude of the tracking error is at the scale of 5×10^{-6}, or equivalently $O(T^2) = O(0.001^2) = O(10^{-6})$. Figure 7.10a shows the sliding function σ at 1 ms sampling period. From Fig. 7.9b we can see that the ISM control can respond very fast when encountering a discontinuity generated by the static friction at around 1s. This fast control response is owing to the incorporation of the disturbance observer which can effectively estimate uncertain changes. Figure 7.10b shows the sliding function $\sigma_{\rm d}$ which has a magnitude of 20×10^{-6} which is quite small.

Finally, to illustrate the robustness of ISM control, an extra load of 2.5 kg is added to the piezo-motor driven linear motion stage, which is 250 % of the original motor mass of 1 kg, meanwhile the parameters of the controller and observers remain unchanged. Figure 7.11 shows the responses with and without the extra load. It can be seen that the tracking error remains at a low level despite the extra load.

7.4 Discrete-Time Terminal Sliding Mode Control

In this study, the TSM control is designed in discrete-time, but the real plant is analog in nature. TSM control is applied with the sampling period of 1 ms. The tracking errors of both TSM and PI controllers are shown in Fig. 7.12. It can be seen that the tracking performance of the TSM control is far better. Figure 7.13 shows the control signals of TSM and PI. In Fig. 7.14 the reference velocity and the state x_2 are plotted. It is clearly seen that the velocity tracking is very good with minimal chattering. It is well known that sliding mode control requires a fast switching frequency in order to maintain the sliding motion. The proposed TSM control however does not produce much chattering though with the limited sampling frequency.

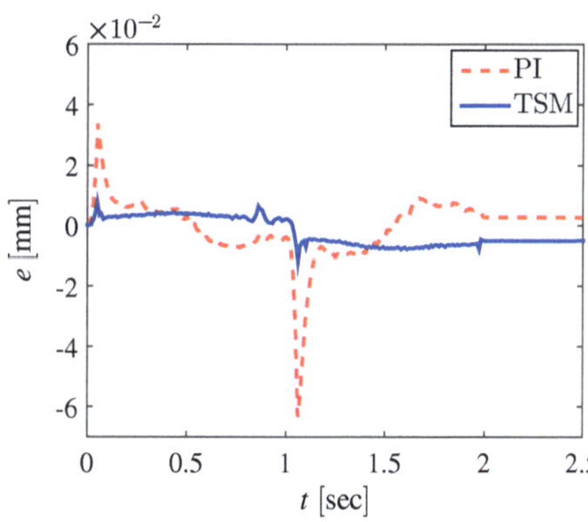

Fig. 7.12 Tracking error of TSM control and PI control

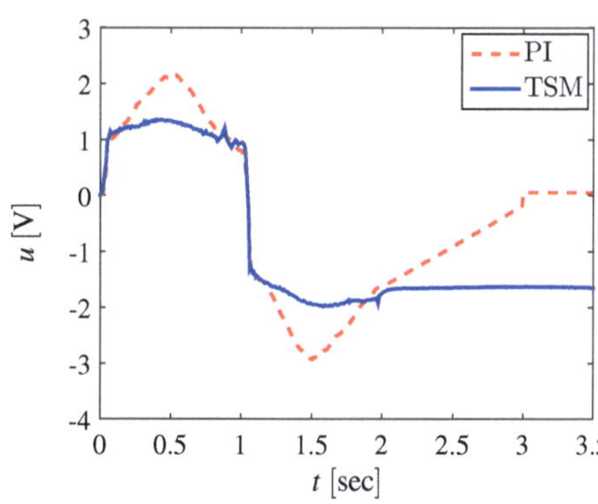

Fig. 7.13 Comparison of the control inputs of TSM and PI controllers

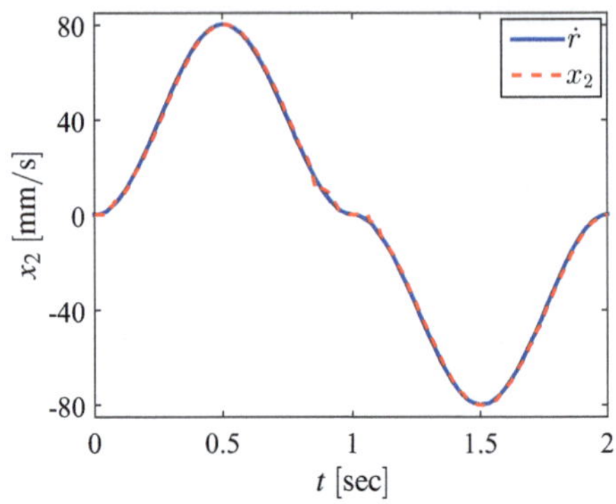

Fig. 7.14 State x_2 and reference velocity $\dot{r}$

Finally, it is also interesting to check the tracking error bound according to the theoretical analysis and experimental result. Through analysis we have shown the tracking error bound is proportional to the size of the sampling period T. Therefore, we can expect a smooth control response and low tracking error when the sampling period is 1ms. The magnitude of the tracking error obtained in the experiment, as shown in Fig. 7.12, is at the scale of 8×10^{-6}, which is consistent with the theoretical error bound of $O(T^2) = O(0.001^2) = O(10^{-6})$.

7.5 Sampled-Data ILC Design

In this section we shall show the ILC design for the piezo-motor stage. Unlike the ISM controller design where the controller structure is designed from the model, the ILC controller has a standard control structure and the design factors $Q(z)$ and $L(z)$ are determined from the nominal model or from experimentally obtained frequency response data.

7.5.1 Controller Parameter Design and Experimental Results

The objective of the ILC design is to achieve as precisely as possible motion control after the smallest number of iterations. Due to the existence of uncertainties and other unmodelled disturbances the most suitable selection would be the current-cycle iterative learning control where the iterative controller would act as an add-on to the feedback controller. Experiments conducted on the system have shown that PI control works quite well and, so, it shall be used as the feedback control law. The optimum PI gains found for this system are $k_\mathrm{p} = 6$ and $k_\mathrm{i} = 10$, [161]. The resulting closed-loop system is given by

$$P_{CL}(z) = \frac{2.826 \times 10^{-6} z^2 - 5.124 \times 10^{-8} z - 2.771 \times 10^{-6}}{z^3 - 2.944 z^2 + 2.888 z - 0.944} \tag{7.46}$$

This system is stable and so it will be possible for us to use the frequency domain tools for the design of the ILC controller.

Since, we want to achieve the best possible tracking performance we will not retune the sampling-time according to Table 5.1, instead we will use the other design factors $Q(z)$ and $L(z)$. Before we proceed with the design of the function $Q(z)$ and $L(z)$ we plot the phase and magnitude diagram for $zP_{CL}(z)$ in order to decide on what type of functions $Q(z)$ and $L(z)$ should be. According to Table 5.1, $L(z)$ cannot be selected as P-type, D-type, or D^2-type as the order of $P(z)$ is 3 and it has no integrators. Therefore, $L(z)$ will be selected as a lead compensator. From Fig. 7.15 we see that the phase falls below $-\frac{\pi}{2}$ and so we need $L(z)$ to have a leading phase of no more than 90°. We select the following simple function $L(s) = 0.1(s+1)$ which would be $L(z) = \frac{z-0.9996}{0.0004}$ in discrete-time, we can also plot $L(z)$ in Fig. 7.16. If we

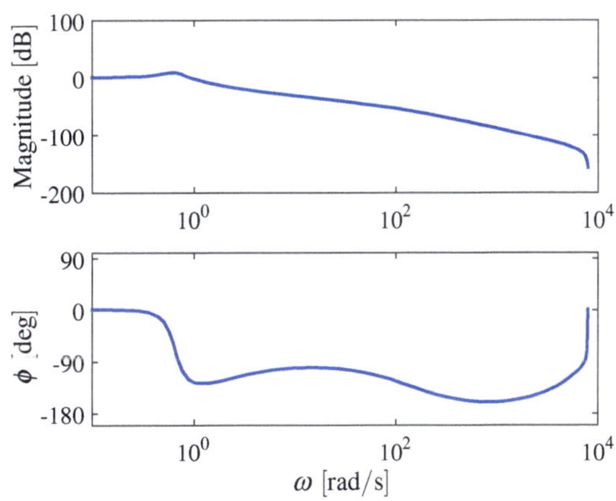

Fig. 7.15 Phase and Magnitude for $zP_{CL}(z)$

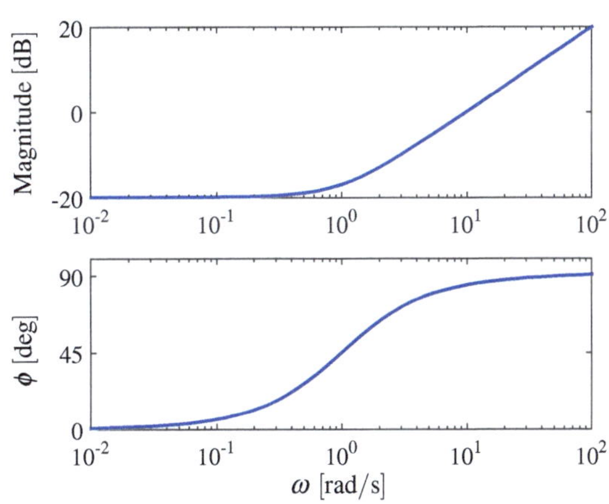

Fig. 7.16 Phase and Magnitude for $L(z)$

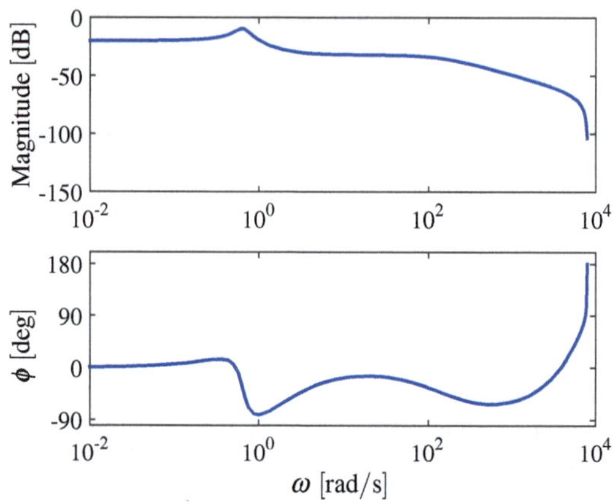

Fig. 7.17 Phase and Magnitude for $zL(z)P_{\mathrm{CL}}(z)$

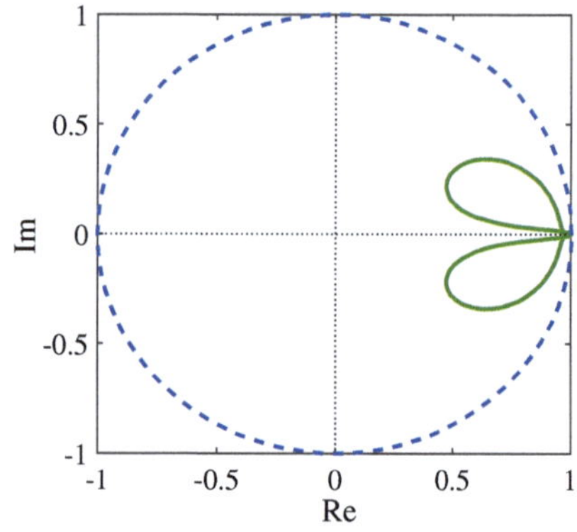

Fig. 7.18 Nyquist diagram for $1 - zL(z)P_{\mathrm{CL}}(z)$

now combine $L(z)$ and $zP_{\mathrm{CL}}(z)$ we obtain the frequency response in Fig. 7.17 which satisfies our requirements. We can also plot the Nyquist diagram for $1-zL(z)P_{\mathrm{CL}}(z)$ to confirm if $|1-zL(z)P_{\mathrm{CL}}(z)| < 1$. This can be seen from Fig. 7.18. Since, stability is achieved the function $Q(z)$ is selected as unity.

We are now ready to implement the ILC control law with the designed parameters. The reference trajectory of the system is shown in Fig. 7.19. Figure 7.20 shows the output tracking error of the system at the 0th and the 15th iterations. It is easily seen that the tracking error is greatly reduced by the 15th iteration and is of a magnitude of 2 μm at the transient and 0.6 μm at steady state. Finally, Fig. 7.21 shows the control effort at the 0th and the 15th iteration.The above results show the exceptional performance of the ILC laws as add-ons to existed feedback control. The rather straight forward design also shows that the method has a lot of promise for practical applications.

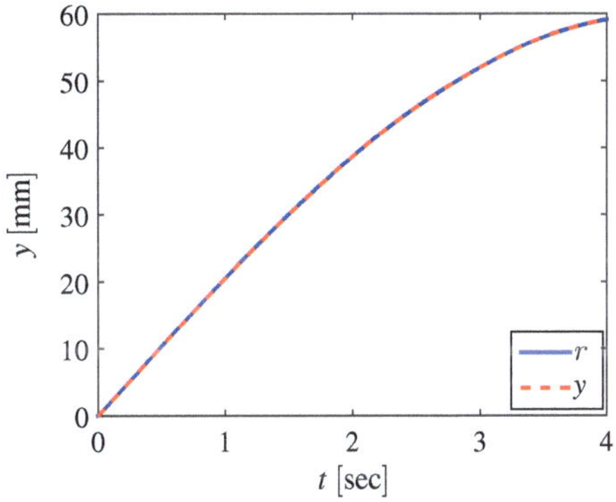

Fig. 7.19 Desired and actual output of the system

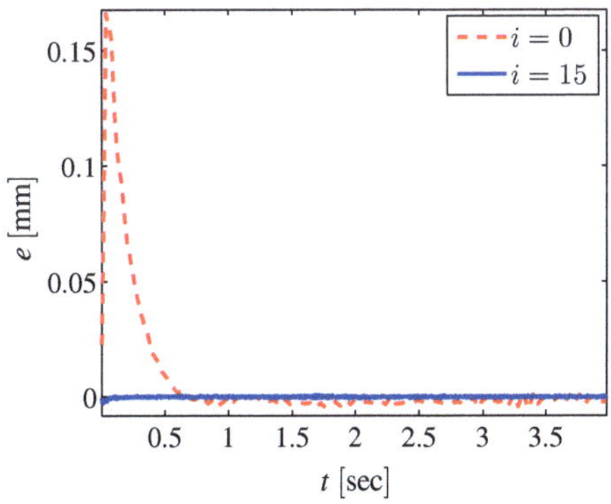

Fig. 7.20 Ouput tracking error of the system at the 0th and the 15th iteration

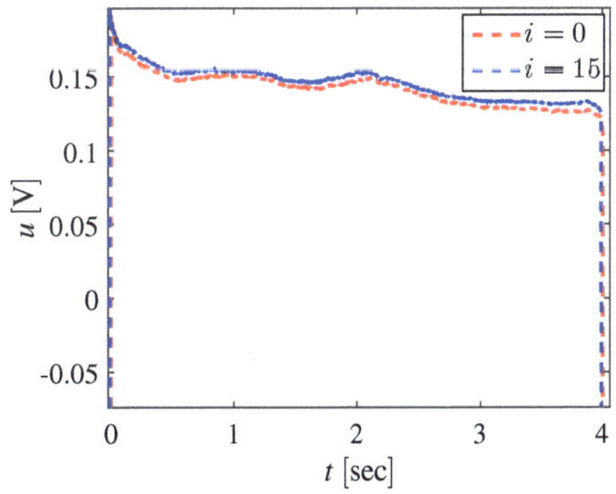

Fig. 7.21 Control input of the system at the 0th and the 15th iteration

7.6 Conclusion

This chapter presents a few the various controller designs for sampled-data systems applied to the tracking control of a piezo-motor driven linear motion stage.

For the ISM control design, proper disturbance and state observers were presented, and in particular the disturbance observer is designed using the idea and method of integral sliding mode to achieve the desired performance. Experimental comparisons with a PI controller evidence the effectiveness of the proposed control method. It is worth to point out that the designs of controller and obervsers are separate, in other words, what we present in this chapter is a modular design approach. Two observers can be added or removed individually according to practical applications. For instance, state observer can be removed if the velocity is accessible. The disturbance observer can also be removed when the disturbance is negligible. In either cases, the ISM control design remain valid and the tracking error bound is guaranteed at least to be $O(T^2)$.

For the ILC design, the parameter and filter selection has been shown. The design procedure has been shown to be straight forward and intuitive. With some basic information about the system it was possible to achieve high-precision motion for a repetitive task.

Chapter 8
Advanced Control for Practical Engineering Applications

Abstract Although the theoretical results for periodic adaptive control are based on purely time-based uncertainties, in practical scenarios, state based periodic uncertainties can also be considered time-based if the system is at steady state. In this study, this is investigated and verified using a PM synchronous motor as a platform. In some systems with low quality components, it is unavoidable to have varying sampling rates at the various A/D ports of the system. Multirate ILC was developed with the intention to, as effectively as possible, attenuate the effects of the lower sampling rate components on the overall system performance. In this study, a Ball-and-Beam apparatus is used to verify that indeed, in spite of having lower sampling rates of certain components, superior performance is achievable. Fuzzy PID has been shown to be effective on highly nonlinear and difficult to model systems. In this study, a coupled tank apparatus is used as a test bed to investigate and verify the effectiveness of fuzzy PID. Experimental results show that even with the uncertain nature of the model, it is possible to achieve high performance of the system. Finally, using the fact that traffic patterns are repeatable, ILC is implemented to improve the flow of traffic in a freeway. It is shown by the simulation results that superior performance can be achieved using a very simple controller structure.

8.1 Introduction

In the previous chapter, the effectiveness of the controller designs for high-precision applications was demonstrated. In this chapter, the application of the controller designs to other engineering applications will be demonstrated. As was discussed previously, a majority of modern controller implementations involve digital microprocessors and, hence, provide the case for discrete-time controllers.

In the first application, periodic adaptive control is implemented on a PM synchronous motor. PM synchronous motors are popular due to high power density, high torque-to-inertia ratio and efficiency. However, control of PM synchronous motors has proven to be challenging due to the state dependent parameters which cause torque ripples. Under steady state operation the PM synchronous motor parameters exhibit periodicity that can be approximated as time dependent. This way it is

K. Abidi and J.-X. Xu, *Advanced Discrete-Time Control*, Studies in Systems, Decision and Control 23, DOI 10.1007/978-981-287-478-8_8

possible to implement periodic adaptive control to deal with the periodicity in the parameters.

In the second application, a Ball-and-Beam apparatus will be used to demonstrate the effectiveness of the multirate ILC. The apparatus used is nonlinear in structure and its A/D port has a low sampling rate (large sampling period) which is ideal for testing the multirate ILC. Multirate ILC has been shown to be an ideal approach to achieve smoother control input and higher control bandwidth under low sampling rates. These aspects will be highlighted in this implementation.

In the third application, a coupled tank apparatus will be used to demonstrate the practical implementation of fuzzy PID. In process industries, control of fluid levels in storage tanks is a common and important control problem. By controlling the fluid level in the tank, material balance can be achieved so that the inflow and outflow are equal in steady state. Nonlinear nature of the system coupled with the difficulties in modeling provide a good opportunity to implement fuzzy PID which is suitable for such systems.

Finally, based on the traffic pattern repeatability, in this work we apply iterative learning control (ILC), [159], aiming at improving the traffic performance in a freeway. The main idea in ILC is to use information from pervious repetitions to come up with new control actions that improve the tracking accuracy as the number of repetitions increases. It has a very simple structure, that is, integration along the iteration axis, and is a memory based learning process. ILC requires much less information of the system variations to yield the desired dynamic behaviors. This is a very desirable feature in traffic control as the traffic model and the exogenous quantities may not be well known in practice. ILC focuses on the input signal update algorithms and differs from most existing control methods in the sense that it exploits every possibility to incorporate past control information such as tracking errors and control input signals into the construction of the present control action.

8.2 Periodic Adaptive Control of a PM Synchronous Motor

8.2.1 Problem Definition

Consider the simplified dynamics of a PM synchronous motor given by

$$\begin{aligned}\dot{\theta}(t) &= \omega(t)\\ \dot{\omega}(t) &= \frac{1}{J}\left[k_t(\theta)u(t) - \tau_l(\theta,\omega)\right]\end{aligned} \tag{8.1}$$

where θ is the angular displacement of the motor, ω is the angular velocity, J is the inertia of the motor shaft, k_t is the torque coefficient and τ_l is the load torque. For a

PM synchronous motor the torque coefficient is given as $k_t(\theta) = \frac{3}{2}\frac{P}{2}\psi_{dm}(\theta)$ where P is the number of poles of the motor and $\psi_{dm}(\theta)$ is given as

$$\psi_{dm}(\theta) = \psi_{d0} + \psi_{d6} + \cos(6\theta)\psi_{d12}\cos(12\theta) + \cdots \tag{8.2}$$

where ψ_{d0}, ψ_{d6} and ψ_{d12} are the DC, 6th and 12th harmonics terms of the flux linkage respectively. The load torque τ_l is assumed to be as a result of an eccentric loading on the shaft as well as the dynamic friction of the shaft bearings. This can be modeled as

$$\tau_l(\theta, \omega) = \mu\cos(\theta) + b\omega \tag{8.3}$$

where μ is a constant coefficient of the eccentric load and b is the damping constant of the shaft bearings.

Consider the discrete-time form of the system model given as

$$\begin{aligned} \theta_{k+1} &= \theta_k + T\omega_k \\ \omega_{k+1} &= \omega_k + \frac{T}{J}\left[k_t(\theta_k)\,u_k - \tau_l(\theta_k, \omega_k)\right] \end{aligned} \tag{8.4}$$

which is an approximate discrete-time model of the PM synchronous motor at a sampling period T.

Remark 8.1 Since the PM synchronous motor parameters are state dependent it is not possible to obtain an exact discrete-time/sampled-data model of the system. Hence, an Euler approximation is used and for sufficiently small T the discretization errors incurred will be insignificant in the overall performance of the system.

Given the system model (8.4), the control objective is to design a controller to achieve as precisely as possible steady speed control of the motor.

8.2.2 Control Strategy and Results

Initially, the effectiveness of a PI speed controller is tested on the system. The PM synchronous motor parameters used for this are: $J = 0.01\,\text{kg}\,\text{m}^2$, $P = 6$, $\psi_{d0} = 2$, $\psi_{d6} = 0.8$, $\psi_{d12} = 0.5$, $\mu = 2.5$ and $b = 0.05\,\text{Nm s}$. The continuous-time PI controller is given as

$$u(t) = 0.5\left[\omega_r - \omega(t)\right] + 2\int\left[\omega_r - \omega(t)\right]dt = 0.5\omega_r + 2\omega_r t - 0.5\omega(t) - 2\theta(t) \tag{8.5}$$

for a constant ω_r. Since the final controller form shown in (8.5) requires no discretization, the PI controller will be used as

$$u_k = 0.15\omega_r + 0.8k\omega_r T - 0.15\omega_k - 0.8\theta_k. \tag{8.6}$$

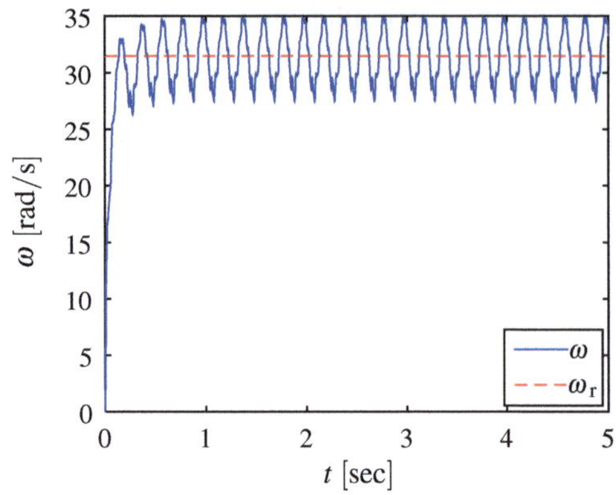

Fig. 8.1 Speed control performance of a PI controller for a target speed of 300 rpm ($\omega_r = 10\pi$ rad/s)

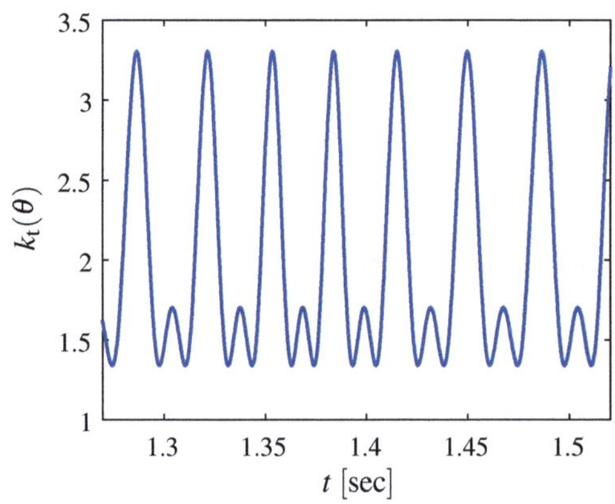

Fig. 8.2 Steady state variation of k_t

The performance of the controller can be seen in Fig. 8.1. As it can be seen from the figure, the average speed of $\omega = 10\pi$ rad/s is achieved, however, there is significant ripple due to the variation in k_t and the load torque τ_l. A closer look at the coefficient k_t and the periodic part of the load torque τ_l shows that, at steady state, the periodicity of the terms becomes time based as seen in Figs. 8.2 and 8.3. It can be seen that the torque coefficient k_t repeats every $1/35$ s while $\mu\cos(\theta)$ repeats every $1/5$ s. This means that there is a possibility to implement periodic adaptive control to handle these variations and produce better speed control performance in comparison to PI control.

The control strategy that will be used will involve using a PI controller to drive the system to steady state after which the periodic adaptive control will be activated to improve the steady state performance as seen in the block diagram in Fig. 8.4. As seen the coefficient ρ will switch between 1 and 0 for either PI control or periodic

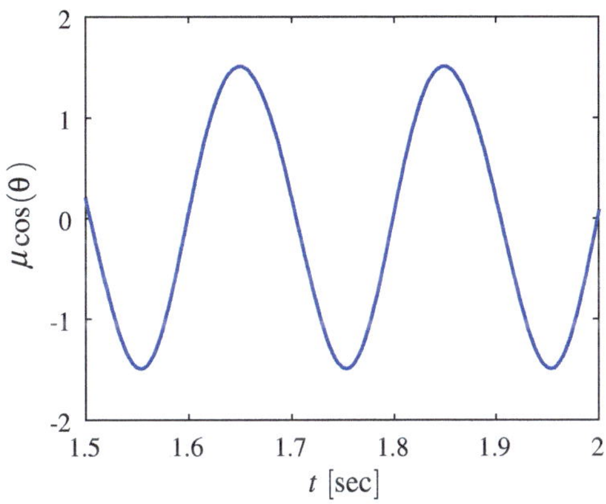

Fig. 8.3 Steady state variation of $\mu\cos(\theta)$

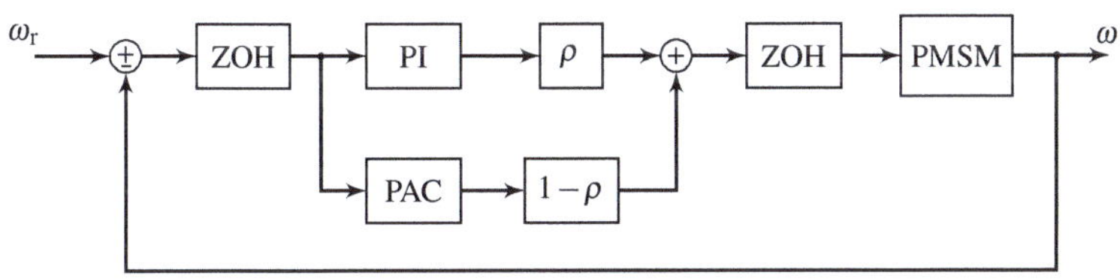

Fig. 8.4 Configuration of the PM synchronous motor speed control system

adaptive control. Initially ρ will be set to 1 until steady state is reached. Once steady state is reached ρ will be set to 0 and periodic adaptive control will replace the PI controller.

In order to design the periodic adaptive control, consider (8.4) and for a constant speed reference ω_r the error dynamics can be written as

$$\omega_r - \omega_{k+1} = \omega_r - \omega_k - \frac{T}{J}\left[k_t(\theta_k)\, u_k - \tau_l(\theta_k, \omega_k)\right]. \tag{8.7}$$

Define $e_k = \omega_r - \omega_k$ and rewrite (8.7) as

$$e_{k+1} = e_k - \phi_{u,k} u_k + \phi_{\mu,k} + \phi_b \omega_k \tag{8.8}$$

where $\phi_{u,k} = \frac{T}{J} k_t(\theta_k)$ and $\phi_{\mu,k} = \frac{T}{J}\mu\cos(\theta_k)$ are uncertain and periodic and $\phi_b = \frac{T}{J} b$ is a constant uncertainty. Let the control law be given as, (3.32),

$$u_k = \hat{\phi}_{u,k}^{-1}\left(e_k + \hat{\phi}_{\mu,k} + \hat{\phi}_{b,k}\omega_k\right) \tag{8.9}$$

where $\hat{\phi}_{u,k}$, $\hat{\phi}_{\mu,k}$ and $\hat{\phi}_{b,k}$ are the estimates of $\phi_{u,k}$, $\phi_{\mu,k}$ and $\hat{\phi}_b$ respectively. The periodic adaptive control problem here clearly fits the mixed parameter case where some uncertain parameters are periodic and the remaining uncertain parameter is constant. The adaptation law is selected as, (3.33),

$$\begin{aligned}\hat{\phi}_{u,k} &= \hat{\phi}_{u,k-N} - q_1 \frac{u_{k-N}}{1+\xi_{k-N}^{\top} Q \xi_{k-N}} e_{k-N+1}\\ \hat{\phi}_{\mu,k} &= \hat{\phi}_{\mu,k-N} + q_2 \frac{1}{1+\xi_{k-N}^{\top} Q \xi_{k-N}} e_{k-N+1}\\ \hat{\phi}_{b,k} &= \hat{\phi}_{b,k-1} + q_3 \frac{\omega_{k-1}}{1+\xi_{k-1}^{\top} Q \xi_{k-1}} e_k \end{aligned} \tag{8.10}$$

where q_1, q_2, q_3 are the adaptation gains, $Q = \text{diag}(q_1, q_2, q_3)$, N is the period in time steps and $\xi_k^{\top} = \left[-u_k \,\vdots\, 1 \,\vdots\, \omega_k\right]$.

Using a sampling period of $T = 1\,\text{ms}$ and adaptation gains $q_1 = q_2 = q_3 = 1 \times 10^{-2}$ the system is simulated. Periodic adaptive control is activated at $t = 2\,\text{s}$ which is roughly when the system achieves steady state operation. The periodicity is selected as $N = 200$ which is the smallest common factor between both periodic terms. Once, periodic adaptive control replaces the PI controller the ripple due to the periodic coefficients vanishes. The result in Figs. 8.5 and 8.6 shows that even in the case when the parameters are periodic with respect to state rather than time, under certain conditions, it is still possible to implement periodic adaptive control to achieve good performance. Further, from Fig. 8.7 it is seen that the strategy of switching from PI control to periodic adaptive control does not result in a large jump in the control input as the transition from PI to periodic adaptive control occurs.

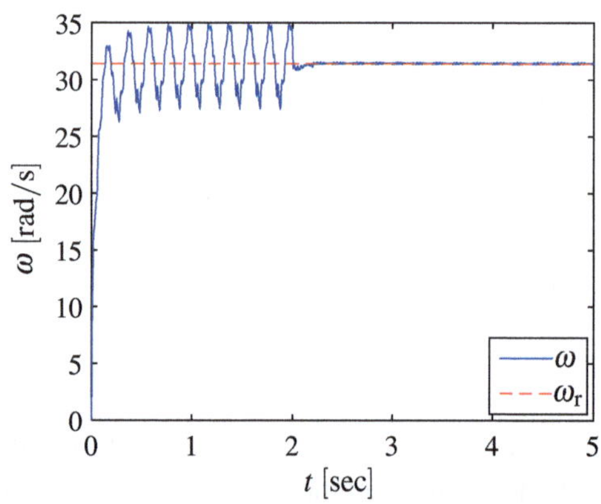

Fig. 8.5 Speed control performance of a PI controller with periodic adaptive control

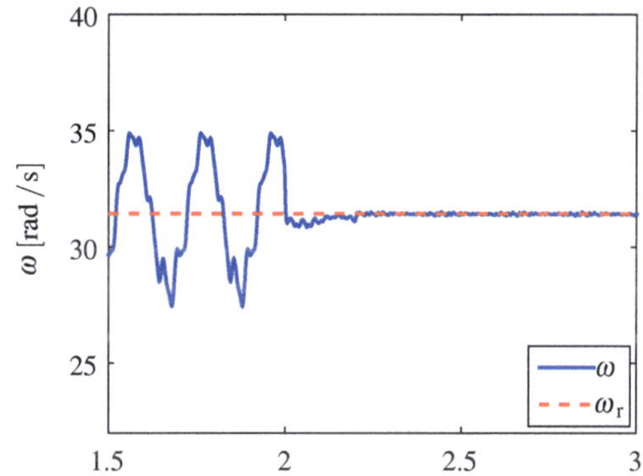

Fig. 8.6 Speed control performance of a PI controller with periodic adaptive control (close-up)

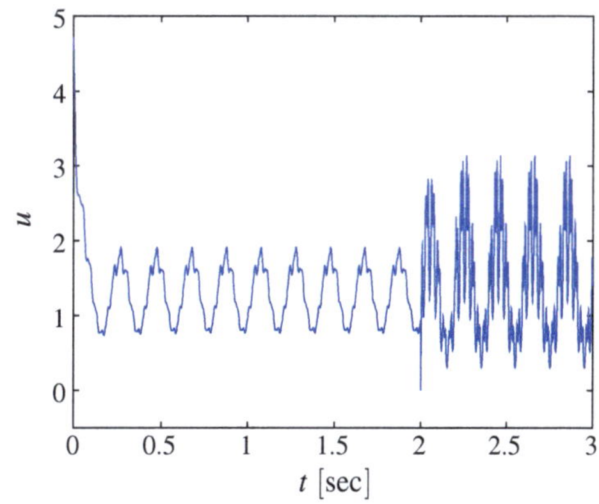

Fig. 8.7 Control input profile of the PI controller with periodic adaptive control

8.3 Multirate ILC of a Ball and Beam System

The algorithm is applied to a CE106 Ball-and-Beam system.

8.3.1 System Model

The block diagram of the Ball-and-Beam system is illustrated in Fig. 8.8. In Fig. 8.8 a simplified linear model is shown, with a double integrator representing the ball-and-beam dynamics, and a simple integrator representing the actuator dynamics, can well fit the real system behaviour when the motion speed is reasonably low, [182]. Consequently the ball-and-beam system model is expressed by

$$G_{\text{bnb}} = \frac{7.79}{s^2}$$

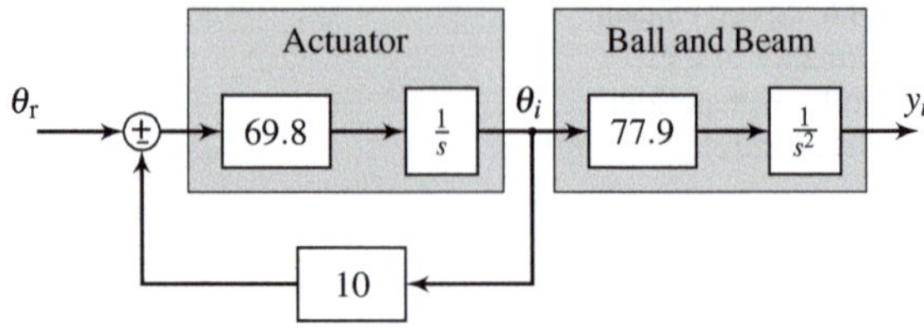

Fig. 8.8 Configuration of the Ball-and-Beam system

where G_{bnb} is the ball-and-beam dynamics. If all the states are obtainable and the ball-and-beam system is sampled at T_c, its state space model is similar to (5.75), with the coefficient matrices

$$\Phi_c = \begin{bmatrix} 1 & T_c \\ 0 & 1 \end{bmatrix}, \quad \Gamma_c = \begin{bmatrix} \frac{T_c^2}{2} \\ T_c \end{bmatrix} \quad \& \quad C = \begin{bmatrix} 1 & 1 \end{bmatrix}$$

and the control gain is given as $\gamma = 7.79$.

8.3.2 Target Trajectory

The ball-and-beam system is required to track a target trajectory as shown in Fig. 8.9. The trajectory is designed as

$$y_d(t) = \begin{cases} -0.00228t^5 + 0.0285t^4 - 0.095t^3 + 0.988 & \forall t \in [0, 5] \\ -0.2 & \forall t \in (5, \infty) \end{cases} \tag{8.11}$$

Note that the trajectory is designed based on a tracking task on the real-time Ball-and-Beam apparatus. From the Fast Fourier transform in Fig. 8.9, we can see that the cut-off frequency of the desired trajectory is about 0.2 Hz.

Fig. 8.9 Target trajectory

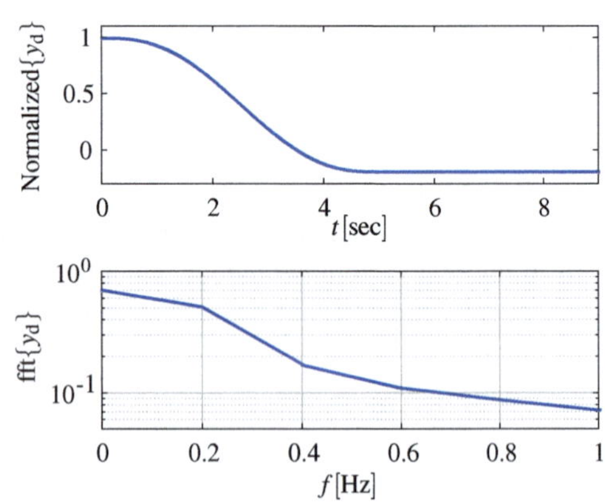

8.3.3 Controller Configurations

Since the target trajectory is a smooth, low cut-off frequency curve and it can be tracked repetitively in an off-line fashion, an appropriate PCCL scheme can perform perfect tracking of this model. The multirate approach improves the limitation of the data acquisition rate. Following the design method in Sect. 5.5.1, we first design a stabilized feedback loop containing the multirate structure, then design a PCCL scheme to minimize the tracking error.

Assume $T_s = 0.04$ s and $T_c = 0.01$ s. The multirate index is $n = 4$. According to Fig. 8.8, the controller gain $\gamma = 7.79$. Recalling the equations (5.77), to place the estimator poles at 0.01, we have

$$L = \left[-22.5225 \;\; 24.5025\right]^{\top}.$$

According to (5.106), the more the denominator of G_{learn} is greater than its numerator, the faster the PCCL scheme may converge. Based on this point of view, the learning gains are chosen such that

$$G_{\text{ff}} = g_{\text{ff}} I \quad \& \quad G_{\text{fb}} = g_{\text{fb}} I$$

where $g_{\text{ff}} = 0.01$ and $g_{\text{fb}} = 1.5$. The convergence condition (5.106) is obviously complicated to analyze in this case. We may verify it numerically by substituting all coefficients. Since the cut-off frequency of the desired trajectory is at about 0.2 Hz, the bandwidth we are interested in is at a low frequency band, say up to 10 Hz.

In Fig. 8.10, the absolute values of the eigenvalues of the learning convergence matrix G_{learn} at the frequency band (0, 10] Hz are shown. For the chosen coefficients, only the absolute value of the first eigenvalue (eigenvalue 1) is strictly less than 1. The rest of the eigenvalues are very closed to the unit circle. However, from the projections of the tracking error to the corresponding eigenvectors, as shown in Fig. 8.11, the first eigenvalue plays a decisive role. The first eigenvalue in Fig. 8.10 dominates the characteristics of the overall system and, thus, makes the system converge. In other words, such a design of multirate and PCCL is asymptotically stable.

8.3.4 System Verifications

In this section we shall verify the design presented in Sect. 8.3.3 using a simulation example. In Fig. 8.12 the comparison of the maximum error amplitudes of multirate and single rate systems at T_s is shown. The maximum error of the multirate PCCL scheme reduces from 2×10^{-2} to 1.5×10^{-5} after 7 iterations. In other words, the learning scheme reduces the maximum error to less than $1/1000$ of its original amplitude in 7 learning cycles. This design performs perfect tracking.

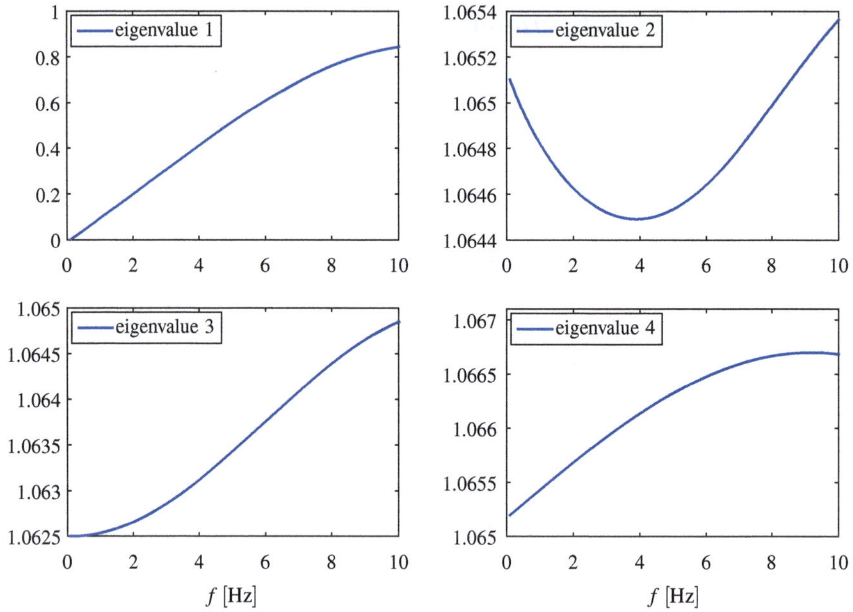

Fig. 8.10 Eigenvalues of G_{learn} at frequency band (0, 10] Hz

In comparison to the single rate at T_s, the multirate approach overcomes the problem of disparity between the sampling rates of the data acquisition and the controller output and improves the efficiency of the measured data and controller signals. By employing this approach, the maximum error was further reduced to half or even 30 % of a single rate system (as shown in Fig. 8.12). In Fig. 8.13 the comparison of the tracking errors at the 8th iteration is shown. The 3 curves in Fig. 8.13 are for a single rate system sampled at T_c, a multirate system sampled at T_s and a single rate system sampled at T_s, respectively. Because the errors of the single rate system sampled at T_c and the multirate system sampled at T_s are too close (the difference is about 3×10^{-7}), the error curves are almost overlapping and could not be distinguished in Fig. 8.13. It shows that, by employing the multirate approach, a performance that is as good as a conventional system at a sampling rate that is 4 times faster can be achieved.

In Fig. 8.13 two oscillations are shown: one is after 0 s and the other is after 5 s. As the computer control system is a digital system by nature, the controller designs, such as the multirate estimator and the PCCL scheme, are based on the discretized system model (5.75). Initial errors may occur when such a discrete controller performs a tracking task in continuous time domain. The target trajectory shown in Fig. 8.9 is a piecewise function with intervals [0, 5] and (5, 9], respectively. Thus, the errors may occur at 0 s and 5 s respectively, although it might be very small. Recalling the eigenvalues of the learning convergence matrix G_{learn} in Fig. 8.10, although the first eigenvalue dominates the system dynamics, the other 3 eigenvalues are

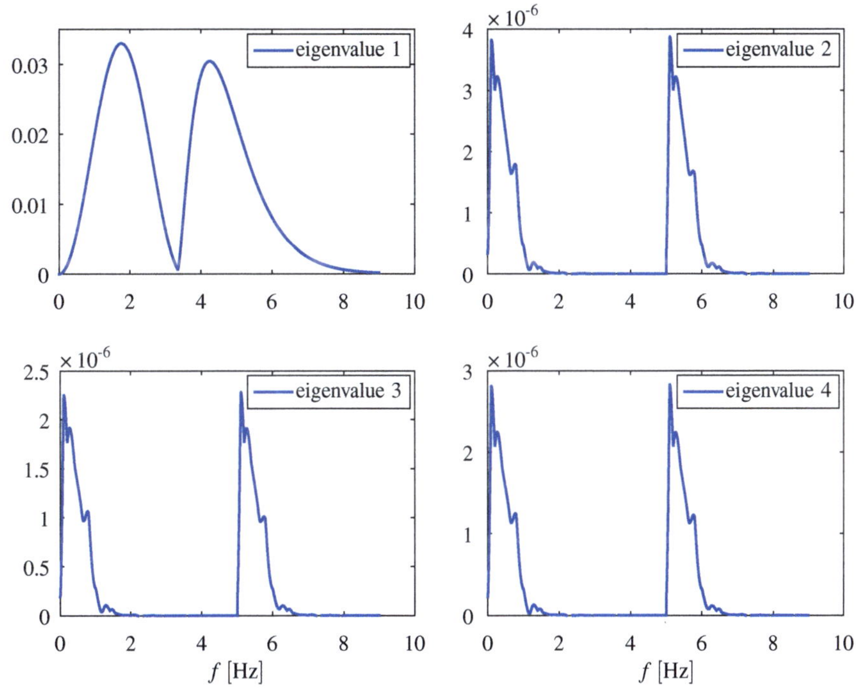

Fig. 8.11 Projections of e to eigenvectors of G_{learn}

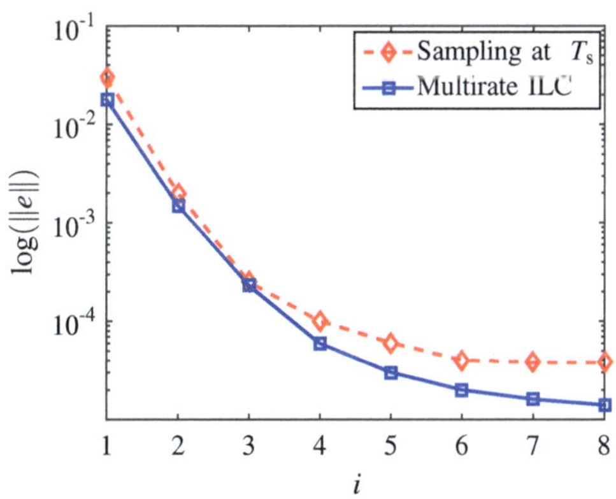

Fig. 8.12 Comparison of $\|e\|_{\max}$

actually located slightly outside the unit circle, which may generate oscillations and cause the tracking error to diverge gradually. Furthermore, the integrable property of the learning scheme accumulates tracking errors iteration after iteration. These will slow the convergence speed after several learning iterations as shown in Fig. 8.12.

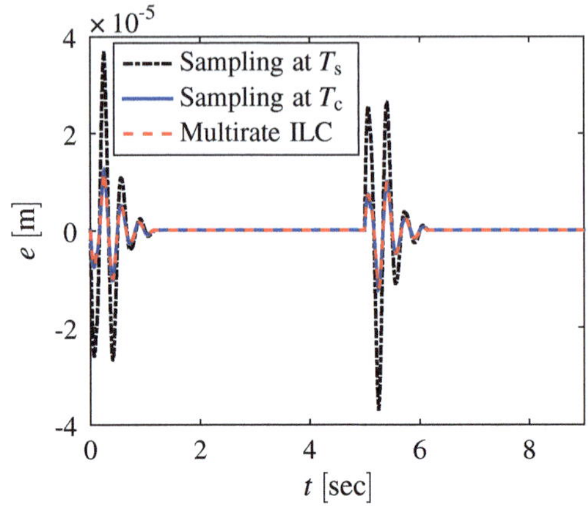

Fig. 8.13 Comparisons of tracking errors at 8th iteration

Overall, though, as the first eigenvalue plays a dominant role the asymptotic stability of the system is guaranteed. In practice the learning scheme is disabled once the performance specification is satisfied, or no further improvement can be observed.

8.4 Discrete-Time Fuzzy PID of a Coupled Tank System

To further compare the fuzzy and conventional PID controllers, an experiment is carried out in which a nonlinear system is used. The system employed in the experiment is a coupled-tank system shown in the picture (Fig. 8.14) and the purpose of the experiment is to control the fluid level in the second tank.

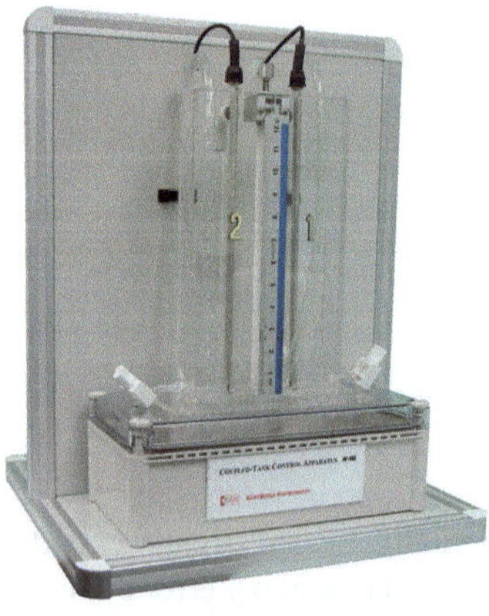

Fig. 8.14 Coupled tanks apparatus

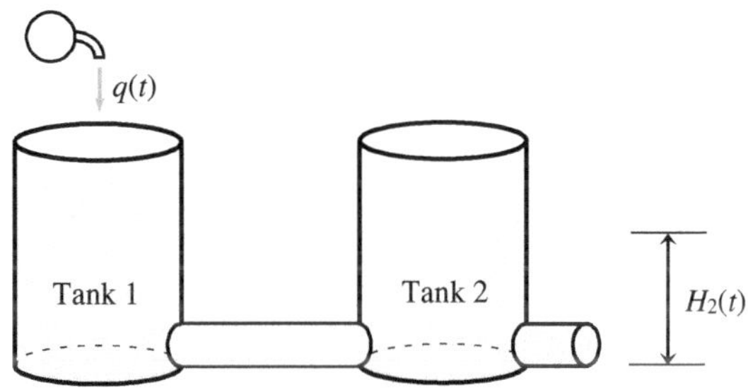

Fig. 8.15 Coupled tanks system

8.4.1 System Description

The basic experimental system which consists of two hold-up tanks coupled by an orifice can be simplified as shown in Fig. 8.15. The input $q(t)$ is supplied by the variable speed pump, which pumps water to the first tank. The orifice between two tanks allows the water to flow into the second tank and hence out as a outflow. The basic control problem is to control the water level in the second tank $H_2(t)$ by varying the speed of the pump. The measurement voltage for water level is read in and the control signal for the pump is written out by a computer through A/D and D/A interfaces. The control algorithm is realized by computer programming. Since two tanks are coupled by an orifice, the system is a second order system. Moreover, the outflow rate is determined by the square root of the fluid level in second tank, $H_2(t)$, thus the system is essentially nonlinear. For the design of conventional and fuzzy PID controllers using gain and phase margin specifications, the system must be linearized, simplified and modeled by a second order plus dead time structure. After performing a relay feedback experiment, the ultimate gain and period are obtained as 21.21 and 100 respectively. By conducting another set-point changing experiment, the system gain can also be obtained as 0.75. Thus, a simplified system model will be:

$$G(s) = \frac{0.75e^{-8.067s}}{(1 + 61.45s)^2}$$

and the parameters of the conventional and fuzzy PID controllers are determined based on this model and formulas (6.13), (6.14) and (6.15).

8.4.2 Experiment

The experiments are carried out during a time period from 0 to 4000 s. The sampling period is 1 s. First, the system is settled at 10 cm. There are set-point changes from 10 to 16 cm, 16 to 11 cm and 11 to 13 cm at time instants of 100, 1100, 2000 respectively. Moreover, there is a load disturbance at time instant 2800, which is a 20 cm^3/s flow introduced by a pump to the first tank to emulate the change in inflow. The specified

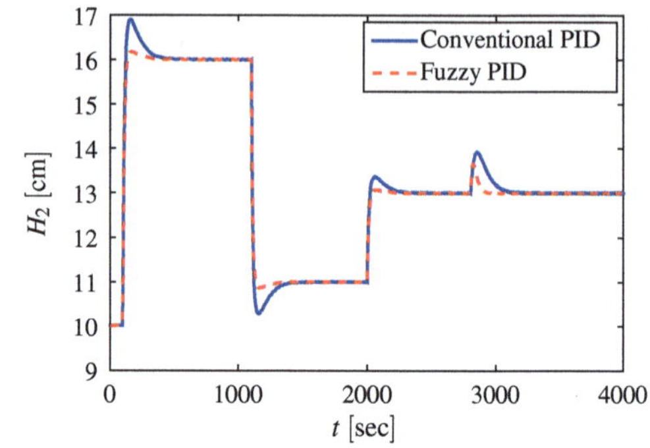

Fig. 8.16 Tank 2 water level

gain and phase margins are 3 and 45° respectively. The size of orifice is set to $0.396\,cm^2$. The sampling interval is 1 s and the experimental results are shown in Fig. 8.16. In the figure, the solid and dotted lines are the water levels of the second tank controlled by the fuzzy and conventional, [80], PID controllers respectively. After the water level is settled at 13 cm (time instant 2500), the parameters of fuzzy PID controller are re-calculated by increasing χ in (6.15) from 0.2 to 0.5. From the experimental data, we can find that the closed-loop system performance of fuzzy PID controller is evidently better than that of conventional PID controller. The fuzzy PID controller gives only slight overshoot to set-point changes with almost same response speed as its conventional counterpart. On the other hand, the conventional PID controller gives large overshoots. The magnitude of the system response to load disturbance is relatively smaller and the convergence is faster when using fuzzy PID controller. Experimental results confirm again the advantage of the proposed fuzzy PID controller.

8.5 Iterative Learning Control for Freeway Traffic Control

It is worth to take note that the macroscopic traffic flow patterns are in general repeated everyday. For example, the traffic flow will start from a very low level in the midnight, and increase gradually up to the first peak during morning rush hour, which is often from 7 to 9 A.M., and the second one from 5 to 7 P.M. This repeatability in macroscopic level implies two features on them: (1) the macroscopic traffic model is invariant in day/week or month axis and (2) the macroscopic exogenous quantities (independent of the freeway states, off-ramp flow, initial inflow on the mainstream, etc.) are invariant along the day/week or month axis too.

Based on the traffic pattern repeatability, in this work we apply iterative learning control (ILC) [159] aiming at to improving the traffic performance in a freeway. The main idea in ILC is to use information from pervious repetitions to come up with new control actions that improve the tracking accuracy as the number of repetitions increases. It has a very simple structure, that is, integration along the iteration axis,

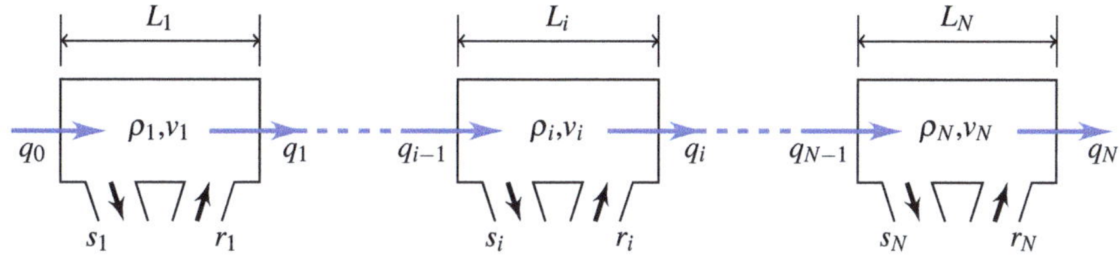

Fig. 8.17 Segments on a freeway with on/off ramp

and is a memory based learning process. ILC requires much less information of the system variations to yield the desired dynamic behaviors. This is a very desirable feature in traffic control as the traffic model and the exogenous quantities may not be well known in practice. ILC focuses on the input signal update algorithms and differs from most existing control methods in the sense that it exploits every possibility to incorporate past control information such as tracking errors and control input signals into the construction of the present control action.

8.5.1 Traffic Model and Analysis

8.5.1.1 Traffic Model

The space and time-discretized traffic flow model for a single freeway with one on-ramp and one off-ramp is given as follows and its diagram shown in Fig. 8.17.

$$\rho_{i,k+1} = \rho_{i,k} + \frac{T}{L_i}\left[q_{i-1,k} - q_{i,k} + r_{i,k} - s_{i,k}\right] \tag{8.12}$$

$$q_{i,k} = \omega\rho_{i,k}\cdot v_{i,k} + (1-\omega)\,\rho_{i+1,k}\cdot v_{i+1,k} \tag{8.13}$$

$$\begin{aligned} v_{i,k+1} = {} & v_{i,k} + \frac{T}{\tau}\left[V\left(\rho_{i,k} - v_{i,k}\right)\right] + \frac{T}{L_i} v_{i,k}\left[v_{i-1,k} - v_{i,k}\right] \\ & - \frac{\nu T}{\tau l_i}\left[\frac{\rho_{i+1,k} - \rho_{i,k}}{\rho_{i,k} + \kappa}\right] - \delta T\left[\frac{q_{i,k}v_{i,k}\left(\lambda_i - \lambda_{i+1}\right)}{\lambda_i L_i\left(\rho_{i,k} + \kappa\right)}\right] \\ & - \frac{\phi T}{L_i}\left(\frac{\rho_{i,k}}{\rho_{cr}\lambda_i}\right)\left[\frac{\lambda_i - \lambda_{i+1}}{\lambda_i}\right] v_{i,k}^2 \end{aligned} \tag{8.14}$$

$$V\left(\rho_{i,k}\right) = v_{\text{free}}\left(1 - \left[\frac{\rho_{i,k}}{\rho_{\text{jam}}}\right]^l\right)^m \tag{8.15}$$

where T is the sampling interval in hours, $k = [1, 2, \ldots, K]$ is the time instant, and $i = [1, 2, \ldots, I]$ is the number of sections. Other variables are listed as follows:

$\rho_{i,k}$: Density in section i at time kT (veh/lane/km);
$v_{i,k}$: Space mean speed in section i at time kT (km/h);
$q_{i,k}$: Traffic flow leaving section i entering section $i+1$ at time kT (veh/h);
$r_{i,k}$: On-ramp traffic volume for section i at time kT (veh/h);
$s_{i,k}$: Off-ramp traffic volume for section i at time kT (veh/h), which is regarded as an unknown disturbance;
$L_{i,k}$: Length of freeway in section i (km);
λ_i: Number of lanes in section i.

The constants v_{free} and ρ_{jam} are the free speed and maximum possible density per lane, respectively. The coefficients τ, ν, κ, l, m, ϕ, δ are constant parameters which reflect particular characteristics of a given traffic system and depend upon street geometry, vehicle characteristics, drivers behaviors, etc. For a real life network these parameters are determined by a validation procedure. A validated model, however, that is accurate for one place at a particular time of the day may not hold in another place at another time.

Equation (8.13) is the well-known conservation equation, (8.14) is the flow equation, (8.15) is the empirical dynamic speed equation, and (8.15) represents the density-dependent equilibrium speed. The mathematical representations show highly nonlinear property of the traffic flow model, which leads to sophisticated problems in controlling process.

8.5.1.2 Traffic Conditions

We assume that the traffic flow rate entering section 1 during the time period kT and $(k+1)T$ is $q_{0,k}$ and the mean speed of the traffic entering section 1 is equal to the mean speed of section 1, i.e. $v_{0,k} = v_{1,k}$. We also assume that the mean speed and traffic density of the traffic exiting section $N+1$ are equal to those of section N, i.e. $v_{N+1,k} = v_{N,k}$, $\rho_{N+1,k} = \rho_{N,k}$. Boundary conditions can be summarized as follows:

$$\rho_{0,k} = \frac{q_{0,k}}{v_{1,k}} \tag{8.16}$$

$$v_{0,k} = v_{1,k} \tag{8.17}$$

$$\rho_{N+1,k} = \rho_{N,k} \tag{8.18}$$

$$v_{N+1,k} = v_{N,k} \tag{8.19}$$

In this study, a freeway traffic flow process around the desired density $y_{\text{d}} = 30\,\text{veh/km}$ is simulated in the presence of a large exogenous disturbance.

Consider a long segment of freeway that is subdivided into $I = 12$ sections. The length of each section is $L_i = 0.5\,\text{km}$. The initial traffic volume entering section

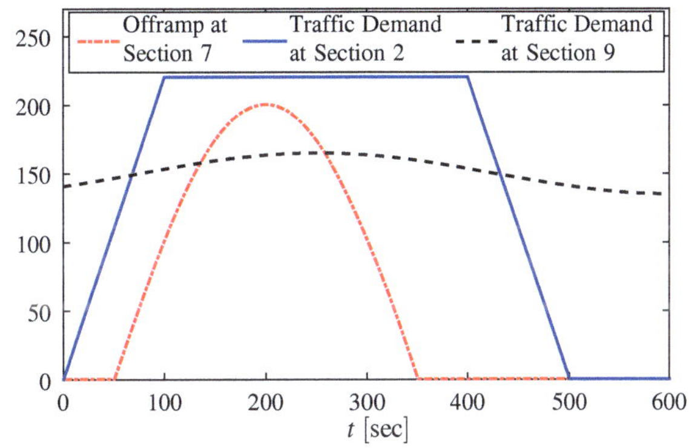

Fig. 8.18 Traffic demand constraints and disturbance (offramp)

1 is 1500 vehicles per hour. The initial densities is set as 30 veh/lane/km for each section and the parameters used in this model are set as $v_{\text{free}} = 80\,\text{km/h}$, $\rho_{\text{jam}} = 80\,\text{km/h}$, $l = 1.8$, $m = 1.7$, $\kappa = 13\,\text{veh/km}$, $\tau = 0.01\,\text{h}$, $T = 0.00417\,\text{h}$, $\nu = 35\,\text{km}^2/\text{h}$, $q_{0,k} = 1500\,\text{veh/h}$, $r_{i,0} = 0\,\text{veh/h}$, $\omega = 0.95$. Furthermore, in order to examine the influence of insufficient traffic supply on ramp metering case, situations under demand constraint have also been tested to provide a comparison. The detailed constraints and disturbances information is shown in Fig. 8.18.

The demands impose a constraint on the control inputs of ramp metering in the sense that the on ramp volumes cannot exceed the current demands plus the existing waiting queue at on ramps at time k. In mathematical representation, it is written as

$$r_{i,k} \leq d_{i,k} + \frac{l_{i,k}}{T} \tag{8.20}$$

where $l_{i,k}$ denotes the length (in veh) of a possibly existing waiting queue at time kT at ith on ramp, $d_{i,k}$ is the demand flow at time kT at the ith on ramp (veh/h). In this case, i belongs to the set $\{2, 9\}$, where an on ramp exists. On the other hand, the waiting queue is the accumulation of the difference between the demand and the actual on ramp, i.e.,

$$l_{i,k+1} = l_{i,k} + T\left[d_{i,k} - r_{i,k}\right] \tag{8.21}$$

8.5.1.3 Equilibrium Analysis

A systematic model analysis is done in order to study and develop effective control algorithms based on the traffic model. The most important feature of the nonlinear traffic model is the density and speed equilibrium points, which can be found with respect to a constant entering flow, $q_{0,k}$. Theoretically, the conditions for searching the equilibrium are $r_{i,k} = 0$, $s_{i,k} = 0$, $\rho_{i,k} = \rho_{i-1,k}$, $v_{i,k} = v_{i-1,k}$, $q_{i,k} = q_{0,k}$, $\lambda_i = \lambda_{i-1}$ and $k = [1, 2, \ldots, K]$, $i = [1, 2, \ldots, I]$.

Fig. 8.19 Density versus speed at equilibrium points

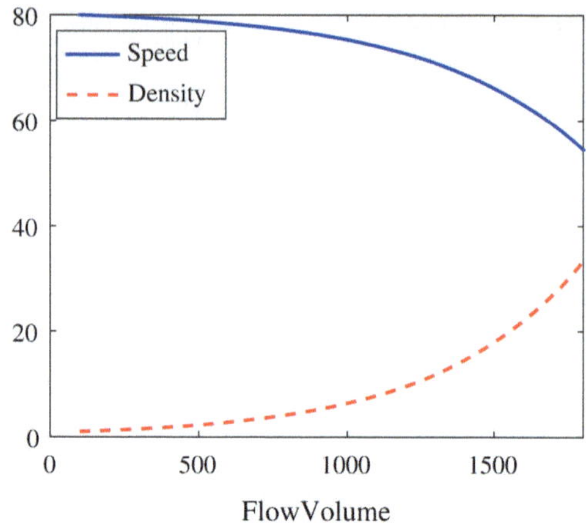

According to (8.13), $\rho_{i,k} = \rho$, $v_{i,k} = v$, thus, from (8.14), it is obtained that $q = \rho v$, $\rho = \frac{q}{v} = \frac{q_0}{v}$ and from (8.15), it is obtained that $V(\rho) - v = 0$. Substitute (8.15) for $V(\rho)$, $v_{\text{free}}\left(1 - \left[\frac{\rho}{\rho_{\text{jam}}}\right]^l\right)^m = v$, further, $v_{\text{free}}\left(1 - \left[\frac{\rho}{\rho_{\text{jam}}}\right]^l\right)^m = \frac{q_0}{\rho}$, from which the equilibrium density value could be acquired. The equilibrium speed value could then be obtained by $v = \frac{q}{\rho}$. The density and speed relationship under different entering flow volume values is indicated in Fig. 8.19.

With a specific flow value $q_{0,k}$, if the initial density value is different from the equilibrium value, the density curve will quickly drop down or rises up to the equilibrium point, see Fig. 8.20. A similar phenomenon can be observed in the speed response.

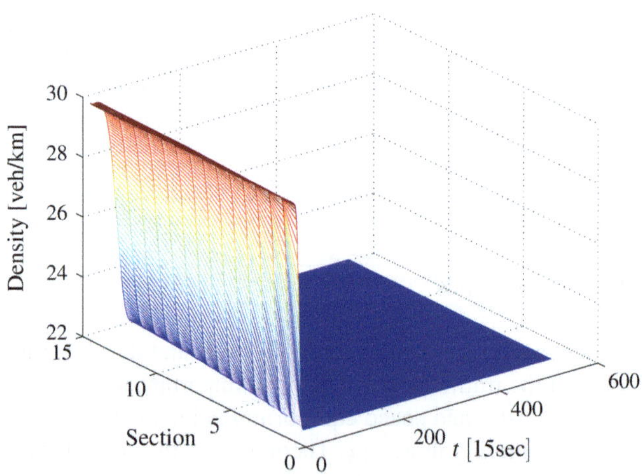

Fig. 8.20 Density drops to the equilibrium point entry flow is 1500 veh/h, initial density is 30 veh/km, equilibrium density is 22.5 veh/km

8.5.2 Density Control

In this section, the control objective is to track the desired density control.

8.5.2.1 On-Ramp Metering

$$u_{n+1,k} = u_{n,k} + \beta e_{n,k+1} \tag{8.22}$$

The control vector used in this methodology is $r_{i,k}$ and $i = \{2, 9\}$. The feedback gain is chosen to be 40, and the ILC learning gain is chosen to be $\beta = 20$.

Without any external control action, the equilibrium may differ from the set-point, hence there exists a large deviation from the desired density value. Traffic jam may occur due to the traffic inputs in sections 2 and 9. By adding in pure error feedback control, traffic jam has been reduced, but deviation still exists. In terms of ILC control, without the traffic demand constraint, it gives the perfect tracking after 20 iterations, [84, 85]. However, as a result of the demand constraint, the system could not offer enough traffic flow than it requires (calculated in theory), and thus, there is deviation (compared with desired density value) exists.

The simulation result in Fig. 8.26 shows that there is large initial error in pure ILC control (first iteration $n = 1$). This is because in pure ILC control, it learns from information obtained in previous iterations. As there is no information provided in the first iteration, the traffic flow is under no control.

ILC combined with error feedback control is proved to be an effective way to overcome this setback, [84, 85]. The reason is that error feedback control covers the control action in first iteration, whereas ILC takes care of the rest of iteration rounds.

Based on the observation that there is a sudden change in gradient in Section 9, see Fig. 8.26, two gain tuning methods have been proposed and proved to be able to decrease the initial error and smoothen the density error curve (Fig. 8.21).

(1) Set two separated learning gain values, $\beta = 20$ for section 2 and $\beta = 10$ for section 9.
(2) Define the learning gain, β, as a function of iteration number, some examples are given below.

Applied to both sections 2 and 9:

$$\beta_n = \beta_{n-1} - 0.5\beta_{n-1}, \text{ or}$$
$$\beta_n = \beta_{n-1} - (15 - n).$$

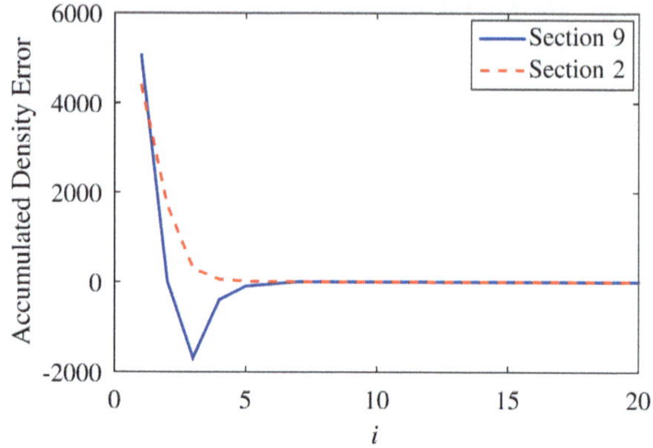

Fig. 8.21 Accumulated error under pure ILC algorithm

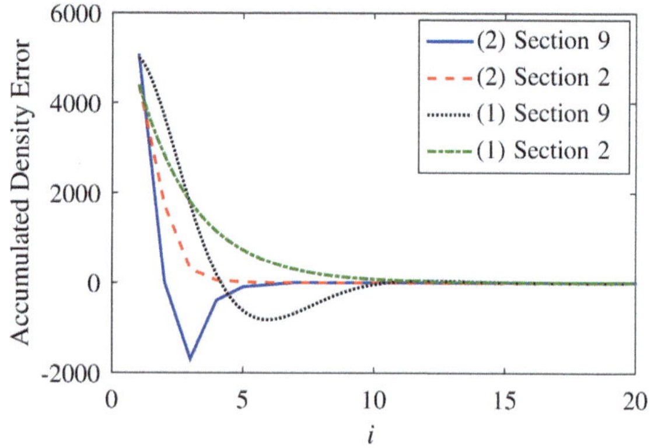

Fig. 8.22 Accumulated error under pure ILC algorithm (modified)

Applied to section 9:

$$\beta_n = \beta_{n-1} e^{1-n}, \text{ or}$$

$$\beta_n = \beta_{n-1} - 1.33\beta_{n-1}\left(\frac{\beta_{n-1}}{n}\right)$$

where β_n for section 2 remains at a value of 20 and n represents how many iteration rounds have gone through. The results by applying the first method, and the last mentioned formula in second method are shown in Fig. 8.22.

However, both of the approaches compromise the converging speed (to the steady state) to the improvement of the two problems addressed above. Next, consider a stochastic environment, a random system disturbance $m_{1,k}$ which is uniformly distributed on interval $(-1, 1)$ is added to the right hand side of the dynamic speed equation, and an output measurement noise $m_2(k)$ which is uniformly distributed on interval $(-0.5, 0.5)$ is added to the density output. The resulting control signal is shown in Fig. 8.23.

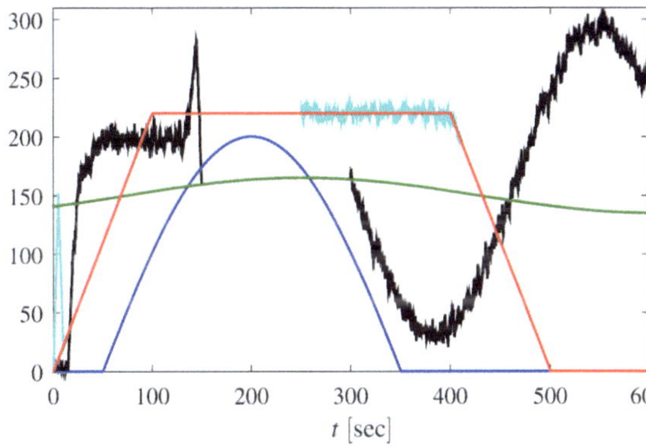

Fig. 8.23 Unfiltered control signal

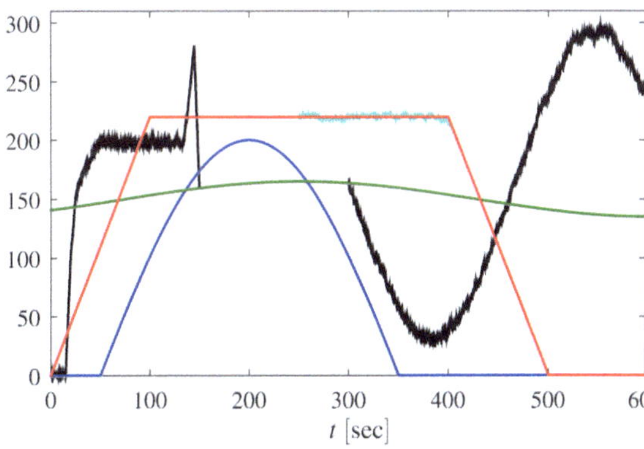

Fig. 8.24 Filtered control signal

One proposed solution to reduce the influence of the random perturbations is to make use of filtering approach. From the power spectrum of the two control signal (on-ramp metering in sections 2 and 9, it is observed they are at low frequency, hence, butterworth filter with order 2 is chosen. Mathematically, it is represented as

$$u_{n+1,k} = F[u_{n,k}, \omega_0] + \beta e_{n,k+1} \tag{8.23}$$

where ω_0 is the cutoff frequency of the filter $F[\cdot, \cdot]$ and is chosen to be 0.01 Hz. The obtained result is shown in Fig. 8.24.

8.5.2.2 Speed Regulation

Speed regulation aims at constraining drivers' behavior by suggesting proper travel speeds, and the control is applied to all sections.

This is realized by adding in a control input to the right hand side of (3), dynamic speed equation. From (1) to (4), it is observed there are two steps delay from the control signal to the output signal. Therefore, pure error feedback control is hardly

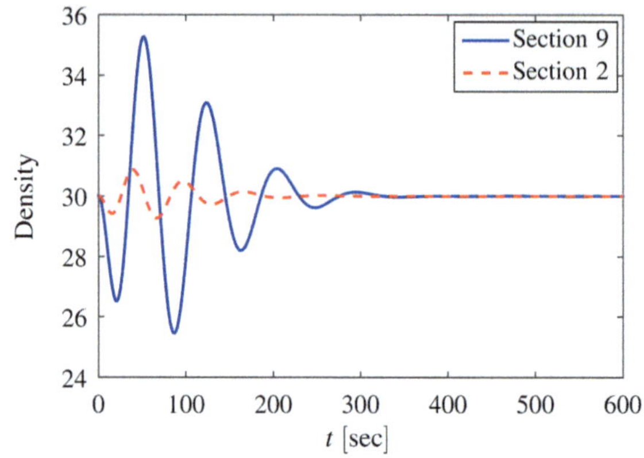

Fig. 8.25 Density in sections 2 and 9

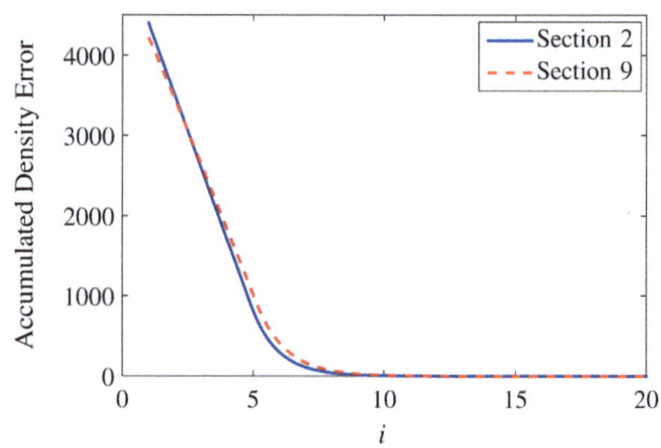

Fig. 8.26 Accumulated error in each iteration

applicable in this case because the system relative degree is two. ILC, however, is able to perform well because it retrieves information from pervious iteration, where information at all time instances is available. The learning gain is chosen to be 0.15.

In the simulations, the traffic demands in sections 2 and 9 shown in Fig. 8.18 are considered as on-ramp disturbance. The density value and speed control signal at sections 2 and 9, after learning 20 iterations, are shown in Fig. 8.25 while Fig. 8.26 shows the accumulated error in each iteration round.

8.5.3 Flow Control

Here the control objective is to track the desired flow value. Three methods have been proposed. The control in the second and third method is applied to all sections.

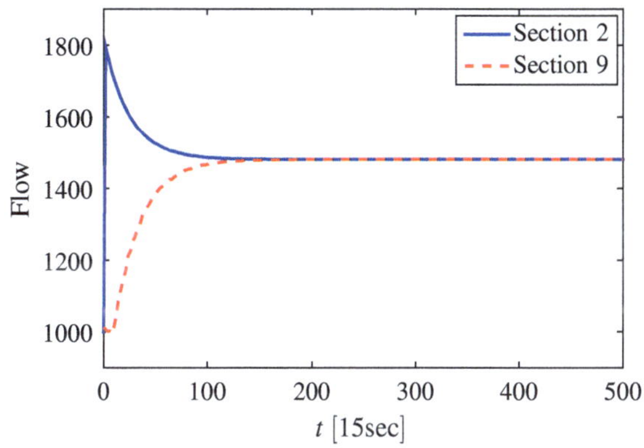

Fig. 8.27 Flow in sections 2 and 9

8.5.3.1 Freeway Input Flow Volume Control

The input flow volume is set at 2020 veh/h. From Fig. 8.27, it is observed though there is initial fluctuation, it becomes smooth around $k = 50$, and converge around $k = 150$.

8.5.3.2 Flow Tracking

A controllable term $\beta \frac{Q_k - Q_{d,k}}{\lambda_i \rho_{i,k}} \cdot \frac{Q_{i,k}}{Q_k}$ is added to the right hand side of (8.15), where β is tuned to obtain a satisfactory result, $Q_{i,k} = Q_{i,k-1} + q_{i-1,k} - q_{i,k}$ is the flow volume in section i at time kT, Q_k is the sum of the flow volume in all N sections at time kT, and is the desired flow volume in all N sections at time kT. The term $\frac{Q_{i,k}}{Q_k}$ is the percentage of flow volume in section i to the entire N sections. Therefore, $\frac{Q_k - Q_{d,k}}{\lambda_i} \cdot \frac{Q_{i,k}}{Q_k}$ is the difference of flow volume in each lane contributed by section i. It is then divided by $\rho_{i,k}$ to get the corresponding speed value. After $Q_{i,k}$ is obtained for all sections at each time instant in the first iteration, it is then available at time kT in the dynamic speed equation for the next iteration. This provides instantaneous feedback though the information is from the previous iteration.

Besides, a constraint $\left|Q_{k+1} - Q_{d,k+1}\right| < \textit{Threshold}$ is added in each time instant kT. The variable Q_{k+1} is available at time kT because $\rho_{i,k+1}$ and $v_{i,k+1}$ are both accessible then. This constraint means the variable speed will be continuously adjusted until Q_{k+1} fulfills the condition. The value for β is chosen to be 0.1. The result shows the accuracy in Fig. 8.28. It is implemented in a double lane freeway link, where the desired flow volume is 3000 veh/h. The program needs to go through a large number of iterations to make the Q_{k+1} value settled within the desired range.

Fig. 8.28 Flow at all sections

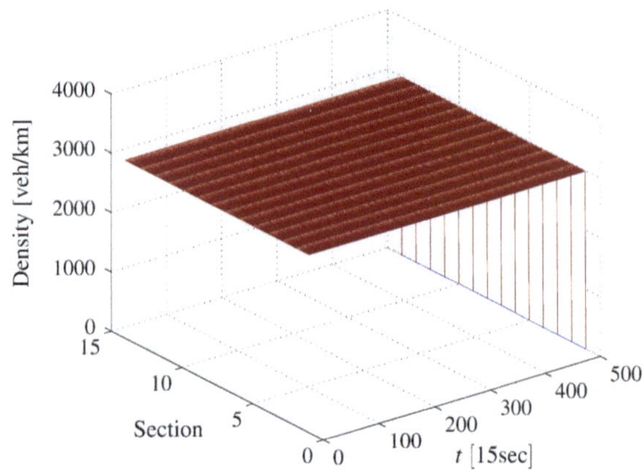

8.5.3.3 Flow Learning

It uses similar algorithm as the case of speed regulating, but different tracking objectives. The learning gain is chosen to be 0.025. Besides, the input flow volume is set at the desired value. The results of the density and control signal are shown in Figs. 8.29 and 8.30. Because of the short transient period, the initial fluctuation could be fairly ignored.

After combining with pure error feedback control, the results are about the same. This is because in flow learning, it needs information about the error message two steps ahead (same as speed regulating case). Therefore, even when the feedback control is added in, the information is with a two-step delay, resulting in not much improvement.

Fig. 8.29 Flow value

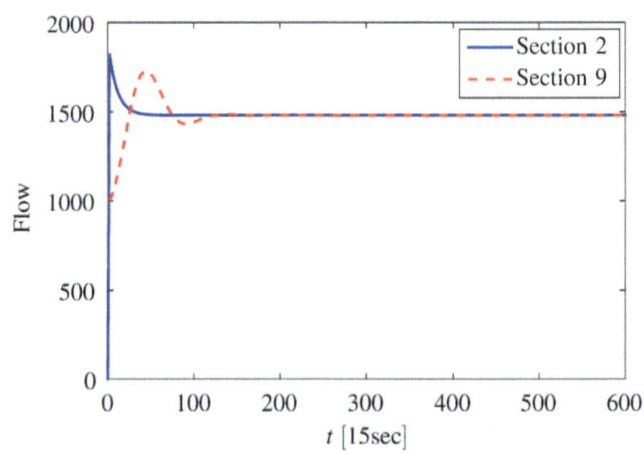

Fig. 8.30 Speed commands

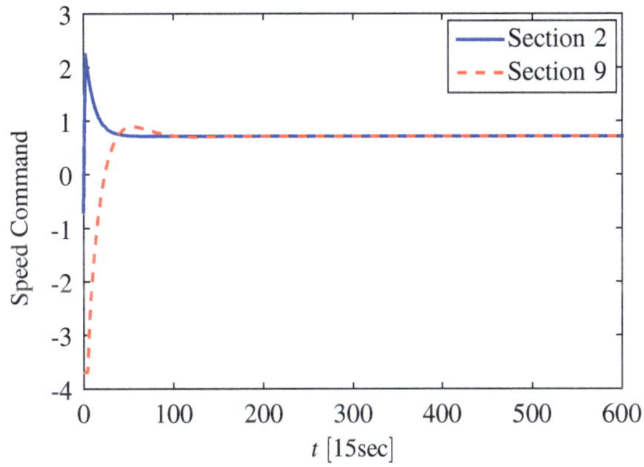

8.6 Conclusion

In this chapter different control approaches for practical engineering applications were presented. The intent was to provide the reader with a practical and systematic design procedure for the many control approaches that were discussed earlier in the book.

In the first example, periodic adaptive control was implemented on a PM synchronous motor. It is shown that although the uncertain parameters are state dependent, under certain conditions it is possible to assume time based periodicity which can be attenuated by periodic adaptive control. The control strategy used involved the implementation of a PI controller to drive the system to steady state and then replace the PI controller with periodic adaptive control to further improve the steady state performance. From the results presented it is shown that such a strategy is very effective and produces good performance.

In the second example, multirate ILC is implemented on a Ball-and-Beam apparatus with a slow A/D port while the controller is set at a higher sampling rate. It is shown that the multirate ILC overcomes the problem of different sampling rate between the data acquisition and the controller output. It estimates inter-sample states so as to improve the efficiency of the measured data and controller signals. The learning scheme keeps the system convergent along the iteration axis to achieve nearly perfect tracking performance. This algorithm reduces maximum tracking error to less than 1/1000 of its original amplitude after 7 iterations of learning. By employing multirate ILC, it can achieve as good a result as a conventional system that is set at a 4 times faster sampling rate.

In the third example, fuzzy PID controller is implemented on a coupled tanks apparatus. The theoretically validated fuzzy PID controller with gain and phase margin based tuning formula were confirmed by the experimental results. The experimental results show that the fuzzy PID controller has the nonlinear properties of (1) higher control gains when the system is away of its steady states; and (2) lower control profile when set-point changes occur. As a result, these nonlinear properties provide the fuzzy PID control system with a superior performance over the conventional PID control system.

In the last example, ILC was implemented on a traffic control problem. It was shown in the example that the on-ramp metering method proved to be effective, but the implementation could be costly in changing freeway infrastructure. Furthermore, when there is insufficient traffic supply, the controller performance will be strongly affected. Hence, the development of ramp metering is limited. Speed control and flow control generally give a more satisfactory performance and could avoid congestion if drivers can respond and obey the signals. For both of the approaches, the control to the first segment is critical because it is the basis for regulating the entire freeway system. Thus, combined with control of input flow volume, the approaches would provide a more accurate solution to either the underflow or the overflow problem. Moreover, these approaches could be used to control individual lanes as well.

Appendix
Derivation of BIBO Stability Condition of Linear PID Control System

Suppose a nonlinear process N is controlled by a linear PID controller. In the computerized implementation, the PID controller is discretized by using a zero order holder where s in the transfer function of PID controller is substituted by $\frac{1-z^{-1}}{\Delta t}$ and for a linear PID controller in the series form, we have:

$$
\begin{aligned}
\frac{u_k - u_{k-1}}{\Delta t} = & \frac{K_c}{T_i}\left(1 + \frac{T_i}{\Delta t} + \frac{T_d}{\Delta t} + \frac{T_i T_d}{\Delta t^2}\right) e_k \\
& - \frac{K_c}{T_i}\left(\frac{T_i}{\Delta t} + \frac{T_d}{\Delta t} + \frac{2T_i T_d}{\Delta t^2}\right) e_{k-1} + \frac{K_c T_d}{\Delta t^2} e_{k-2}.
\end{aligned}
$$

Referring to Fig. 6.4, if we define:

$$
\begin{cases}
e_{1,k} = e_k = r_k - y_k \\
e_{2,k} = u_k \\
u_{1,k} = r_k \\
u_{2,k} = u_{k-1} \\
H_1(e_{1,k}) = \Delta u_k = u_k - u_{k-1} \\
H_2(e_{2,k}) = N(e_{2,k}) = y_k
\end{cases}
$$

Applying the small gain theorem, we can obtain the following sufficient condition for the BIBO stability of the linear PID controlled system, as:

$$
K_c\left(1 + \frac{T_d}{T_i} + \frac{\Delta t}{T_i} + \frac{T_d}{\Delta t}\right) \cdot \|H_2\| < 1
$$

K. Abidi and J.-X. Xu, *Advanced Discrete-Time Control*, Studies in Systems, Decision and Control 23, DOI 10.1007/978-981-287-478-8

References

1. Abidi, K., & Xu, X.-J. (2007). On the discrete-time integral sliding-mode control. *IEEE Transactions on Automatic Control*, *52*(4), 709–715.
2. Abidi, K., & Yildiz, Y. (2011). Discrete-time adaptive posicast controller for uncertain time-delay systems. In *Proceedings of the AIAA Guidance, Navigation and Control Conference*, AIAA Paper 2011–5788. Portland, Oregon.
3. Ackermann, J., & Utkin, V. (1998). Sliding mode control design based on Ackermann's formula. *IEEE Transactions on Automatic Control*, *43*(2), 234–237.
4. Ahn, H.-S., & Bristow, D. A. (2011). Iterative learning control. *Asian Journal of Control*, *13*(1), 1–2.
5. Akhtar, S., & Bernstein, D. S. (2005). Lyapunov-stable discrete-time model reference adaptive control. *International Journal of Adaptive Control and Signal Processing*, *19*(10), 745–767.
6. Akhtar, S., & Bernstein, D. S. (2004). Logarithmic Lyapunov functions for direct adaptive stabilization with normalized adaptive laws. *International Journal of Control*, *77*(7), 630–638.
7. Al-Rahmani, H. M., & Franklin, G. F. (1990). A new optimal multirate control of linear periodic and time-invariant systems. *IEEE Transactions on Automatic Control*, *35*(4), 406–415.
8. Annaswamy, A. M., Jang, J., & Lavretsky, E. (2008). Adaptive gain-scheduled controller in the presence of actuator anomalies. In *Proceedings of the AIAA Guidance, Navigation and Control Conference*, AIAA Paper 2008–7285. Huntington Beach, CA.
9. Annaswamy, A. M., & Narendra, K. S. (1989). Adaptive control of simple time-varying systems. In *Proceedings of the IEEE Conference on Decision and Control*. Tampa, FL.
10. Araki, M., & Yamamoto, K. (1986). Multivariable multirate sampled-data systems: State-space description, transfer characteristics and Nyquist criterion. *IEEE Transactions on Automatic Control*, *31*(2), 145–154.
11. Arimoto, S., Kawamura, S., & Miyazaki, F. (1984). Bettering operation of robots by learning. *Journal of Robotic Systems*, *1*, 123–140.
12. Armstrong-Helouvry, B., Dupont, P., & De Wit, C. C. (1994). A survey of models, analysis tools and compensation methods for the control of machines with friction. *Automatica*, *30*(7), 1083–1138.
13. Åström, K. J. (1996). Tuning and Adaptation. In *IFAC World Congress* (pp. 1–18).
14. Åström, K. J. (1992). Intelligent tuning. In L. Dugard, M. M'Saad & I. D. Landau, (Eds.), *Adaptive systems in control and signal processing* (pp. 360–370). Oxford, UK: Pergamon.
15. Åström, K. J., Hang, C. C., Persson, P., & Ho, W. K. (1992). Toward intelligent PID control. *Automatica*, *28*, 19.

K. Abidi and J.-X. Xu, *Advanced Discrete-Time Control*, Studies in Systems, Decision and Control 23, DOI 10.1007/978-981-287-478-8

16. Åström, K. J., Hang, C. C., & Lim, B. C. (1994). A new Smith predictor for controlling a process with an integrator and long dead-time. *IEEE Transactions on Automatic Control, 39*(2), 343–345.
17. Åström, K. J., & Wittenmark, B. (1997). *Computer-controller systems*. Upper Saddle River, NJ: Prentice Hall.
18. Åström, K. J., & Hägglund, T. (1993). Automatic tuning of PID controllers. In *Instrument Society of America*. North Carolina: Research Triangle Park.
19. Åström, K. J., Hägglund, T., Hang, C. C., & Ho, W. K. (1992). Automatic tuning and adaptation for PID controllers—a survey. In L. Dugard, M. M'Saad & I. D. Landau (Eds.), *Adaptive Systems in Control and Signal Processing*(pp. 371–376). Oxford, U.K.: Pergamon.
20. Åström, K. J., Hang, C. C., Persson, P., & Ho, W. K. (1992). Towards intelligent PID control. *Automatica, 28*(1), 1–9.
21. Baek, S.-E., & Lee, S.-H. (1999). Design of a multi-rate estimator and its application to a disk drive servo system. In *Proceedings of the American Control Conference* (pp. 3640–3644).
22. Becker, K. (1990). Closed-form solution of pure proportional navigation. *IEEE Transactions on Aerospace and Electronic Systems, 26*(3), 526–533.
23. Bekiaris-Liberis, N., & Krstic, M. (2010). Stabilization of strict-feedback linear systems with delayed integrators. *Automatica, 46*(11), 1902–1910.
24. Bekiaris-Liberis, N., & Krstic, M. (2010). Delay-adaptive feedback for linear feedforward systems. *Systems and Control Letters, 59*, 277–283.
25. Bekiaris-Liberis, N., & Krstic, M. (2011). Compensation of time-varying input and state delays for nonlinear systems. *ASME Journal of Dynamic Systems Measurement and Control, 134*(1), 1100–1123.
26. Bekiaris-Liberis, N., & Krstic, M. (2013). Compensation of state-dependent input delay for nonlinear systems. *IEEE Transactions on Automatic Control, 58*(2), 275–289.
27. Bekiaris-Liberis, N., & Krstic, M. (2013). Robustness of nonlinear predictor feedback laws to time-and state-dependent delay perturbations. *Automatica, 49*(6), 1576–1590.
28. Bellemans, T., de Schutter, B., & de Moor, B. (2002). Models for traffic control. *Journal A, 43*(3–4), 13–22.
29. Ben-Asher, J., & Yaesh, I. (1998). Advance in missile guidance theory. In *Progress in Astronautics and Aeronautics*, AIAA paper (Vol. 180).
30. Berg, M. C., Amit, N., & Powell, J. D. (1988). Multirate digital control system design. *IEEE Transations Automatic Control, 33*(12), 1139–1150.
31. Bien, Z., & Xu, J.-X. (1998). *Iterative learning control: Analysis, design, integration and applications*. Boston: Kluwer.
32. Blakelock, J. H. (1991). *Automatic control of aircraft and missiles* (2nd ed.). Englewood Cliffs: Wiley.
33. Bodson, M., & Groszkiewicz, J. E. (1994). Multivariable adaptive algorithms for reconfigurable flight control. In *Proceedings of the Conference on Decision and Control* (pp. 3330–3335). Lake Buena Vista, FL.
34. Boskovic, J. D., & Mehra, R. K. (2002). Multiple-model adaptive flight control scheme for accommodation of actuator failures. *Journal of Guidance, Control and Dynamics, 25*(4), 712–724.
35. Bresch-Pietri, D., & Krstic, M. (2009). Adaptive trajectory tracking despite unknown input delay and plant parameters. *Automatica, 45*(3), 2074–2081.
36. Bristow, D. A., & Alleyne, A. G. (2003). A manufacturing system for microscale robotic deposition. In *Proceedings of the American Control Conference* (pp. 2620–2625).
37. Bristow, D. A., Tharayil, M., & Alleyne, A. G. (2006). A survey of iterative learning control. *IEEE Control Systems Magazine, 26*(3), 96–114.
38. Buckley, J. J., & Ying, H. (1989). Fuzzy controllers theory: Limit theorems for linear fuzzy control rules. *Automatica, 25*(3), 469–472.
39. Calise, A., Lee, S., & Sharma, M. (1998). Direct adaptive reconfigurable control of a tailless fighter aircraft. In *Proceedings of the AIAA Guidance, Navigation and Control Conference*, AIAA Paper 98–4108.

40. Cao, W. J., & Xu, J.-X. (2004). Eigenvalue assignment in full-order sliding mode using integral type sliding surface. *IEEE Transactions on Automatic Control*, *49*(8), 1355–1360.
41. Carrasco, J. M., Galvn, E., Valderrama, G. E., Ortega, R., & Stankovic, A. M. (2000). Analysis and experimentation of nonlinear adaptive controllers for the series resonant converter. *IEEE Transactions on Power Electronics*, *15*(3), 536–544.
42. Casalino, G., & Bartolini, G. (1984). A learning procedure for the control of movements of robotic manipulators. In *Proceedings of the IASTED Symposium on Robotics and Automation* (pp. 108–111).
43. Chapra, S. C., & Canale, R. P. (1998). *Numerical methods for engineers*. Singapore: McGraw Hill.
44. Chen, G., & Ying, H. (1993). Stability analysis of nonlinear fuzzy PI control system. In *Proceedings of the International Conference on Fuzzy Logic Applications* (pp. 128–133).
45. Chen, Y., & Wen, C. (1999). *Iterative learning control: Convergence, robustness and applications*. London: Springer.
46. Chiang, W.-W. (1990). Multirate state-space digital controller for sector servo systems. In *Proceedings of the IEEE Conference on Decision and Control* (pp. 1902–1907).
47. Coffey, T. C., & Williams, I. J. (1966). Stability analysis of multiloop, multirate sampled systems. *AIAA Journal*, *4*(12), 2178–2190.
48. Craig, J. J. (1984). Adaptive control of manipulators through repeated trials. In *Proceedings of the American Control Conference* (pp. 1566–1573).
49. Crespo, L., Matsutani, M., Jang, J., Gibson, T., & Annaswamy, A. M. (2009). Design and verification of an adaptive controller for the generic transport model. In *Proceedings of the AIAA Guidance, Navigation and Control Conference*, AIAA Paper 2009–5618. Chicago, IL.
50. Desoer, C. A., & Vidyasagar, M. (1975). *Feedback system: Input-output properties*. New York: Academic Press Inc.
51. Dydek, Z. T., Annaswamy, A. M., & Lavretsky, E. (2008). Adaptive control and the NASA X-15 program: A concise history, lessons learned and a provably correct design. In *Proceedings of the American Control Conference* (pp. 2957–2962).
52. Dydek, Z. T., Annaswamy, A. M., Slotine, J. E., & Lavretsky, E. (2010). High performance adaptive control in the presence of time delays. In *American Control Conference* (pp. 880–885).
53. Dydek, Z. T., Jain, H., Jang, J., & Annaswamy, A. M. (2006). Theoretically verifiable stability margins for an adaptive controller. In *Proceedings of the AIAA Guidance, Navigation and Control Conference*, AIAA Paper 2006–6416. Keystone, CO.
54. Edwards, C., & Spurgeon, S. K. (1996). Robust output tracking using a sliding-mode controller/observer scheme. *International Journal of Control*, *64*(5), 967–983.
55. El-Khazali, R., & DeCarlo, R. (1995). Output feedback variable structure control design. *Automatica*, *31*(4), 805–816.
56. Elci, H., Longman, R. W., Phan, M., Juang, J.-N., & Ugoletti, R. (1994). Discrete frequency based learning control for precision motion control. In *Proceedings of the IEEE International Conference on Systems, Man, Cybernetics* (pp. 2767–2773).
57. de Figueiredo, R. J. P., & Chen, G. (1993). *Nonlinear feedback control systems: An operator theory approach*. New York: Academic Press Inc.
58. Feng, G., & Lozano, R. (1999). *Adaptive control systems*. Oxford, U.K.: Newnes.
59. Franklin, G. F., & Powell, J. (1980). *Digital control of dynamic systems*. Reading, MA: Addison-Wesley.
60. Fridman, L., Castaños, F., M'Sirdi, N., & Kharfa, N. (2004). Decomposition and robustness properties of integral sliding mode controllers. In *Proceedings of the IEEE Variable Structure Systems workshop*. Spain.
61. Gafvurt, M. (1999). Dynamic model based friction compensation on the furuta pendulum. In *Proceedings of the IEEE International Conference on Control Applications* (pp. 1260–1265).
62. Gao, W., Wang, Y., & Homaifa, A. (1995). Discrete-time variable structure control systems. *IEEE Transactions on Industrial Electronics*, *42*(2), 117–122.

63. Gao, F., Yang, Y., & Shao, C. (2001). Robust iterative learning control with applications to injection molding process. *Chemical Engineering Science*, *56*(24), 7025–7034.
64. Garden, M. (1971). *Learning control of actuators in control systems*. U.S. Patent 3555252.
65. Gibson, T., Crespo, L., & Annaswamy, A. M. (2009). Adaptive control of hypersonic vehicles in the presence of modeling uncertainties. In *Proceedings of the American Control Conference* (pp. 3178–3183). St. Louis, MO.
66. Giri, F. M. M., Dugard, L., & Dion, J. M. (1990). Pole placement direct adaptive control for time-varying ill-modeled plants. *IEEE Transactions on Automatic Control*, *35*, 723–726.
67. Golo, G., & Milosavljevic, C. (2000). Robust discrete-time chattering free sliding mode control. *Systems and Control Letters*, *41*(1), 19–28.
68. Goodwin, G. C., Ramadge, P., & Caines, P. (1980). Discrete-time multivariable adaptive control. *IEEE Transactions on Automatic Control*, *25*(3), 449–456.
69. Goodwin, G. C., & Sin, K. S. (1984). *Adaptive filtering, prediction and control*. Englewood Cliffs, NJ: Prentice-Hall.
70. Gregory, I. M., Cao, C., Xargay, E., Hovakimyan, N., & Zou, X. (2009). L_1 Adaptive control design for NASA AirSTAR flight test vehicle. In *Proceedings of the AIAA Guidance, Navigation and Control Conference, AIAA Paper* (pp. 2009–5738).
71. Gu, K., & Niculescu, S.-I. (2003). Survey on recent results in the stability and control of time-delay systems. *Journal of Dynamic Systems, Measurement and Control*, *125*(2), 158–165.
72. Ha, I.-J., Hur, J.-S., Ko, M.-S., & Song, T.-L. (1990). Performance analysis of PNG laws for randomly maneuvering targets. *IEEE Transactions on Aerospace and Electronic Systems*, *26*(5), 713–721.
73. Hablani, H. (2006). Endgame guidance and relative navigation of strategic interceptors with delays. *Journal of Guidance, Control and Dynamics*, *29*(1), 82–94.
74. Hang, C. C., Ho, W. K., & Cao, L. S. (1994). A comparison of two design methods for PID controllers. *ISA Transactions*, *33*, 147–151.
75. Hang, C. C., Lee, T. H., & Ho, W. K. (1993). Adaptive control. In *Instrument Society of America*. NC: Research Triangle Park.
76. Hara, T., & Tomizuka, M. (1998). Multi-rate controller for hard disk drive with redesign of state estimator. In *Proceedings of the American Control Conference* (pp. 3033–3037).
77. Havlicsek, H., & Alleyne, A. (1999). Nonlinear control of an electrohydraulic injection molding machine via iterative adaptive learning. *IEEE/ASME Transactions on Mechatronics*, *4*(3), 312–323.
78. Hildebrand, F. B. (1987). *Introduction to numerical analysis*. New York: Dover Publications.
79. Hillerstrom, G., & Walgama, K. (1997). Repetitive control theory and applications: A survey. In *Proceedings of the World Congress Volume D: Control Design II, Optimization* (pp. 1–6).
80. Ho, W. K., Hang, C. C., & Cao, L. S. (1994). Tuning of PID controllers based on gain and phase margins specifications. *Automatica*, *31*, 497–502.
81. Ho, W. K., Gan, O. P., Tay, E. B., & Ang, E. L. (1996). Performance and gain and phase margins of well-known PID tuning formulas. *IEEE Transactions on Control Systems Technology*, *4*(4), 473–477.
82. Holder, E., & Sylvester, V. (1990). An analysis of modern versus classical homing guidance. *IEEE Transactions on Aerospace and Electronic Systems*, *26*(4), 599–606.
83. Horowitz, R. (1993). Learning control of robot manipulators. *ASME Journal of Dynamic Systems Measurement and Control*, *115*(2B), 402–411.
84. Hou, Z., & Xu, J.-X. (2003). Freeway traffic density control using iterative learning control approach. In *Proceedings of the Intelligent Transportation Systems Conference*. Shanghai.
85. Hou, Z., Zhong, H., & Xu, J.-X. (2004). Iterative Learning Approach for Local Ramp Metering. In *Proceedings of the IEEE Conference on Control Applications*.
86. Hunt, K. J., Sbarbaro, D., Zbikowski, R., & Gawthrop, P. J. (1992). Neural networks for control systems—a survey. *Automatica*, *28*(6), 1083–1112.
87. Ioannou, P. A., & Sun, J. (1969). *Robust adaptive control*. Upper Saddle River, NJ: Prentice-Hall.

88. Jang, J. (2008). *Adaptive control design with guaranteed margins for nonlinear plants*. PhD Thesis, Massachusetts Institute of Technology.
89. Jankovic, M. (2009). Cross-term forwarding for systems with time delay. *IEEE Transactions on Automatic Control*, *54*(3), 498–511.
90. Kaneko, K., & Horowitz, R. (1997). Repetitive and adaptive control of robot manipulators with velocity estimation. *IEEE Transactions on Robotics and Automation*, *13*(2), 204–217.
91. Kanellakopoulos, I. (1994). A discrete-time adaptive nonlinear system. *IEEE Transactions on Automatic Control*, *39*(11), 2362–2365.
92. Kawamura, S., Miyazaki, F., & Arimoto, S. (1984). Iterative learning control for robotic systems. In *Proceedings of the International Conference on Industrial Electronics, Control and Instrumentation* (pp. 393–398).
93. Khan, B. Z., & Lehman, B. (1996). Setpoint PI controllers for systems with large normalized dead time. *IEEE Transactions on Control Systems Technology*, *4*(4), 459–466.
94. Kim, D.-I., & Kim, S. (1996). An iterative learning control method with application for CNC machine tools. *IEEE Transactions Industrial Applications*, *32*(1), 66–72.
95. Kim, K.-S., Lee, K.-J., & Kim, Y. (2003). Reconfigurable flight control system design using direct adaptive method. *Journal of Guidance, Control and Dynamics*, *26*(4), 543–550.
96. Kokotovic, K. P. V. (1991). *Foundations of adaptive control*. New York: Springer.
97. Krstic, M. (2010). Input delay compensation for forward complete and feedforward nonlinear systems. *IEEE Transactions on Automatic Control*, *55*(2), 287–303.
98. Krstic, M. (2009). *Delay compensation for nonlinear, adaptive and PDE systems*. Boston, MA: Birkhauser.
99. Kurek, J. E., & Zaremba, M. B. (1993). Iterative learning control synthesis based on 2-d system theory. *IEEE Transactions on Automatic Control*, *38*(1), 121–125.
100. Kurz, H., & Goedecke, W. (1981). Digital parameter-adaptive control of processes with unknown dead time. *Automatica*, *17*(1), 245–252.
101. Lai, N. O., Edwards, C., & Spurgeon, S. K. (2003). On discrete time output feedback sliding-mode like control for non-minimum-phase systems. In *Proceddings of the IEEE Conference on Decision and Control* (pp. 1374–1379). Hawaii.
102. Lai, N. O., Edwards, C., & Spurgeon, S. K. (2004). On discrete time output feedback min-max controller. *International Journal of Control*, *77*(2), 554–561.
103. Lavretsky, E. (2009). Combined/composite model reference adaptive control. In *Proceedings of the AIAA Guidance, Navigation and Control Conference*, AIAA Paper 2009–6065. Chicago, IL.
104. Longman, R. W. (2000). Iterative learning control and repetitive control for engineering practice. *International Journal of Control*, *73*(10), 930–954.
105. Lum, K.-Y. (2003). Infinite-dimensional linearization and extended adjoint method for non-linear homing loop analysis. In *Proceedings of the AIAA Guidance, Navigation and Control Conference* (pp. 2003–5449). Austin, TX.
106. Majhi, S., & Atherton, D. (1998). A new Smith predictor and controller for unstable and integrating processes with time delay. In *Proceedings of the Conference on Decision and Control*. Tampa, FL.
107. Mahut, M. (2001). A multi-lane link model of traffic dynamics based on the Space-Time queue. In *Proceedings of the Intelligent Transportation Systems Conference*. Oakland, CA.
108. Malki, H. A., Li, H. D., & Chen, G. R. (1994). New design and stability analysis of fuzzy proportional-derivative control system. *IEEE Transactions on Fuzzy Systems*, *2*(4), 245–254.
109. Manitius, A. Z., & Olbrot, A. W. (1979). Finite spectrum assignement problem for systems with delays. *IEEE Transactions on Automatic Control*, *24*(4), 541–553.
110. Marino, R., & Tomei, P. (2003). Adaptive control of linear time-varying systems. *Automatica*, *39*, 651–659.
111. Mazenc, F., Malisoff, M., & Lin, Z. (2008). Further results on input-to-state stability for nonlinear systems with delayed feedbacks. *Automatica*, *44*(9), 2415–2421.
112. Mazenc, F., & Niculescu, S.-I. (2011). Generating positive and stable solutions through delayed state feedback. *Automatica*, *47*(3), 525–533.

113. Mazenc, F., Niculescu, S.-I., & Bekaik, M. (2011). Backstepping for nonlinear systems with delay in the input revisited. *SIAM Journal on Control and Optimization*, *49*, 2263–2278.
114. Messner, W., Horowitz, R., Kao, W.-W., & Boals, M. (1991). A new adaptive learning rule. *IEEE Transactions on Automatic Control*, *36*(2), 188–197.
115. Middleton, R. H., & Goodwin, G. G. C. (1988). Adaptive control of time-varying linear systems. *IEEE Transactions on Automatic Control*, *33*, 150–155.
116. Mondie, S., & Michiels, W. (2003). Finite spectrum assignment of unstable time-delay systems with a safe implementation. *IEEE Transactions on Automatic Control*, *48*(12), 2207–2212.
117. Moon, J., & Kim, Y. (2001). Design of missile guidance law via variable structure control. *Journal of Guidance, Control, and Dynamics*, *24*(4), 659–664.
118. Moore, K. L. (1993). *Iterative learning control for deterministic systems*. London: Springer.
119. Moore, K. L., Dahleh, M., & Bhattacharyya, S. P. (1992). Iterative learning control: A survey and new results. *Journal of Robotic Systems*, *9*(5), 563–594.
120. Moore, K. L., & Xu, J.-X. (2000). Editorial: Iterative learning control. *International Journal of Control*, *73*(10), 930–954.
121. Narendra, K. S., & Annaswamy, A. M. (1989). *Stable adaptive systems* (Vol. 3). Prentice-Hall: Upper Saddle River, NJ.
122. Nesline, F. W., & Zarchan, P. (1981). A new look at classical vs modern homing missile guidance. *Journal of Guidance, Control and Dynamics*, *4*(1), 78–85.
123. Nguyen, N., Krishnakumar, K., Kaneshige, J., & Nespeca, P. (2006). Dynamics and adaptive control for stability recovery of damaged asymmetric aircraft. In *Proceedings of the AIAA Guidance, Navigation and Control Conference*, AIAA paper 2006–6049.
124. Niculescu, S. I. (2001). *Delay effects on stability: A robust control approach*. Heidelberg, Germany: Springer.
125. Niculescu, S.-I., & Annaswamy, A. M. (2003). An adaptive Smith-controller for time-delay systems with relative degree $n^* \leq 2$. *Systems and control letters*, *49*, 347–358.
126. Norrlof, M. (2002). An adaptive iterative learning control algorithm with experiments on an industrial robot. *IEEE Transactions on Robotics and Automation*, *18*(2), 245–251.
127. Norrlof, M., & Gunnarsson, S. (2002). Time and frequency domain convergence properties in iterative learning control. *International Journal of Control*, *75*(14), 1114–1126.
128. Norrlof, M., & Gunnarsson, S. (2001). Disturbance aspects of iterative learning control. *Engineering Applications of Artificial Intelligence*, *14*(1), 87–94.
129. Ortega, R., & Lozano, R. (1988). Globally stable adaptive controller for systems with delay. *International Journal of Control*, *47*(1), 17–23.
130. Pagilla, P. R., Yu, B., & Pau, K. L. (2000). Adaptive control of time-varying mechanical systems: Analysis and experiments. *IEEE/ASME Transactions on Mechatronics*, *5*(4), 410–418.
131. Palmor, Z. J. (1996). Time-delay compensation-Smith predictor and its odifications. In W. Levine (Ed.), *The Control Handbook* (pp. 224–237). Boca Raton, FL, USA: CRSC Press.
132. Papageorgiou, M., & Kotsialos, A. (2002). Freeway ramp metering: An overview. *IEEE Transactions on Intelligent Transportation Systems*, *3*(4), 271–281.
133. Papageorgiou, M., Blosseville, J.-M., & Hadj-Salem, H. (1989). Macroscopic modeling of traffic flow on the boulevard Peripherique in Paris. *Transportation Research B*, *23B*(1), 29–47.
134. Papageorgiou, M., Blosseville, J.-M., & Hadj-Salem, H. (1990). Modeling and real-time control of traffic flow on the southern part of boulevard Peripherique in Paris, Part I: Modeling. *Transportation Research A*, *24A*(5), 345–359.
135. Pepe, P., & Jiang, Z.-P. (2006). A Lyapunov-Krasovskii methodology for ISS and iISS of time-delay systems. *Systems and Control Letters*, *55*, 1006–1014.
136. Poulin, È., Pomerleau, A., Desbiens, A., & Hodouin, D. (1996). Development and evaluation of an auto-tuning and adaptive PID controller. *Automatica*, *32*(1), 71–82.
137. Qin, S. J., & Borders, G. (1994). A multiregion fuzzy logic controller for nonlinear process control. *IEEE Transactions on Fuzzy Systems*, *2*(1), 74–81.
138. Richard, J.-P. (2003). Time-delay systems: An overview of some recent advances and open problems. *Automatica*, *39*, 1667–1694.

139. de Roover, D., & Bosgra, O. H. (2000). Synthesis of robust multivariable iterative learning controllers with application to a wafer stage motion system. *International Journal of Control, 73*(10), 968–979.
140. Saab, S. S. (1995). A discrete-time learning control algorithm for a class of linear time-invariant systems. *IEEE Transactions on Automatic Control, 40*(6), 1138–1141.
141. Sastry, S. S., & Boston, M. (1989). *Adaptive control: Stability, convergence and robustness*. Upper Saddle River, NJ: Prentice-Hall.
142. Seborg, D. E., Edgar, T. F., & Mellichamp, D. A. (1989). *Process dynamics and control*. New York: Wiley.
143. Shima, T., Shinar, H., & Weiss, H. (2003). New interceptor guidance law integrating time-varying and estimation-delay models. *Journal of Guidance, Control and Dynamics, 26*(2), 295–303.
144. Shima, T., Idan, M., & Golan, O. M. (2006). Sliding-mode control for integrated missile autopilot guidance. *Journal of Guidance, Control and Dynamics, 29*(2), 250–260.
145. Shtessel, Y., & Buffington, J. (1998). Continuous sliding mode control. In *Proceedings of the American Control Conference* (Vol. 1, pp. 562–563). Philadelphia, PA.
146. Slotine, J. J. E., Hedrick, J. K., & Misawa, E. A. (1987). On sliding observers for nonlinear systems. *Transactions of the ASME: Journal of Dynamic Systems, Measurement and Control, 109*(1), 245–252.
147. Smith, O. J. (1959). A controller to overcome dead time. *ISA Journal, 6*, 28–33.
148. Smith, O. J. (1976). *Feedback control systems*. New York: McGraw-Hill Inc.
149. Stepanyan, V., Krishnakumar, K., & Nguyen, N. (2009). Adaptive control of a transport aircraft using differential thrust. In *AIAA Guidance, Navigation and Control Conference*, AIAA Paper 2009–5741. Chicago, IL.
150. Su, W. C., Drakunov, S., & Ozguner, U. (2000). An $O(T^2)$ boundary layer in sliding mode for sampled-data systems. *IEEE Transactions on Automatic Control, 45*(3), 482–485.
151. Taoa, G., Chena, S., & Joshi, S. M. (2000). An adaptive control scheme for systems with unknown actuator failures. *Automatica, 38*, 1027–1034.
152. Tsakalis, K. S., & Ioannou, P. A. (1993). *Linear time-varying systems—control and adaptation*. Upper Saddle River, NJ: Prentice-Hall.
153. Uchiyama, M. (1978). Formation of high-speed motion pattern of a mechanical arm by trial. *Transactions of the Society of Instrument and Control Engineers, 14*(6), 706–712.
154. Utkin, V. (1994). Sliding mode control in discrete-time and difference systems. In A.S.I. Zinober (Ed.), *Variable Structure and Lyapunov Control*, Chap. 5 (Vol. 193, pp. 87–107). Springer.
155. Utkin, V., & Shi, J. (1996). Integral sliding mode in systems operating under uncertainty conditions. In *Proceedings of the Conference on Decision and Control* (pp. 4591–4596). Kobe, Japan.
156. Wang, Q.-G., Lee, T. H., & Tan, K. K. (1999). Finite spectrum assignment for time-delay systems. In *Lecture Notes in Control and Information Sciences* (Vol. 239). New York: Springer.
157. Wang, Z., Polycarpou, M. M., Uber, J. G., & Shang, F. (2006). Adaptive control of water quality in water distribution networks. *IEEE Transactions on Control Systems Technology, 14*(1), 149–156.
158. Wen, C., Soh, Y. C., & Zhang, Y. (2000). Adaptive control of linear systems with unknown time delay. In G. Tao & F. Lewis (Eds.) *Adaptive control of nonsmooth dynamic systems*. UK: Springer.
159. Xu, J.-X. (1997). Analysis of iterative learning control for a class of nonlinear discrete-time systems. *Automatica, 33*, 1905–1907.
160. Xu, J.-X. (2004). A new adaptive control approach for time-varying parameters with known periodicity. *IEEE Transactions on Automatic Control, 49*(4), 579–583.
161. Xu, J.-X., & Abidi, K. (2008). Discrete-time output integral sliding-mode control for a piezomotor-driven linear motion stage. *IEEE Transactions Industrial Applications, 55*(11), 3917–3927.

162. Xu, J.-X., Liu, C., & Hang, C. C. (1996). Tuning of fuzzy PI controllers based on gain/phase margin specifications and ITAE index. *ISA Transactions*, *35*, 79–91.
163. J. Xu, Lum, K.-Y., & Xu, J.-X. (2007). Analysis of PNG laws with LOS angular rate delay. In *Proceedings of the AIAA Guidance, Navigation, and Control Conference*, AIAA Paper 2007–6788. Hiltonhead Island, SC, USA.
164. Xu, J.-X., Pan, Y. J., & Lee, T. H. (2004). Sliding mode control with closed-loop filtering architecture for a class of nonlinear systems. *IEEE Transactions on Circuits and Systems II: Express Briefs*, *51*(4), 168–173.
165. Xu, J.-X., Panda, S. K., Pan, Y. J., Lee, T. H., & Lam, B. H. (2004). A modular control scheme for PMSM speed control with pulsating torque minimization. *IEEE Transactions on Industrial Electronics*, *51*(3), 526–536.
166. Xu, J.-X., & Tan, Y. (2003). *Linear and nonlinear iterative learning control*. Berlin: Springer.
167. X.-G. Yan, Edwards, C., & Spurgeon, S. K. (2004). Output feedback sliding mode control for non-minimum phase systems with non-linear disturbances. *International Journal of Control*, *77*(15), pp. 1353–1361.
168. Yang, C.-D., & Yang, C.-C. (1996). Analytical solution of 3D true proportional navigation. *IEEE Transactions on Aerospace and Electronic Systems*, *4*, 1509–1522.
169. Yildiz, Y. (2010). Adaptive control for time delay systems applied to flight control. In *Proceedings of the AIAA Guidance, Navigation and Control Conference*, AIAA Paper 2010–7576. Keystone, CO.
170. Yildiz, Y. Y., Annaswamy, A. M., Kolmanovsky, I., & Yanakiev, D. (2010). Adaptive posicast controller for time-delay systems with relative degree $n^* \leq 2$. *Automatica*, *46*, 279–289.
171. Yildiz, Y., Y., Annaswamy, A. M., Yanakiev, D., & Kolmanovsky, I. (2007). Adaptive idle speed control for internal combustion engines. In *Proceedings of the American Control Conference* (pp. 3700–3705). New York City.
172. Yildiz, Y., Y., Annaswamy, A. M., Yanakiev, D., & Kolmanovsky, I. Spark ignition engine idle speed control: an adaptive control approach. *IEEE Transactions on Control Systems Technology*, in press.
173. Yildiz, Y., Y., Annaswamy, A. M., Yanakiev, D., & Kolmanovsky, I. (2008). Adaptive air fuel ratio control for internal combustion engines. In *Proceedings of the American Control Conference* (pp. 2058–2063). Seattle, Washington.
174. Yildiz, Y. Y., Annaswamy, A. M., Yanakiev, D., & Kolmanovsky, I. (2010). Spark ignition engine fuel-to-air ratio control: an Adaptive control approach. *Control Engineering Practice*, *18*(12), 1369–1378.
175. Yildiz, Y., Y., Annaswamy, A. M., Yanakiev, D., Kolmanovsky, I. (2008). Automotive powertrain control problems involving time delay: an adaptive control approach. In *Proceedings of the ASME Dynamic Systems and Control Conference* (pp. 1421–1428). Ann Arbor, Michigan.
176. Ying, H. (1993). The simplest fuzzy controllers using different inference methods are different nonlinear proportional-integral controllers with variable gains. *Automatica*, *29*(6), 1579–1589.
177. Zak, S. H., & Hui, S. (1993). On variable structure output feedback controllers for uncertain dynamical systems. *IEEE Transactions on Automatic Control*, *38*(8), 1509–1512.
178. Zarchan, P. (1990). Tactical and strategic missile guidance. In *Progress in Astronautics and Aeronautics*, AIAA Paper (Vol. 124).
179. Zarchan, P. (1979). Complete statistical analysis of nonlinear missile guidancesystems SLAM. *Journal of Guidance and Control*, *2*(1), 71–78.
180. Zhao, Z. Y., Tomizuka, M., & Isaka, S. (1993). Fuzzy gain scheduling of PID controllers. *IEEE Transactions on SMC*, *23*(5), 1392–1398.
181. Zhou, D., Mu, C., Ling, Q., & Xu, W. (1999). Optimal sliding-mode guidance of a homing missile. In *Proceedings of the Conference on Decision and Control* (pp. 2131–5136).
182. *CE106 Ball and Beam Apparatus Manual* (1999). Distributed by TecQuipment Education and Training, Bonsall Street, Long Eaton, Nottingham, BG10 2AN, England, http://www.tq.com.

MIX
Papier aus verantwortungsvollen Quellen
Paper from responsible sources
FSC® C105338

If you have any concerns about our products,
you can contact us on
ProductSafety@springernature.com

In case Publisher is established outside the EU,
the EU authorized representative is:
Springer Nature Customer Service Center GmbH
Europaplatz 3, 69115 Heidelberg, Germany

Printed by Libri Plureos GmbH
in Hamburg, Germany